Tec
Bioproduct Analysis

BOOKS IN THE BIOTOL SERIES

The Molecular Fabric of Cells
Infrastructure and Activities of Cells

Techniques used in Bioproduct Analysis
Analysis of Amino Acids, Proteins and Nucleic Acids
Analysis of Carbohydrates and Lipids

Principles of Cell Energetics
Energy Sources for Cells
Biosynthesis and the Integration of Cell Metabolism

Genome Management in Prokaryotes
Genome Management in Eukaryotes

Crop Physiology
Crop Productivity

Functional Physiology
Cellular Interactions and Immunobiology
Defence Mechanisms

Bioprocess Technology: Modelling and Transport Phenomena
Operational Modes of Bioreactors

In vitro Cultivation of Micro-organisms
In vitro Cultivation of Plant Cells
In vitro Cultivation of Animal Cells

Bioreactor Design and Product Yield
Product Recovery in Bioprocess Technology

Techniques for Engineering Genes
Strategies for Engineering Organisms

Principles of Enzymology for Technological Applications
Technological Applications of Biocatalysts
Technological Applications of Immunochemicals

Biotechnological Innovations in Health Care

Biotechnological Innovations in Crop Improvement
Biotechnological Innovations in Animal Productivity

Biotechnological Innovations in Energy and Environmental Management

Biotechnological Innovations in Chemical Synthesis

Biotechnological Innovations in Food Processing

Biotechnology Source Book: Safety, Good Practice and Regulatory Affairs

BIOTECHNOLOGY BY OPEN LEARNING

Techniques Used in Bioproduct Analysis

PUBLISHED ON BEHALF OF :

Open universiteit and **University of Greenwich (formerly Thames Polytechnic)**

Valkenburgerweg 167
6401 DL Heerlen
Nederland

Avery Hill Road
Eltham, London SE9 2HB
United Kingdom

Butterworth-Heinemann Ltd
Linacre House, Jordan Hill, Oxford OX2 8DP

 PART OF REED INTERNATIONAL BOOKS

OXFORD LONDON BOSTON
MUNICH NEW DELHI SINGAPORE SYDNEY
TOKYO TORONTO WELLINGTON

First published 1992

British Library Cataloguing in Publication Data
A catalogue record for this book is
available from the British Library

Library of Congress Cataloguing in Publication Data
A catalogue record for this book is
available from the Library of Congress

ISBN 0 7506 1501 X ✓

Composition by University of Greenwich
(formerly Thames Polytechnic)
Printed and Bound in Great Britain by
Thomson Litho, East Kilbride, Scotland

The Biotol Project

The BIOTOL team

OPEN UNIVERSITEIT, THE NETHERLANDS
Prof M. C. E. van Dam-Mieras
Prof W. H. de Jeu
Prof J. de Vries

UNIVERSITY OF GREENWICH (FORMERLY THAMES POLYTECHNIC), UK
Prof B. R. Currell
Dr J. W. James
Dr C. K. Leach
Mr R. A. Patmore

This series of books has been developed through a collaboration between the Open universiteit of the Netherlands and University of Greenwich (formerly Thames Polytechnic) to provide a whole library of advanced level flexible learning materials including books, computer and video programmes. The series will be of particular value to those working in the chemical, pharmaceutical, health care, food and drinks, agriculture, and environmental, manufacturing and service industries. These industries will be increasingly faced with training problems as the use of biologically based techniques replaces or enhances chemical ones or indeed allows the development of products previously impossible.

The BIOTOL books may be studied privately, but specifically they provide a cost-effective major resource for in-house company training and are the basis for a wider range of courses (open, distance or traditional) from universities which, with practical and tutorial support, lead to recognised qualifications. There is a developing network of institutions throughout Europe to offer tutorial and practical support and courses based on BIOTOL both for those newly entering the field of biotechnology and for graduates looking for more advanced training. BIOTOL is for any one wishing to know about and use the principles and techniques of modern biotechnology whether they are technicians needing further education, new graduates wishing to extend their knowledge, mature staff faced with changing work or a new career, managers unfamiliar with the new technology or those returning to work after a career break.

Our learning texts, written in an informal and friendly style, embody the best characteristics of both open and distance learning to provide a flexible resource for individuals, training organisations, polytechnics and universities, and professional bodies. The content of each book has been carefully worked out between teachers and industry to lead students through a programme of work so that they may achieve clearly stated learning objectives. There are activities and exercises throughout the books, and self assessment questions that allow students to check their own progress and receive any necessary remedial help.

The books, within the series, are modular allowing students to select their own entry point depending on their knowledge and previous experience. These texts therefore remove the necessity for students to attend institution based lectures at specific times and places, bringing a new freedom to study their chosen subject at the time they need and a pace and place to suit them. This same freedom is highly beneficial to industry since staff can receive training without spending significant periods away from the workplace attending lectures and courses, and without altering work patterns.

Contributors

AUTHORS

Dr R. D. J. Barker, De Montfort University, Leicester, UK

Dr M. D. Brownleader, Royal Holloway and Bedford New College, Egham, Surrey, UK

Dr R. Cannell, Glaxo Group Research, Greenford, Middlesex, UK

Dr T. G. Cartledge, Nottingham Polytechnic, Nottingham, UK

Dr H. B. C. M. Haaker, Agricultural University, Wageningen, The Netherlands

Professor A. M. James, Tooting, London, UK

Dr R. O. Jenkins, De Montfort University, Leicester, UK

Dr C. K. Leach, De Montfort University, Leicester, UK

Dr I. Simpkin, Hatfield Polytechnic, Hatfield, UK

Professor Dr K. van Dam, University of Amsterdam, The Netherlands

Dr J. M. Walker, Hatfield Polytechnic, Hatfield, UK

EDITOR

Professor A. M. James, Tooting, London, UK

SCIENTIFIC AND COURSE ADVISORS

Professor M. C. E. van Dam-Mieras, Open universiteit, Heerlen, The Netherlands

Dr C. K. Leach, De Montfort University, Leicester, UK

ACKNOWLEDGEMENTS

Grateful thanks are extended, not only to the authors, editors and course advisors, but to all those who have contributed to the development and production of this book. They include Mrs S. Connor, Miss J. Skelton and Professor R. Spier.

The development of this BIOTOL text has been funded by **COMETT, The European Community Action Programme for Education and Training for Technology**. Additional support was received from the Open universiteit of The Netherlands and by University of Greenwich (formerly Thames Polytechnic).

Contents

How to use an open learning text

An open learning text presents to you a very carefully thought out programme of study to achieve stated learning objectives, just as a lecturer does. Rather than just listening to a lecture once, and trying to make notes at the same time, you can with a BIOTOL text study it at your own pace, go back over bits you are unsure about and study wherever you choose. Of great importance are the self assessment questions (SAQs) which challenge your understanding and progress and the responses which provide some help if you have had difficulty. These SAQs are carefully thought out to check that you are indeed achieving the set objectives and therefore are a very important part of your study. Every so often in the text you will find the symbol Π, our open door to learning, which indicates an activity for you to do. You will probably find that this participation is a great help to learning so it is important not to skip it.

Whilst you can, as an open learner, study where and when you want, do try to find a place where you can work without disturbance. Most students aim to study a certain number of hours each day or each weekend. If you decide to study for several hours at once, take short breaks of five to ten minutes regularly as it helps to maintain a higher level of overall concentration.

Before you begin a detailed reading of the text, familiarise yourself with the general layout of the material. Have a look at the contents of the various chapters and flip through the pages to get a general impression of the way the subject is dealt with. Forget the old taboo of not writing in books. There is room for your comments, notes and answers; use it and make the book your own personal study record for future revision and reference.

At intervals you will find a summary and list of objectives. The summary will emphasise the important points covered by the material that you have read and the objectives will give you a check list of the things you should then be able to achieve. There are notes in the left hand margin, to help orientate you and emphasise new and important messages.

BIOTOL will be used by universities, polytechnics and colleges as well as industrial training organisations and professional bodies. The texts will form a basis for flexible courses of all types leading to certificates, diplomas and degrees often through credit accumulation and transfer arrangements. In future there will be additional resources available including videos and computer based training programmes.

Preface

Advances of the biological sciences per sé and the application of biological knowledge depend upon a variety of analytical techniques. Those whose chosen field is to investigate and use biological systems are faced with a particularly daunting task. The variety and complexity of organisms is well known. Each cell contains many thousands of different molecules all performing their own specialised functions in a controlled and orderly manner. Furthermore, many of these molecules are large and complex and their functional activities depend upon their compositions, conformations and the environment in which they are found. To analyse such systems calls upon the development of sound strategies and the employment of a wide variety of techniques. This text has been prepared to provide those embarking on, or developing careers in the molecularly - orientated biological sciences and their application to biotechnological enterprises, with appropriate knowledge of these essential techniques.

The enormous range of techniques now available has, inevitably meant that we had to make some selection of material. We had chosen a 'strategic' approach used with special emphasis on key areas and the principles which underpin the techniques of greatest general value. The text, however, is not simply a description of the mechanistic background to laboratory techniques. It provides a discussion of strategies that can be used to achieve particular objectives and many specific examples are included. On completion of this text, the reader will be able to contribute to the development of strategies for analysing biological materials and understand the potential and limitations of a wide range of techniques.

We start by explaining the general strategy for measuring biomass with particular emphasis placed on unicellular systems. We then turn our attention to describing the techniques used to disrupt cells and to separate cell extracts into specific sub-cellular fractions. This naturally leads to an examination of the techniques that may be used to purify particular molecular species from these sub-cellular fractions.

The detection and measurement of specific biomolecules are important when purifying and studying biological materials. In order to follow separation performance, the biomolecules we wish to purify must be made 'visible'. We focus on this aspect in the next two chapters. In these we explain how biological (enzymatic) activity can be used to visualise and measure biomolecules and how immune assays and molecular hybridisation may be used to detect and quantify biomolecules.

In Chapter 8, we concentrate on spectroscopic techniques using electromagnetic radiation (UV, visible, IR, X-ray) or nuclear magnetic resonance and on mass spectroscopy as techniques for quantifying and determining the structure of biomolecules. In Chapter 9 we examine the use of radioactive isotopes. These may be used as 'labels' to measure particular molecules but also are used to monitor the movement and transformation of molecules within intact systems. We complete this transition back to the whole cell in the final two chapters where we examine microcalorimetry and electrometric techniques. Microcalorimetry enables us to detect and study heat effects accompanying metabolic conversions, whilst electrometric techniques are widely used in research, process automation and environmental monitoring.

We were faced with one further dilemma, what system of units to use. Although the Systeme Internationale de Unites are widely accepted in the sciences, this system is not universally used in the biological sciences. Thus litres (l) and millilitres (ml) are still in common usage. It is therefore necessary to know both the SI units and the other systems and to be able to interconvert between them. The BIOTOL series predominantly uses SI units. Nevertheless in areas of activities where their use is not common, other units have been used. Some of the alternative methods of expressing various physical quantities are described in an appendix.

One final point we would like to make. 'Hands on' experience is a vital pre-request to becoming a practising scientist or technologist. Nevertheless, there is much that can, and should, be done before setting foot into the laboratory. Understanding the principle upon which laboratory techniques are based and being aware of the range of techniques that are available, allows for sound experimental design which in turn facilitates effective use of laboratory time and resources. This text provides an effective way of gaining knowledge of techniques which are applied in a large number of areas including biological research, process control and monitoring, product purification and quality control, health care, food processing, agricultural production and environmental monitoring. We thank the author: editor team for their contributions to such a useful learning resource.

Scientific and Course Advisors: Professor M.C.E van Dam-Mieras

Dr C. K. Leach

An introduction to the techniques used in bioproduct analysis

An introduction to the techniques used in bioproduct analysis

1.1 The purpose of this chapter

The aims of this brief introductory chapter are to:

- explain the importance of being able to extract, purify and analyse the molecules which are found in, or are produced by, living systems;
- describe the layout of the chapters found within this texts and explain their significance;
- explain the relationship between this text and other BIOTOL text on the analytical procedures which are applied to living systems at a molecular level.

1.2 Why study the molecules of living systems?

The aim of biochemists and molecular biologists is to understand the molecular and chemical basis of biological phenomena. The success of their efforts, thus far, is reflected by the wealth of knowledge we now have of the molecular composition of, and the chemical transformations which occur in, living systems. There is, however, still much of which we have little knowledge and understanding. We have, for example, studied only a tiny fraction of the many millions of biological species and in those which have received attention, only a small proportion of the molecules found in these organisms have been studied to any great extent. One reason, therefore, for developing the skills needed to study these molecules, is to contribute to the general knowledge base we have of biochemistry and molecular biology.

There are many other compelling reasons for studying living systems at this level. Mankind is dependent upon living systems and their products to fulfil an enormous number of his needs. We are dependant upon them for food, for medicines, building materials, for specific chemicals and so on. History has shown us that increased knowledge of the molecules and chemical process of living systems greatly enhances our abilities to harness the capabilities of biological systems to carry out the processes, and make the products, we need or desire.

The whole of contemporary biotechnology has been dependent upon our expanding knowledge of the molecular activities of cellular material. To achieve all that biotechnology potentially offers us requires further extension of our knowledge of biological phenomena at a molecular level.

Perhaps the most compelling reason of all for studying the molecular events of living material is the notion that the very survival of the Earth as a place fit for habitation may depend upon it. Increasing pressure for food arising from expansion of the human population; pollution arising from chemically-based technologies; energy consumption; global warming; destruction of rain forests are all seen as threats. The development of cleaner technologies and better environmental management are essential for global health. Increased knowledge of the molecular biology of living systems leading to new biologically-based processes and products is recognised as providing opportunities for developing strategies for overcoming or avoiding the worst excesses of our current behaviour. Biochemistry and molecular biology have already made enormous, practical contributions to improving health, providing food and in cleaning the environment. But, these contributions still remain of minor significance to what potentially such studies will enable us to achieve in future decades. The study of bio-molecules is, and will remain, of vital importance to the biosphere and human society.

1.3 Organisms as chemical systems

A typical response to the question, 'what are the characteristic properties of living systems?' usually encompasses ideas about reproduction and development. Although these ideas are true, they are underpinned by much more fundamental characteristics. All living systems are highly organised entities, which use an external energy source and chemicals to create new highly ordered and complex structures. To achieve this, living systems must bring about a wide variety of chemical changes, transforming one set of chemicals (nutrients) into the building blocks that make up cells. Some molecules (DNA) store the information that is needed to carry out these changes, others (enzymes) actually carry out those required changes while yet others perform structural functions. Cells are indeed extremely complex and refined machines, carrying out a vast array of chemical changes. Some would regard cells as the ultimate chemical factories. But there is more to cells than being merely chemical factories. We know that many can adapt to changing circumstances and can defend themselves against some chemical and biological threats. We also know that cells can communicate, usually using chemical signals, with each other. This is especially true in multicellular systems where one set of cells can regulate the behaviour of others through the production of hormones, growth factors etc.

If we focus onto particular groups of chemicals within living systems we must also realise that, apart from the great diversity of types of molecules that are produced, many of the molecules that we find are very complex and large. To be functional, many must have highly specific structures. This is especially true of the macromolecules involved in inheritance (DNA) and gene expression (RNA) and in the workers of cells (enzymes).

We conclude that cells carry out a most elaborate array of chemical transformations and can achieve these transformations while controlling and maintaining their internal conditions.

1.4 How are we to cope with the diversity of living systems?

The diversity of biologically produced chemicals and the complexity of cellular structures may appear very daunting. Fortunately we have two main allies to help us.

Studies have revealed a great deal of unity between cells from a wide range of sources. We know for example that genetic information in all organisms is stored within the nucleotide sequences of their nucleic acids and that this genetic information is expressed through the processes of transcription and translation which show great similarities between cells. Likewise the basic energy harnessing processes of cells are limited in number and these common processes are encountered in many organisms from quite different environments. Thus our studies of new systems can start from the knowledge that these core mechanisms and processes will be operational and we can therefore focus on detecting more subtle and specific features of the system under study.

The second ally we have is that a wide range of techniques have been developed that enable us to generate and implement strategies for studying the molecules of living systems. The prime purpose of this text is to provide you with knowledge of a range of techniques which can be widely applied to the analysis of bio-molecules. In the following sections, we will explain how we selected the techniques that are described in this text and describe other BIOTOL texts that have been designed to enable you to contribute to biology and biotechnology by increasing your knowledge of essential analytical skills.

1.5 The strategy of this text

As you may well imagine, since organisms contain such a variety of different chemicals, there is a need for a very large variety of analytical techniques. Some of the molecules you will encounter are small and of low relative molecular mass, others are macromolecular and of exceedingly high molecular mass; some are polar carrying charged groups, others are non-polar and are barely soluble in water; some are present in very low concentrations while others are present in greater abundance. To attempt to catalogue and describe all of the techniques that are needed to cover this diversity would be an enormous task and lead to such an enormous volume that many would be put off by its size, never mind its content!

Here we have adopted a different approach. In this text we have first of all selected techniques that have very general application. For example, in order to study the molecules within cells, it is usually necessary to first break cells open in order to release the molecules. Cell disruption techniques therefore are an essential pre-requisite to biochemical analyses. Likewise producing concentrated suspensions of sub-cellular comments is also a commonly employed as a means of localising the occurrence of particular biomolecules.

A second aspect of the strategy we have adopted is to discuss the principles behind particular techniques rather than their specific application. For example chromatography encompasses a wide variety of techniques that have widespread application in the separation and purification of bio-molecules. Clearly chromatographic systems designed to separate for example non-polar fatty acids is unlikely to be satisfactory for the separation of highly polar nucleotides. Our approach is to attempt to make you aware of the principles employed in chromatography and the diversity of forms in which chromatographic systems can be used. This will enable you to select and use systems appropriate for the particular separation you seek to achieve. We have, therefore, attempted to cover a very wide range of underpinning principles but avoid using a recipe approach to the purification and analysis of bioproducts. Such specific recipes are readily available within the literature. Thus, here we have provided you with the opportunity to understand how a wide variety of techniques work and to give you some feel for the diversity of techniques that are available. The specific details of procedures that are appropriate to achieve a particular objective with particular systems you will acquire through experience.

One final point we would like to make is that although the techniques we have described have widespread application, the techniques we may employ at the outset (for example cell disruption) are much more dependant upon the nature of the starting material. For example the disruption of bacterial cells poses quite different problems from the disruption of animal cells. The former are, of course, thick walled and unicellular, the latter are thin walled and often aggregated. At later stages in the purification of a particular molecule or group of molecules (for example RNA), the character of the samples from the two sources becomes much more similar, RNAs from the two sources are, after all, chemically remarkably similar. Where the techniques are dependent upon the nature of the starting material, we have attempted to keep a broad-based approach to include plant, animal and microbial sources. Nevertheless some emphasis has been placed on microbial (especially bacteria and yeast) systems because they are frequently chosen as preferred systems for study and because they are, until now, often the system of choice for biotechnological application.

1.6 The structure of the text

To some extent, this text covers techniques in the chronological order that they are employed in the laboratory.

We begin, in Chapter 2, by describing how to measure biomass itself. In the study of molecular biology, it is often essential to quantify the amounts of the relevant compounds in terms of quantities found per unit of biomass. This is necessary to compare system in terms of yield or capacity to bring about a desired (or undesired) change. The estimation of cell growth and biomass may also be essential to judge the best time for harvesting and processing a system in order to extract desired molecules. The estimation of biomass therefore frequently underpins molecular and biochemical studies. The estimation of biomass is not, however, in all instances straightforward. In Chapter 2 you will be exposed to a wide variety of techniques and will learn of their advantages and limitations. Once we have our starting material, our next task is to disrupt the cellular material that is present. This is followed by a desire to separate the extract from the broken cell debris and to concentrate the released cellular components into sub-cellular fractions. This is most frequently achieved by centrifugation. Thus Chapters 3 and 4 examine cell disruption techniques and centrifugation. Centrifugation is not however limited to its application as a device for fractionating cells into their sub-cellular constituents. It can also be used as a method of purifying and physically analysing chemical entities, especially macro-molecular components. These applications of centrifugation are described in Chapter 4.

Once the cells have been disrupted and the unwanted debris removed, much of our efforts are usually aimed at separating the desired molecules from other contaminating compounds. Chapter 5 explores the strategies and techniques that may be employed in the separation and purification of bio-molecules. It begins by considering the preliminary purification steps and then moves on to the powerful techniques of chromatography, solvent extractions and those based on the charged nature of biomolecules. Amongst this later group of techniques are ion exchange chromatography and a variety of electrophoretic procedures. Towards the end of the chapter, the contemporary and powerful technique of affinity chromatography is explained. Throughout, this discussion of these separation techniques is based on a consideration of the physical and chemical properties of bio-molecules, since it is the differences in these properties which are exploited in the separation techniques.

In Chapter 6, we take what might appear to be something of a diversion by examining the methods available for the assays of enzymes. Of course much of our interest in the chemical transformations which take place within cells depends upon the catalytic activities of enzymes and this is justification in itself to include a discussion of enzyme assay procedures. There are however other reasons.

If we wish to examine the properties of particular enzymes in detail, then we will need to have a quick sensitive and specific test (or assay) for the enzyme. We will need to know at each stage of the purification if the enzyme is present and, if so, how much is there. This is another important reason for including enzyme assay methods at this stage.

One of the features of enzyme purifications is the desirability to have a quick, easy to perform and specific assay for the enzyme. The availability of such assay procedures is the basis of another compelling reason for including enzyme assay methods at this stage. Enzymes are, themselves potent analytical tools. Because of their specificity and sensitivity, enzymes can be employed to detect and measure many of the molecules we find in biological systems. Enzyme-based assays are not only analytical tools of the research laboratory. They find increasing use in monitoring industrial processes and in diagnostic and therapeutic health care. A good example is the detection and measurement of glucose in spent culture fluids and in urine using the enzyme glucose oxidase. Essentially this enzyme catalyses the oxidation of glucose. The technique using this enzyme to assay glucose depends upon measuring the rate or extent of the oxidation process. Thus although we may primarily think of enzymes in their biological roles, enzymes have become an important analytical tool.

The use of enzymes as analytical tools is not however confined to measuring the substrates of enzymes. Enzymes are being increasing used as 'labels' to detect and measure quite complex molecules. An excellent example of this use is illustrated by the enzyme-linked immunosorbent assay (ELISA) technique. In this technique, an enzyme is covalently linked to an antibody. The antibody is used to detect a particular antigen. A simple sequence for such a procedure is as follows.

The antigen may be adsorbed onto a surface (a plastic dish or membrane)

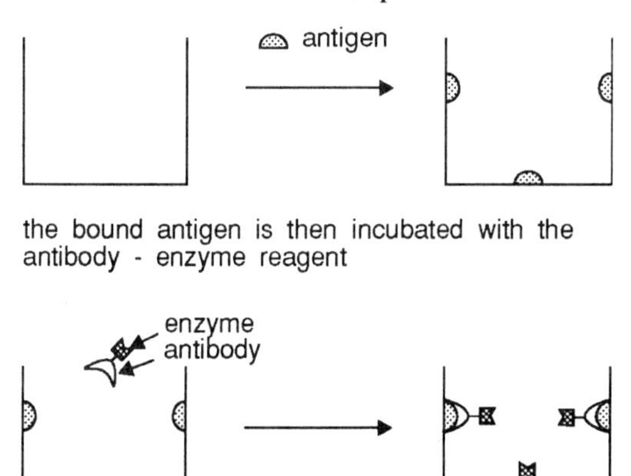

The unbound antibody-enzyme complex is washed out of the dish Then we assay the amount (activity) of the enzyme which has remained in the dish. The activity of the enzyme present in the dish is, of course, a reflection of the amount of antibody that has become attached to the antigen in the dish. Thus by measuring the activity of the enzyme in the dish, we can measure the amount of antigen present. This technique is a potent one for two reasons. It is both specific and sensitive. The specificity of the assay depends upon the specificity of the interaction between the antigen and antibody. The sensitivity of the technique is in part related to the activity of the enzyme. Each enzyme molecule that binds, via the antibody, to the antigen is capable of converting many substrate molecules to product. This in effect acts as a kind of amplification system. The general scheme for ELISA described above is only one of many techniques that have been developed using the specificity of antibodies and the sensitivity of enzymes assays. Other variants are competitive ELISA, double antibody (sandwich) ELISA and indirect ELISA. We will not describe these at this stage. The important point to realise is that the ability to quickly and properly assay enzymes is essential not only in the study of biochemistry but also as analytical tool.

By now you should have realised that enzyme assays techniques are important weapons in the hands of the investigators of the molecular constituents of cells. In this text, we are not proposing to examine all of the scientific and applied uses of enzymes. These are better discussed within the framework of specific issues. A quick glance through other BIOTOL texts will reveal how extensive the application of, and the need to assay, enzymes is in the biological sciences and biotechnologies. The importance of enzymes to technology is reflected by the production two BIOTOL texts ('Principles of Enzymology for Technological Application' and 'Technological Application of Biocatalysts') devoted to this topic.The application of enzymes in immune-assay procedures is described in depth in the BIOTOL text 'Technological Application of Immunochemicals'. In Chapter 6 we have described the strategies and principles which unpin enzyme assay procedures. We have chosen not to discuss enzyme kinetics in the body of the text since it is assumed that the reader has met the important relationships (Michaelis Menton Kinetics) in previous studies. However, we have provided discussion of these kinetics, together with some practical considerations in the assay of enzymes, in the form of an appendix for those who require some revision of this topic.

In Chapter 7 we explore the use of antibodies as analytical tools. The potency of analyses based upon antibodies depends upon the sepcificy of antibody binding and upon our ability to detect the interaction of antibodies with their target molecules. This Chapter, however, also deals with techniques based upon nucleic acid hybridisation, Nucleic acid hybridisation depends upon the ability of single stranded nucleic acids to form double stranded molecules if they are incubated under conditions where hydrogen bands may form between complementary nucleotide bases. Chapter 7, therefore is primarily concerned with techniques that enable us to identify, purify and measure specific macromolecules.

Chapter 8 deals with the physical methods which enable us to determine the structural configuration of biomolecules. Here emphasis is placed on the methods that most workers will encounter within the laboratory. These include a variety of spectrophotometric, (ultraviolet, visible, infrared) techniques, nuclear magnetic resonance spectroscopy and mass spectrometry.

In Chapter 9 we examine the use of radioactive isotopes in biochemical analyses. You will learn that we may use such isotopes to trace the occurence and movement of molecules and to monitor chemical transformations. The focus of Chapter 9 is mainly on the techniques required to measure radioactivity, but the chapter also includes some examples and discussion on matters of safety.

In Chapter 10 we concentrate on the techniques of the measurement of thermal changes associated with molecular (metabolic) transformations. Energetic considerations are vitally important in understanding the behaviour and properties of living systems. Thermodynamics are not only essential to considerations of metabolism but also to understanding the preferred conformation and stability of the major macromolecules of cells. They are also an important factor in designing biological production processes. There are inevitably energetic changes associated with the chemical transformations which occur during metabolism and with the structural changes which occur when the conformation of molecules is modified. In part these energetic changes are internal (thermodynamic) energy changes but, in many instances, energy in the form of heat is either released or absorbed. A discussion of the principles of thermodynamics is provided in the BIOTOL text "Principles of Cell Energetics". Here we focus attention on the sensitive measurement of heat evolution/adsorption as a method of monitoring changes at the molecular (metabolic) level.

In the final chapter (Chapter 11) we examine measurements based on electrochemical methods. These methods are of great value for measuring and monitoring a wide variety of parameters and chemicals of importance to molecular biology and biochemistry. Again the approach has been to examine the underpinning principles of electrometric methods, rather than attempt to cover the whole spectrum of specific examples. We do, however, describe the wide variety of electrode types that are available. Electrometric methods have widespread application ranging from the preparation of reagents through to monitoring culture performance. It is not surprising therefore that electrochemical methods are encountered in many BIOTOL texts. The description provided in Chapter 11 provides an understanding of the operation and limitations of these methods.

1.7 The relationship of this text to other BIOTOL texts

The BIOTOL series of texts should be viewed as a matrix of learning materials. This text provides an opportunity to learn about a wide variety of techniques used to extract, purify and characterise biological products. This text is 'generic' in nature, by which we mean it has widespread application. The principles and techniques described have not only application in the study of living systems at the molecular level, but are also applied in the monitoring of large scale operations, downstream processing (purification) of bioproducts and in the quality control of processes and products. Thus this text may be regarded as a discussion of some of the central issues essential to becoming a practising biotechnologist.

The immediate BIOTOL partner texts to this one are:

- Analysis of Amino Acids, Proteins and Nucleic Acids;
- Analysis of Carbohydrates and Lipids.

In the former of these two texts, the techniques described here are extended and applied to the key molecules of inheritance and metabolism namely the nucleic acids and proteins. The study of these molecules is, of course, vital to understanding how organisms function. In addition the development of the associated enabling technologies (genetic engineering and protein technology) demand that we are equipped with the techniques needed to isolate and analyse the relevant molecular species.

In 'Analysis of Carbohydrates and Lipids' an analogous approach is used. Thus we extend and apply the techniques described in this text to these groups of molecules. These molecules are of fundamental structural and metabolic importance to living systems. The production of these molecules also have great practical significance. For example specific carbohydrates and lipids have particular value in food production and formulation, some are useful as therapeutics while others have properties such as the ability to form gels, which make them useful in a wide variety of industrial processes. There are, therefore compelling academic and practical reasons for acquiring and developing techniques for isolating and characterising these important molecules.

1.8 Pre-requisite knowledge

The text has been written on the assumption that the reader has encountered the major molecular species found in biological systems and has some general knowledge about the structure of cellular material and the occurrence of the major molecular species within sub-cellular structures. Although these assumptions have been made, many helpful reminders have been incorporated into the text. Nevertheless the central theme of this text remains a practical one. If your knowledge of biological chemistry and cell biology is scant we recommend you read the BIOTOL texts 'The Molecular Fabric of Cells' and 'Infrastructure and Activities of Cells' before embarking on this text.

You will have realised from reading this chapter that this text predominantly focusses onto the principles underpinning techniques. We have not included detailed discussion of general laboratory 'manners' and safety or explored the principles underpinning the design of experiments or examined the statistical evaluation of data derived from laboratory measurements. These aspects of laboratory performace are, in the view of the author: editor team, of great importance but fall outside of the scope of this text. For further details of particular techniques we refer you to the suggestions for further reading provided at the end of this text.

Summary and objectives

This chapter has been designed to orientate the reader. It begins by explaining the academic and practical reasons for studying biomolecules and indicated that the study of these entities depended on the availability of suitable preparative and analytical techniques. We then described the strategy and structure adopted in this text and explained the reasons for inclusion of particular sections. In the final part of the chapter we explained the relationship of this text with other BIOTOL books and indicated the nature of the knowledge-base readers are assumed to have before studying this text. Now that you have completed this chapter you should be able to:

- judge whether or not you will be able to cope with the text;
- recognise the value of understanding a wide variety of techniques to being able to select and use these techniques in an extensive range of circumstances;
- understand the sequential approach that has been applied to the material covered by this text.

Measuring biomass and cell growth

Measuring biomass and cell growth

2.1 Introduction

The estimation of biomass is of fundamental importance to a wide variety of scientific and commercial activities and processes. In circumstances where biomass itself represents the desired end product of the process (for example in the production of single celled protein and in agricultural production), it is clearly essential to be able to determine the amount of biomass produced. In a wider sense, the measurement of the amount of biomass present is necessary to quantify physiological studies in biology. For example, it provides a reference criterion by which different biological systems can be compared. It is also essential in monitoring the progress of cultures used for the production of a wide variety of end products such as antibiotics and enzymes.

There are a great variety of methods available for determination of biomass. The choice of methods is influenced by a number of factors. These include:

- the accuracy of the data that is required;
- the speed with which the result is needed. For example measurement might be used for on-line control of media addition to continuous cultures in which case the result would be required almost instantaneously. In contrast, comparing biomass yields of different cultures of organisms might be conducted over a longer time scale;
- the use the measurement data will be put to. For example, in genetic studies with micro-organisms, the number of cells rather than the mass of the biological material is often required. Thus in such cases the estimation of biomass in terms of cell numbers is important. In contrast, in the production of single celled protein, the mass of the material present and more specifically, the mass of protein present are the key features that are measured.

We can, of course, add to this list the other more general factors, such as the cost and availability of equipment, which influence the choice of methods used in any determination.

In this chapter we will describe different approaches used to estimate biomass and cell growth and discuss the advantages and limitations of these methods. You will learn that the choice of method adopted for a particular study is a crucial decision to make and that the limitations of the methods available make the decision less than clear cut. Nevertheless, the discussion presented in this chapter should enable you to make sensible choices.

The chapter begins by discussing general issues concerning precision and accuracy and of sampling. It then examines a variety of methods. We have not, however, attempted to include all of the methods of biomass estimation. With experience, you will learn that there are many variations based on the main methods. For example, the temperature selected to dry material in dry weight determinations is often different in different laboratories. We will, however, give some guidance as to the factors which influence the specifics of the procedures described. We will predominantly focus on the methods which have been applied to unicellular systems grown in suspension because of the predominant importance of these systems in biotechnology. Many of these methods, however, can be adapted for the determination of biomass of filamentous and multicellular systems. We also describe briefly techniques that can be employed with animal and plant cell cultures especially when we only have limited amounts of material. These techniques have applications for tissues as well as with simple cells and can be applied in histopathology (for example for measuring tumour growth).

2.2 General problems of sampling, accuracy and precision

2.2.1 Sampling

representative samples

As with all measurements, it is essential to take representative samples of the material under study. Biomass is particulate and even with the smallest unicellular systems in suspension, there is a tendency for the cells to settle out. This means, of course, that if suspensions are not well mixed just prior to taking the samples to be assayed, there will be uneven distributions of material within the suspensions. The consequence is that, irrespective of the method used for the assay, the result will not accurately reflect the true biomass concentration in the suspension. A key step is therefore to ensure that the suspension is thoroughly mixed prior to sampling. This problem is of course even more acute with suspensions of large cells and flocs. It is preferable to take multiple samples which will enable detection of uneven sampling and reflect the precision of the method being used to determine biomass.

Essentially, the same principle applies to multicellular and filamentous systems. For example, determining the mass of a single plant as an estimation of the mass of a population of plants begs the question, does the plant selected for the estimation represent the mean of the plant population under study? We will not specifically examine the estimation of plant biomass here since this is dealt with elsewhere in the BIOTOL series ('Crop Productivity').

sample size

A second key issue in sampling is the size of the sample to be taken. In practice this largely depends upon the sensitivity and precision of the method being used to determine biomass. Some methods, such as cell counting, require only very small samples. Others, such as methods relying on dry weight determination, require significantly larger samples. The size of the sample required is also influenced by the concentration of biomass present. For example, estimations of biomass based on cell counts of a culture of yeast requires only small samples. In contrast, estimates of cell counts in tap water requires substantially larger samples. Decisions concerning the size of the samples to be taken can only be made, therefore, based on knowledge of the sensitivity of the method, the approximate amount of biomass present and the degree of accuracy required. By the completion of this chapter you should be in a position to make such decisions.

number of samples

The final aspect of sampling we need to consider is the number of samples to be taken. This is influenced by the level of accuracy required. Clearly, the more samples taken and thus the more measurements made, the greater the anticipated accuracy of the determination. However, there is a cost attached to the taking and assaying of each sample. Thus, in practice, the number of samples taken is usually a compromise between the desired accuracy of the determination and the cost incurred in carrying out the determination. Judgements about the number of samples to be taken reflect the actual circumstances. For example, if the estimation of biomass is only needed to determine whether the biomass concentration is above a particular value, a single determination may only be required. On the other hand, if the amount of biomass present needs to be determined accurately then it is common practice to assay at least 3 samples.

2.2.2 Accuracy and precision

There are many common misconceptions about the terms accuracy and precision. Although the terms are related, they are quite distinct and it is worth a few moments thought to ensure that you can distinguish between them. This is particularly important because many of the methods used in biomass estimations are bedeviled by problems of precision and accuracy.

Π For each of the targets shown, describe (good/poor) the accuracy and the precision of the bullet holes.

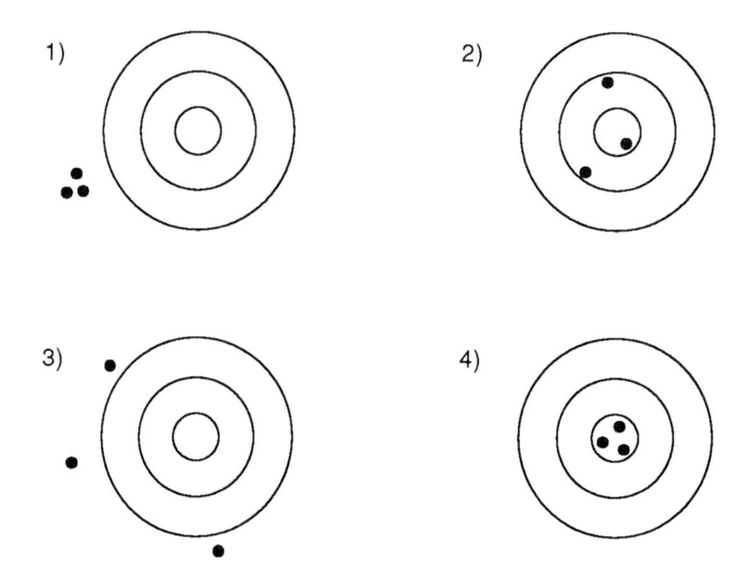

You should have concluded the following:

1) Poor accuracy, good precision. 2) Good accuracy, poor precision. 3) Poor accuracy, poor precision. 4) Good accuracy, good precision.

In principle, therefore it is possible to generate methods which are very precise but quite inaccurate (see example 1 above). Ideally we would prefer methods which are both accurate and precise (see example 4 above) but in practice we often have to settle for methods which may provide good accuracy but rather poor precision (example 2 above). It is with methods which are of rather low precision but potentially good accuracy that we require to assay multiple samples in order to be confident of the accuracy.

Π In which of the situations described in the previous in text activity can we take few (one) sample and still be confident that the result is accurate.

You should have identified situation 4. Any one of the samples (represented by •) gives an accurate measurement. In none of the other situations can we be confident that the result is an accurate one.

2.3 General division of the methods for estimating biomass

We have already mentioned that there is a great diversity of methods available for estimating biomass reflecting the various requirements of the users of the data. Biomass is a complex mixture of biochemicals arranged into cells with a propensity to change. It is these features which have enabled the development of a wide variety of techniques and at the same time, generated uncertainties in the various measurements that are made.

The most commonly used methods for biomass estimation especially for single celled suspensions are the measurement of cell mass, cell number or packed cell volume. We can perhaps regard these as general methods of biomass estimation for they make no reference to the chemical composition of the biomass. Many of the other methods depend upon determining specific components of the biomass, for example its protein content, DNA content or the amount of a specific enzyme. Metabolic activity (eg substrate uptake, CO_2 evolution) can also be used as a method of determining biomass. We can represent this situation as shown in Figure 2.1.

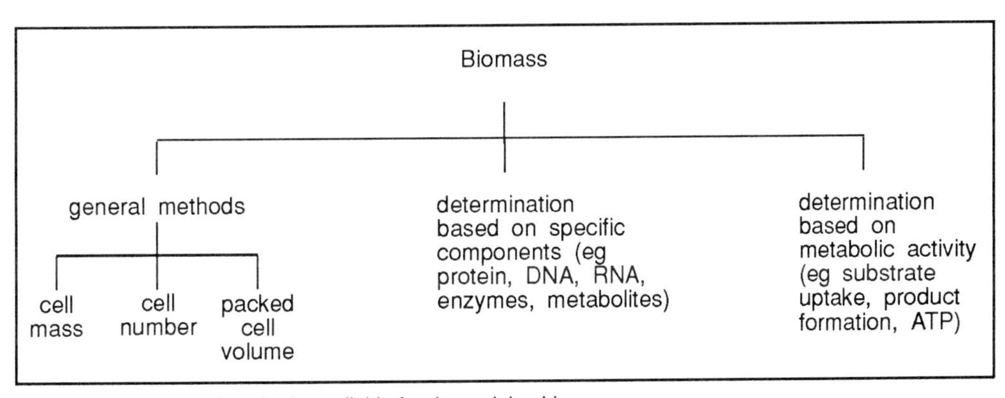

Figure 2.1 Diversity of methods available for determining biomass.

The methods based on the determination of specific components or on metabolic activity often make the assumption that there is a direct correlation between them and the total amount of biomass. This is not always valid, for example, many cells change their chemical composition or metabolic activities in different media. Thus the RNA content of cells as a % of total dry weight can vary dramatically depending on the cultivation conditions. Therefore extrapolation of data on RNA content to calculate total biomass is likely to lead to substantial error unless the exact composition of the cells is known for the various cultivation conditions encountered. For this reason it is safer to report biomass concentrations based on the determination of cell constituents in such a manner that the basis of the estimation is known. For example the biomass concentration could be reported in terms of g.cellular RNA ml^{-1} or as g.cellular protein ml^{-1} (or cm^{-3}). Note that in this text we will use SI Units ($1 cm^3 \equiv 1$ ml and $1 dm^{-3} \equiv 1 l$.)

We can add a further dimension to the methods illustrated in Figure 2.1. The general methods listed in the figure can be further divided into direct and indirect methods. For example, cell mass may measured by directly weighing or indirectly using some property of the cell mass as the method of measurement. For example, cells will scatter certain wavelengths of light (that is suspensions of cells are turbid). The more dense the suspension of cells, the greater the turbidity. We can therefore in principle use turbidity as a measure of cell density and therefore, cell mass.

In the following sections we will first examine the more generally applicable methods of biomass determinations before moving on to the methods based on determination of specific components and determination of metabolic activity.

2.4 General methods of determining biomass

2.4.1 Measurement of cell mass

Cell mass measurements can be performed by direct or indirect techniques. The direct procedures are to determine the dry weight or the wet weight of cell material. The indirect methods usually depend upon light scattering and involve the determination of turbidity or the use of nephelometers.

Dry weight determination

dry weight

Dry weight measurement involves three stages;

- separating the organism from the medium;
- washing the cells;
- drying the biomass.

Organisms may be separated from the medium by filtration or by centrifugation. Washing the biomass should then be carried out in such a way as to prevent lysis of the organism through rupture or osmotic shock. This may occur if the biomass is washed with water, especially when the organisms are taken from a rapidly growing culture. The precaution against lysis is to wash with a near isotonic saline and allow for the dry weight of the salt present after drying. Usually biomass is dried at 80°C for 24 hours or at 110°C for 8 hours. With modern equipment, such as a microwave oven for drying, it is possible to determine the dry weight of the biomass within thirty minutes.

Note however that a wide variety of temperatures have been used. Obviously if a high temperature is used this reduces or prevents continued metabolism by the biomass (and thus loss in mass) during the drying process. This is especially important with biomass with high metabolic rates (for example bacteria). On the other hand, using high temperatures drives off volatile cell products and this results in loss of mass. The choice of temperature used for drying is therefore a compromise and because of the loss of material either through metabolism or by volatilising cell material there are usually some inaccuracies in the determination.

Usually the procedure involves a repeated cycle of weighing and drying until a constant weight is obtained. On removal of the material from the drying cabinet, it is important to cool the sample in a desiccator prior to weighing because the dried biomass is hygroscopic and will begin to take up water from the atmosphere.

The main limitation of a dry weight measurement is that it is relatively insensitive and inaccurate. With routine laboratory equipment it is difficult to weigh with accuracy less that 1 mg, yet this dry weight may represent as many as 5 billion (5×10^9) bacteria. To get a reasonably accurate result we usually need to use about 50mg of cells or more.

Π List circumstances where sample dry weight measurement is not suitable for the estimation of biomass.

The sort of circumstances we hoped you would think of are:

- small sample size;
- medium containing an indeterminate amount of other solids besides the biomass;
- result required in a matter of minutes.

Wet weight determination

wet weight

Wet weight determination also involves the separation of cells from the medium and washing. The weight of the wet biomass is then determined directly. The wet weight will include both intracellular and extracellular water. In the bacterium *Escherichia coli*,for example, the extracelluar water volume of close-packed cells has been estimated to be about 10% of total volume and water represents 75% of the total cell weight. Obviously the proportion of extracelluar water depends on the shape and size of cells.

For wet weight measurements to be of any use the centrifugation or filtration method used to pack down the biomass must be carefully standardised, since this influences the extracellular water content. We discuss centrifugation in greater detail in Chapter 4.

Π List an advantage and a disadvantage of the wet weight method of cell mass determination compared to the dry mass method.

The advantage we hoped you would list is that it is quicker since we do not have to wait until the material has dried out, but it has the disadvantage of being susceptible to greater inaccuracy and poorer precision since the determination depends critically on how the cells were collected.

Turbidity and nephelometry

Techniques of determining cell mass that are more rapid and sensitive are based upon the fact that cells scatter light. Because cells in a population are of a roughly constant size, the amount of scattering is proportional to the concentration of cells present. Using bacteria as an example, when the concentration of bacteria reaches about 10 million (10^7) cells cm^{-3}, the medium appears slightly cloudy or turbid. Further increases in concentration results in greater turbidity and less light is transmitted through the medium. The extent of light scattering can be measured using a spectrophotometer and is almost linearly related to bacterial concentration at low concentration and low absorbance levels. Absorbance (A) is defined as the logarithm of the ratio of intensity of light striking the suspension (Io) to that transmitted by the suspension (I).

$$A = \log \frac{Io}{I}$$

Light scattering by a cell suspension is illustrated in Figure 2.2.

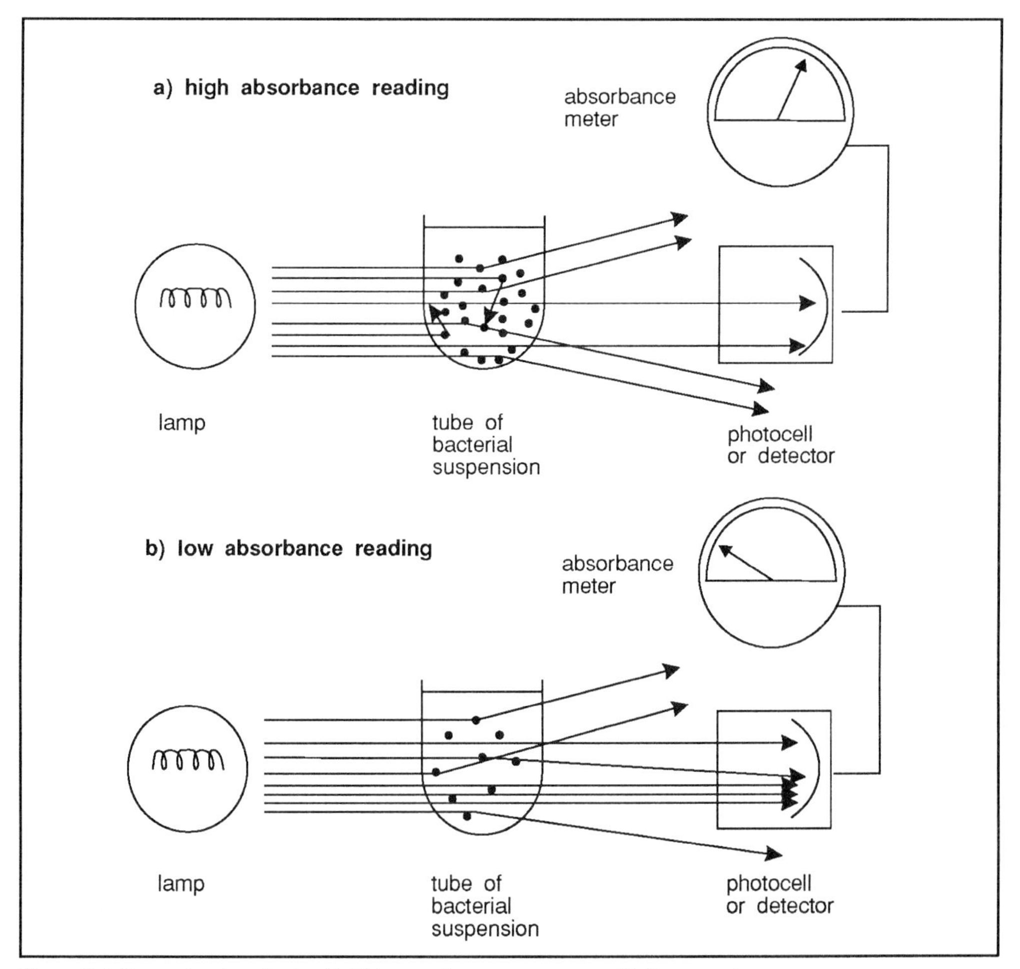

Figure 2.2 Determination of microbial biomass by measurement of light scattering.

In order to use the turbidity of a culture to measure cell mass we need to have a relationship between cell mass and turbidity. This is achieved by preparing a standard curve by measuring the turbidity (absorbance) of several dilutions of culture and determining the dry weight of cell material present. Then a calibration curve similar to that displayed in Figure 2.3 can be prepared.

You should realise that a calibration curve prepared for one organism may not be appropriate for another.

Π List the reasons why a calibration curve for absorbance against dry weight of the type illustrated in Figure 2.3 prepared for the bacterium *Escherichia coli*, may not be appropriate for the yeast *Saccharomyces cerevisiae*.

The amount of light scattered by a cell suspension depends not only on the amount of cell mass present but also on the size and shape of the cells. Yeast and bacteria are quite different in shape and size and therefore their abilities to scatter light are quite different. Thus although both cultures will produce calibrations of the same general shape as that illustrated in Figure 2.3, the actual position of the curves will be different in the two cases.

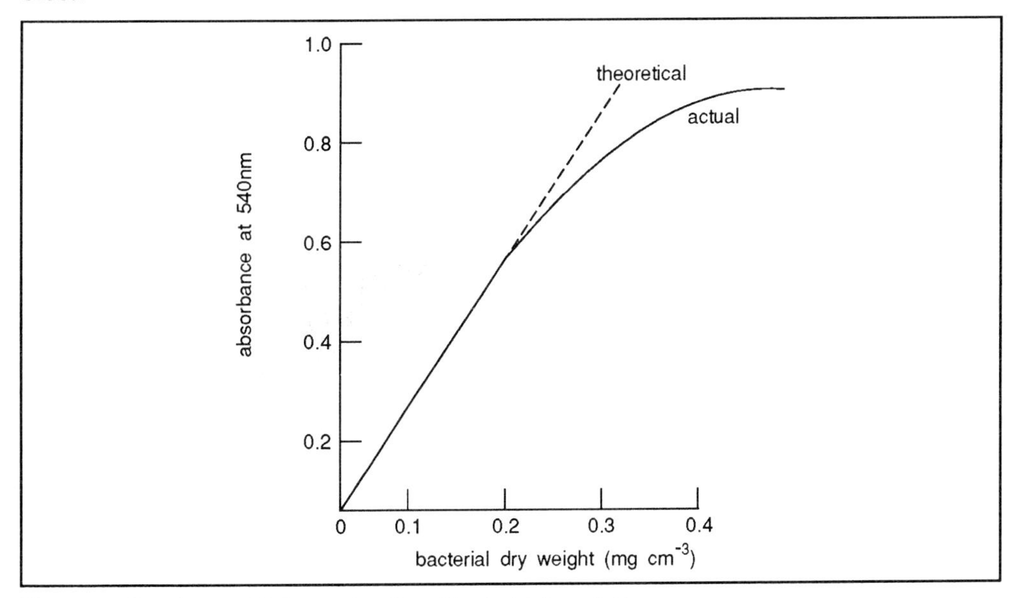

Figure 2.3 Absorbance of light as a function of bacterial dry weight.

Π In Figure 2.3 we show that the relationship between absorbance and dry weight becomes non-linear at relatively high dry weight values. Write down an explanation for this.

As the number of cells in a suspension increases a greater proportion of cells do not contribute to light scattering. We can represent this in the following way:

At low cell density we can represent this as:

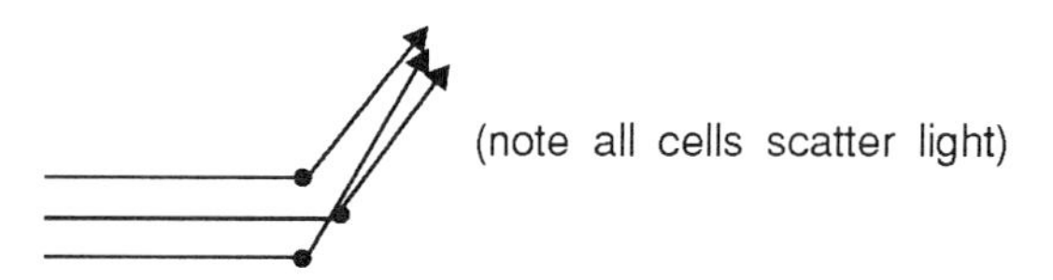

At high cell densities we can represent this as:

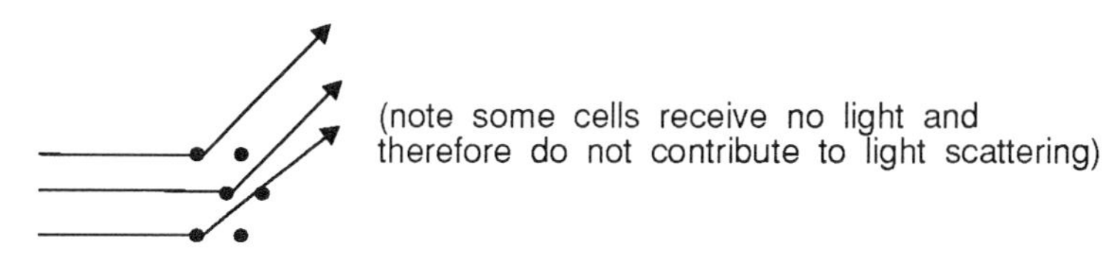

In fact, the relationship between absorbance (turbidity) and cell mass follows the Beer-Lamberts type relationship:

$$A = \alpha [X] - \beta [X^2]$$

where A = absorbance, α and β are constants and [X] = biomass concentration. The values of α and β are characteristic of each culture and α is numerically much larger than β.

You should note that we could also plot a graph of absorbance against cell number. For this we would need to count the number of cells present in a range of dilutions and measure the turbidity of these dilutions. We will deal with counting cells in a later section.

Π Examine the two curves in the following figure. They are plots of turbidity against cell dry weight cm^{-3} for two samples taken from a culture of *Bacillus subtilis*. Sample A was taken from the culture after it had been incubated for 10 hours, after inoculation, sample B was taken from the culture when it had been incubated for 25 hours. Can you explain why the two samples give a different relationship between absorbance and cell dry weight? (Assume that the culture is not contaminated).

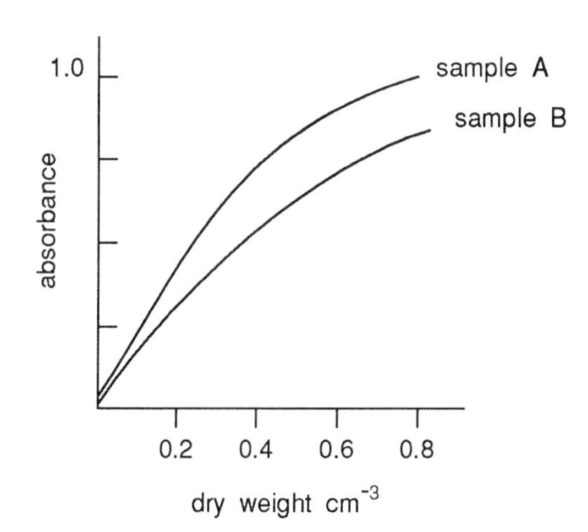

The most likely explanation is as follows.

The absorbance of a suspension depends not only on the amount of biomass present but also on the size and shape of the cells. The explanation for the above result is that the cells in the culture had changed their morphology between 10 and 25 hours in culture. In all probability at 10 hours they were all vegetative cells, while at 25 hours the cells had formed spores.

A key question in using this approach to determining biomass is the choice of wavelength(s) of light to use. The standard curve showing in Figure 2.3 was obtained using a wavelength of 540nm. Usually wavelengths of between 420 and 650nm are used to measure cell density. The sensitivity of the measurement increases sharply if light of a shorter wavelength is used but, at lower wavelengths specific cell components, such as pigments, may absorb light and interfere with cell density measurement.

Another practical consideration is the question of absorbance by the culture medium. This should be measured and subtracted from absorbance values for the sample (culture medium plus cells). Alternatively, the spectrophotometer is calibrated so that it reads zero for medium not containing cells. Note, however, that the intensity of the colour of the medium may change as nutrients are removed from the medium during growth. Thus we must use 'spent' media which truly reflect the situation in the culture if we are not to produce inaccuracies in our procedures. A simple way to achieve this, is to measure the absorbance of the culture sample then quickly centrifuge the sample to remove the cells and then re-measure the absorbance of the clarified media.

Π From the data presented in Figure 2.3, what is the sensitivity of this method of determining biomass?

Essentially this depends on how accurately the spectrophotometer can be read. We would estimate that about 0.1 mg dry weight of cells cm^{-3} are required to obtain a reasonable absorbance. Thus if we use a lml curvette in the spectrophotometer, we need to use about 0.1mg dry weight of cells. This gives some idea of the sensitivity of the method.

A problem with this procedure is that at low cell densities we cannot measure absorbance accurately. For example if 100% of the incident light is transmitted through media containing no cells then, when the cell density is very low then virtually all of the light is still transmitted. Thus the instrument is trying to detect the difference between 100% and almost 100% transmission. A small % error in each of these reading will of course, result in a large error in the measured absorbance.

More sensitive instruments for measuring light scattering at low cell densities are called nephelometers. Look back to Figure 2.2. Nephelometers are similar to spectrophotometers except the photocells (detector) are arranged at an angle to the incident light. Thus when no light is scattered, the photocell receives no signal. This can be illustrated by:

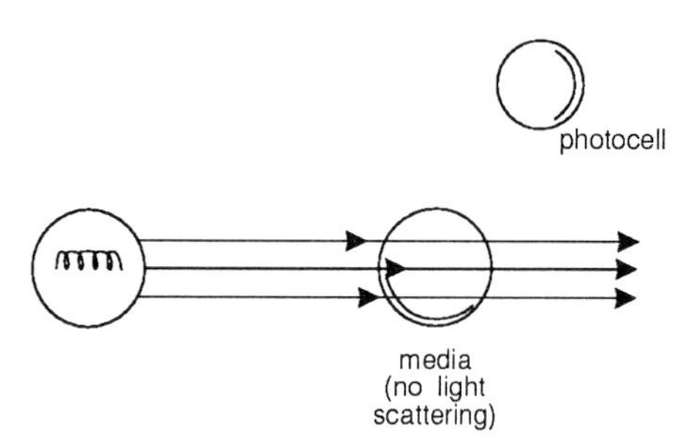

The presence of a few cells and hence some light scattering results in photocell receiving light. Thus

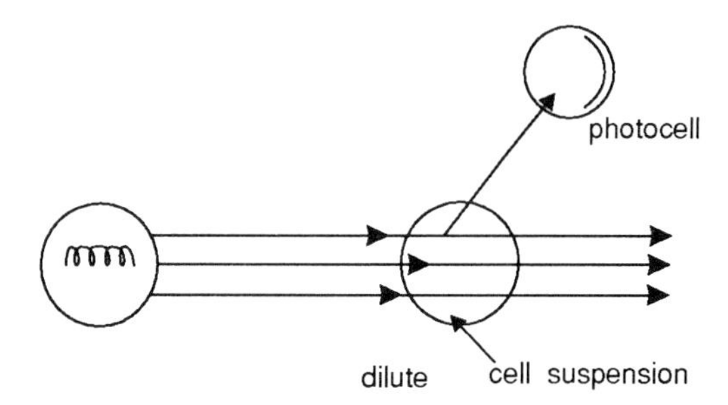

It is easier to measure accurately the difference between a zero and small positive signal than it is to measure the difference between two large values. Nephelometers, therefore, find favour in systems where the cell intensity is low. Of course, in most circumstances we are interested in producing high cell densities and normally spectrophotometer determinations of turbidity are accurate and sensitive enough.

SAQ 2.1

The dry weight of a bacterial culture was measured by weighing. The culture was also diluted in culture medium and absorbances were recorded. The results are shown below:

Dry weight of predried filter = 2.7656g

Dry weight of filter plus cells from $10cm^3$ sample = 2.8456g

	Absorbance at 540nm
Medium minus cells	0.060
Undiluted culture	Infinity
Culture dilution:	
1 in 10	1.050
1 in 20	0.830
1 in 25	0.720
1 in 40	0.470
1 in 70	0.290

1) Plot a graph relating dry weight to absorbance and comment on how this calibration curve could be improved.

2) Determine the dry weight concentration of a culture giving an absorbance (at 540nm) of 0.560 when diluted five-fold in culture medium.

2.4.2 Measurement of cell number

direct microscopic counting

A common way to determine total cell counts is by direct microscopic counting. The small size of most cells and their large population densities make it necessary to use special chambers such as Petroff-Hauser or haemocytometer chambers to count the number of cells in a sample. These specially designed slides have chambers of known depth with an etched grid on the chamber bottom (Figure 2.4). Haemocytometer chambers were designed to count red blood cells. They can, however, be used for a much wider variety of cells.

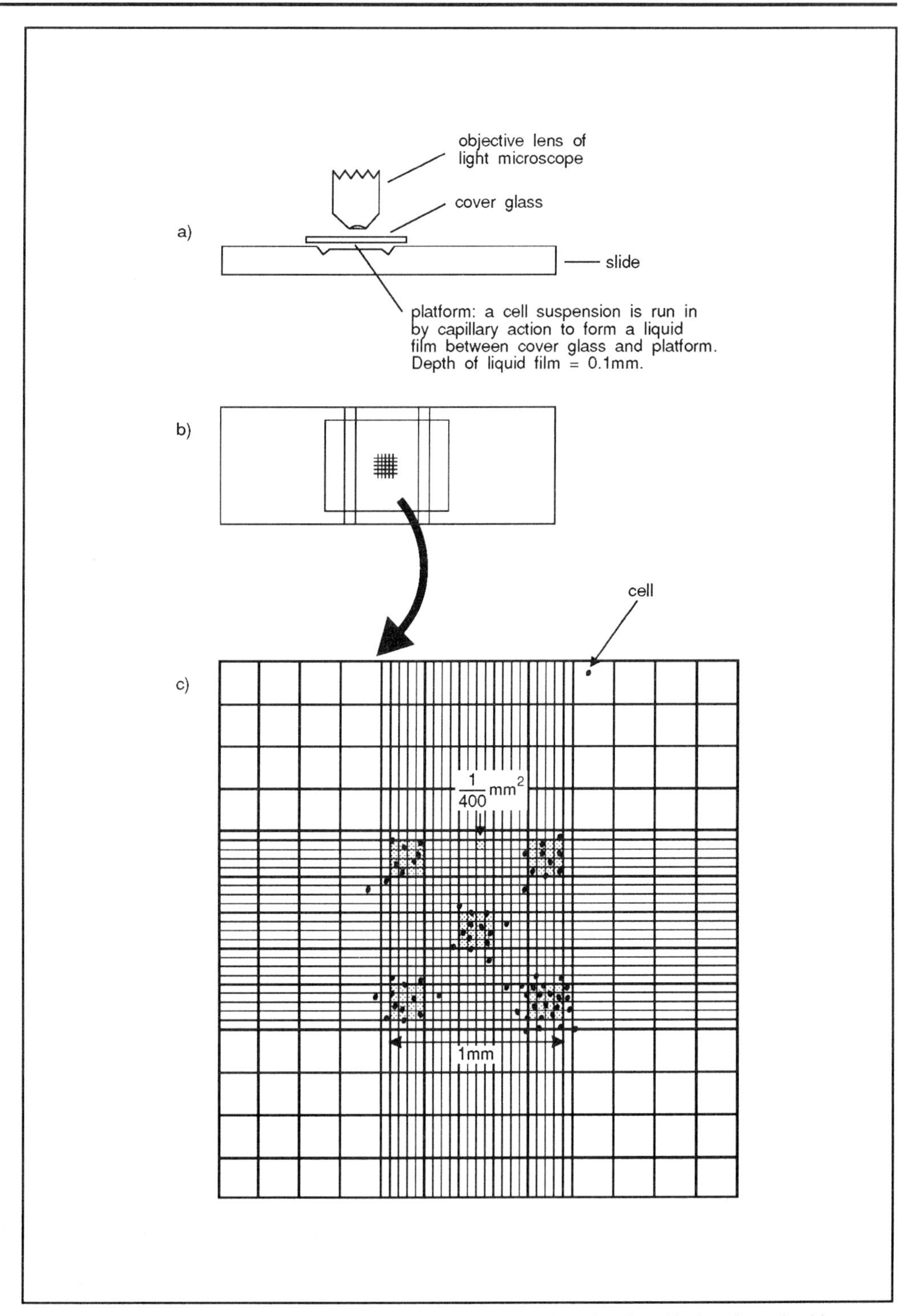

Figure 2.4 A haemocytometer chamber. a) side view of the counting chamber, b) plan view, c) grid magnified. Note we have not put cells evenly over the grid; this is to make an intext activity simpler. Normally, of course, we aim to have cells evenly distributed across the whole grid.

Π Each small square in the grid shown in Figure 2.4 has the dimensions 0.05 x 0.05mm and when the coverslip is applied, the counting chamber has a depth of 0.1mm. What is the number of cells cm^{-3} in suspension in this sample? Hint - we count the cells in the four grids in the corners and the central grid.

An important question to answer in this determination is to decide whether or not to count cells which appear on the lines (edges) of the grid. The easiest way is to decide before you start to count the cells on two of the lines (for example the top and left hand edges) and not on the other two. Thus in counting cells we are effectively removing the problem of deciding whether or not the cells at the edge of the grid are more, or less, than half inside of the grid.

Using this approach our calculation is as follows

number of cells in top left hand grid =	7
number of cells in top right hand grid =	6
number of cells in bottom left hand grid =	6
number of cells in bottom right hand grid =	11
number of cells in central grid =	8
Total cells =	**38**

$$\text{Volume of small square} = \frac{1}{400} \times \frac{1}{10} \text{ mm}^3$$

$$\text{Volume of large square (containing 16 small squares)} = \frac{16}{400} \times \frac{1}{10} \text{ mm}^3$$

$$\text{Volume of 5 large squares} = \frac{16 \times 5}{400 \times 10} \text{ mm}^3$$

$$\text{Therefore } \frac{16 \times 5}{400 \times 10} \text{ mm}^3 \text{ contains 38 cells}$$

$$\text{and} \quad 1 \text{ mm}^3 \quad \text{contains} \quad \frac{38 \times 400 \times 10}{16 \times 5} \text{ cells}$$

$$= 1.9 \times 10^3 \text{ cells}$$

The culture therefore contains 1.9×10^6 cells cm^{-3}.

You may have found it quite difficult to decide if each cell was inside of a grid from our drawing. This reflects the practical situation. Small cells, especially bacteria, are only just visible in the counting chamber under high power magnification and counting is not straightforward and can be tedious. Also the presence of other particulate material can result in counting errors.

How accurate are the data generated using the counting chamber? Let us assume that we can clearly identify cells.

For a reasonable count a minimum of five cells per large square is required. Thus we need a cell density of about 1.2×10^6 cells cm^{-3}.

accuracy of direct counting

The accuracy of this method depends upon producing an evenly distributed suspension of cells. This demands that the chamber is clear and free from contaminants such as oil. Since the distribution of cells is random, obviously the more cells that are counted the more accurate the determination is likely to be. This can of course be determined statistically, but a good simple approximation is that the likely accuracy is reflected by the relationship $N \pm \sqrt{N}$ where N is the number of cells counted. This rule of thumb is quite satisfactory in getting an approximate idea of the likely accuracy of a cell count. Thus if 49 cells are counted the likely accuracy is ± 7 (that is the error could be of the order of 14%).

Π What is the likely error in a similar determination if 4900 cells are counted?

Since $N \pm \sqrt{N}$ then 4900 ± 70, the error is nearer 1.4%.

There are many minor adaptions to the general procedure described above. A common practice is to stain translucent cells prior to placing them in the counter chamber. This makes them easier to see and therefore easier to count. It also often allows us to distinguish between cellular material and non-cellular solids (eg dust particles) suspended in the culture.

In circumstances where cells tend to clump, a variety of procedures and additives can be used to assist with the separation of the cells. These procedures range from simple agitation to the addition of surfactants. In the case of plant cells, dilute chromic acid is frequently used. Care must be taken, however, to select a procedure which leads only to the separation of cells and not to their disruption.

vital staining (viability)

Direct microscopic counting, by the method described, gives a total cell count since it does not distinguish between viable (living) and dead cells. However, if direct microscopic counting is combined with vital (viability) staining it is often possible to determine the proportion of viable cells in a population. There are several types of viable stains which work in different ways. For example, methylene blue is reduced to a colourless form by cells capable of respiring, while dead or non-respiring cells will stain blue. For yeast cell suspensions eosin is often used. Viable yeast cells are impermeable to this dye whereas dead cells stain red.

electronic counters

Larger cells can be counted directly with electronic counters, such as the Coulter counter. Here, the cell suspension is forced through a small hole or orifice which has an electric current flowing across it. Every time a cell passes through the orifice electrical resistance increases (or conductivity drops) and the cell is counted. Obviously to obtain an accurate count the orifice must be small enough to ensure that only one cell passes through at a time. The main advantage of the Coulter counter is that it is not tedious and gives accurate results within a few minutes for larger cells. The instrument has the added benefit of being able to estimate the size of individual cells from the change and duration of the drop in conductivity. However, a limitation of the coulter counter is that it cannot often be used to count bacteria. This is because for bacterial counting problems arise due to the requirement for a small orifice. Small non-biological particles in the samples, cell clumping or chains of bacterial cells can lead to inaccurate cell counts or to even blocking of the orifice. Obviously in cell systems in which the cells have a tendency to clump together or to grow as aggregates there is a likelihood that cell numbers will be underestimated. Because of these difficulties and the cost of the equipment, electronic particle counters have not found widespread application.

Viable cell counts - pour and spread plates

Often we are interested not in determining the total cell mass or number, but need to know the number of living (viable) cells present in a suspension. The determination of cells by viable counting is based on the assumption that each viable cell in a suspension will give rise to a single colony after incubation on a suitable medium under favourable conditions. After incubation the number of colonies formed is counted and used to estimate the number of viable cells in the original suspension. This technique is predominantly applicable to unicellular micro-organism.

The viable count is considered to be a minimum count because the number of colonies on the plate represents only those cells that can multiply under the conditions that have been established. In addition, not all cells give rise to a colony because certain cells have a tendency to clump or aggregate. When plated onto a suitable culture medium a clump will give rise to only one colony, regardless of how many cells are in the clump. For this reason results are usually expressed as colony forming units (CFU) cm^{-3} rather than cells cm^{-3}.

spread-plate
pour-plate

Viable counts are usually carried out by either the spread-plate or the pour-plate technique. In the spread-plate technique, a small sample (usually 0.1 ml) is aseptically spread over the surface of an agar plate containing an appropriate medium. In the pour-plate technique, the sample is mixed with melted agar and the mixture poured into a sterile plate (Figure 2.5). The organisms are thus fixed within the agar gel and form colonies. With pour plates, larger sample volumes can be used, and heavy slurries or suspensions can be used. A potential source of error with pour-plates is that organisms may be inactivated by the brief heating in the melted agar (around 48°C) and will not grow and therefore are not counted.

Usually, only plates containing 30 to 300 colonies are counted because this will enhance the accuracy of the count. If the sample contains too many micro-organisms (many more than 300 per plate) the colonies will be too crowded and colonies will grow into each other to form more-or-less confluent growth. This makes the individual colonies impossible to count. For this reason the samples are usually diluted. If the sample is too diluted, the result will not be statistically valid.

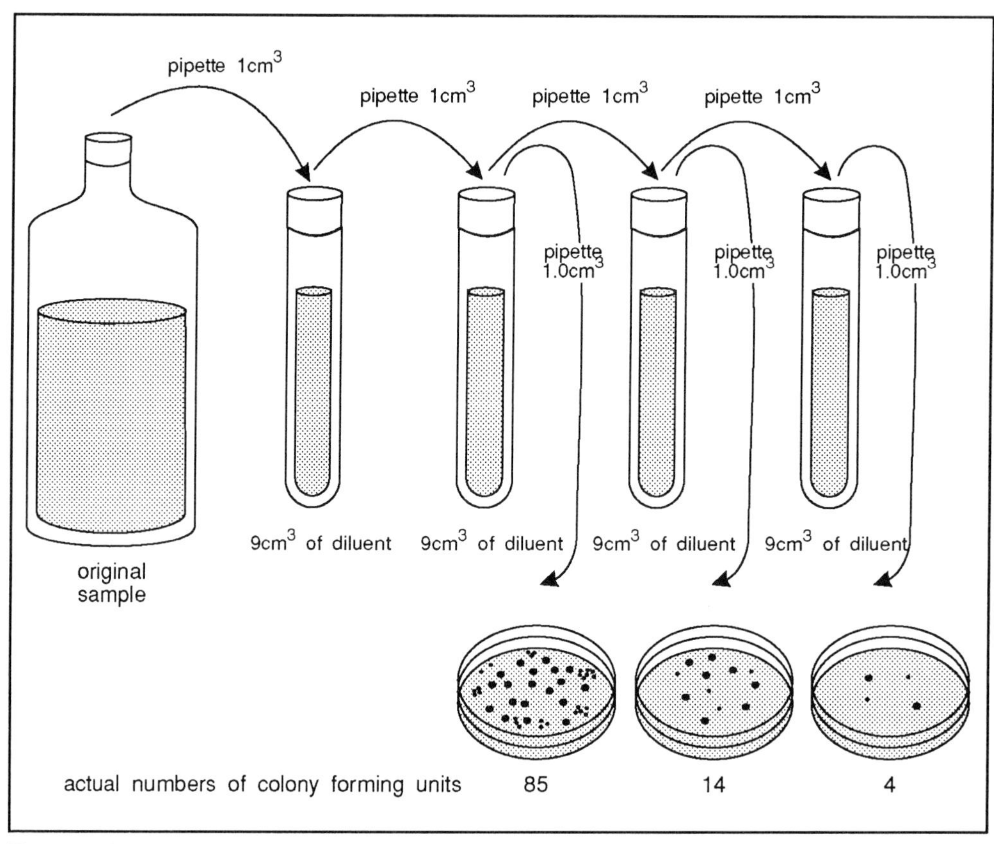

Figure 2.5 The pour-plate technique.

A problem associated with viable counting by the spread-plate and pour-plate technique is the need for a long incubation time. It is usual for plates to be incubated for 1 to 2 days before colonies are counted. There is, however, an adaption of these methods which can speed up the determination using a slide culture. A sample of the population, appropriately diluted, is spread over a thin layer of agar medium on the surface of a sterile glass slide. The inoculated slide is then incubated for a period sufficient for several cell divisions and is examined microscopically. Under these conditions, the viable cells develop into micro-colonies and are readily counted within a few hours. This method is not however widely used.

slide culture

SAQ 2.2

Which of the following are possible sources of error in viable cell counting using the pour-plate technique.

1) pipetting errors;
2) cell clumping;
3) loss of viability due to the temperature of the agar;
4) medium unable to support growth of micro-organisms;
5) loss of viability in diluent;
6) inadequate mixing of dilutions;
7) carry over of cells from low dilution to high dilution.

SAQ 2.3

From the information presented in Figure 2.5, determine the number of CFUs cm^{-3} in the original sample.

SAQ 2.4

The numbers of colony forming units in two growing bacterial cultures were estimated by the spread plate technique using 0.1 cm^3 volumes of suitable dilutions of the cultures. With culture A, aliquots of 10^{-6} dilution of the culture were spread onto 3 plates. After incubation these gave rise to 126, 93 and 81 colonies.

With culture B, aliquots of a 10^{-8} dilution of the culture were spread onto 3 plates. After incubation, these gave rise to 102, 91 and 107 colonies.

Calculate the number of CFUs cm^{-3} in cultures A and B.

Most probable number techniques

This technique is adopted especially for systems in which the biomass (cell number) is extremely low.

most-probable-number

Cultures containing less than 1 cell cm^{-3} can be used. This technique uses only liquid culture for growth and therefore can be used to count cells which will not grow on any agar medium, requiring liquid culture for growth. At these low levels, results are expressed in terms of probability of finding a cell rather than in terms of the cell concentration. For example, if there are 5 cells in 10 cm^3 of liquid sample the concentration could be expressed as 0.5 cells cm^{-3} . However, because cells are individuals we say that there is a probability of 0.5 that each 1 cm^3 portion of the sample would contain a viable cell. Thus if we had 10 cm^{-3} of sample and we divided this up into 1 cm^3 portions, we might anticipate 5 would each contain one viable cell, and 5 would be expected not to contain a cell. Of course, in practice we might find one of the 1 cm^3 portions contains more than one viable cell. The cells are after all randomly distributed in the sample. The presence or absence of viable cells can be analysed statistically. This is the basis of the most probable number (MPN) technique.

In the MPN technique replicate portions of different volumes (eg 0.1, 1.0 and $10cm^3$) are inoculated into separate tubes containing fresh growth medium. After incubation, tubes which received a viable cell will show growth (cloudiness or turbidity) and those which did not, would not show growth. The number of positive tubes (those showing growth) in each replicate can then be used to determine the MPN of cells in the sample by means of MPN tables. There are many versions of these but the most common of these statistical tables are constructed for use with three, five or ten replicate growth tubes. Three replicate tube MPN tables are shown in Table 2.1. Greater reliability is obtained by using a large number of replicate tubes but at the expense of more time and materials in conducting the measurement.

Tubes positive				Tubes positive				Tubes positive			
10 cm^3	1.0 cm^3	0.1 cm^3	MPN	10 cm^3	1.0 cm^3	0.1 cm^3	MPN	10 cm^3	1.0 cm^3	0.1 cm^3	MPN
0	0	1	3	1	2	0	11	2	3	3	53
0	0	2	6	1	2	1	15	3	0	0	23
0	0	3	9	1	2	2	20	3	0	1	39
0	1	0	3	1	2	3	24	3	0	2	64
0	1	1	6	1	3	0	16	3	0	3	95
0	1	2	9	1	3	1	20	3	1	0	43
0	1	3	12	1	3	2	24	3	1	1	75
0	2	0	6	1	3	3	29	3	1	2	120
0	2	1	9	2	0	0	9	3	1	3	160
0	2	2	12	2	0	2	14	3	2	0	93
0	3	0	16	2	0	2	20	3	2	1	150
0	3	0	9	2	0	3	26	3	2	2	210
0	3	1	13	2	1	0	15	3	2	3	290
0	3	2	16	2	1	1	20	3	3	0	240
0	3	3	19	2	1	2	27	3	3	2	460
1	0	0	4	2	1	3	34	3	3	2	1100
1	0	1	7	2	2	0	21	3	3	3	1100+
1	0	2	11	2	2	1	28				
1	0	3	15	2	2	2	35				
1	1	0	7	2	2	3	42				
1	1	1	11	2	3	0	29				
1	1	2	15	2	3	1	36				
1	1	3	19	2	3	2	44				

Table 2.1 Most-probable-number tables.
MPN = most probable number of cells per 100ml using three replicate tubes, each set inoculated with 10, 1.0 or $0.1cm^3$ of sample (see text for details).

Π A MPN assay gave the following results.

Volume of inoculum (ml)	Growth (turbid (+); non-turbid (-))
10	- + -
1	+ + -
0.1	- + +

Use Table 2.1 to determine the most probable number of viable cells per $100ml^{-1}$ in the original sample.

The number of tubes positives in the 10, 1 and 0.1 cm^3 replicate sets of three are 1, 2 and 2 respectively. We can see from Table 2.1 that this combination of positives gives a MPN per 100 cm^{-3} value of 20.

2.4.3 Measurement of cell volume

The measurement of the volume of biomass present in a sample can be used as a rapid method of measuring and comparing the amount of biomass in samples. Usually the measurement is done in a centrifuge tube marked with volume graduations. A known volume of the sample is placed in such a centrifuge tube and then centrifuged at sufficient speed to pellet the cells. The actual speed and duration of centrifugation depends on the sizes and density of the cells as well as on the viscosity of the suspending medium (see Chapter 4). Usually a centrifugal force of about 8000 x g for 5 minutes is sufficient. The volume of the packed cells sedimented by this procedure can be read from the gradations on the centrifuge tube.

packed cell volume

The packed cell volume (PCV) obtained by this procedure depends not only on the amount of biomass present but also on how tightly they become packed. This of course is dependent upon the centrifugal force applied and upon the duration of the centrifugation. It is essential therefore for the centrifugation conditions to be standised and variations in the procedure must be minimised if truly comparative results are to be obtained. If the morphology of the biomass is subject to change, for example as we might encounter with filamentous fungi, then PCV is a rather unreliable method of determining biomass. The morphology of cells greatly influences the way they will pack together and this in turn influences the PCV. Other factors which influence PCV include the osmolarity of the medium which may influence cell swelling and shrinkage thereby affecting the PCV. Nevertheless, the method can be routinely used for cells with rigid cell walls (eg bacteria and plant cells) and since biomass density varies little and assuming that the cell is 20% dry solids, a rough estimate of the cell mass can be made by dividing the cell volume by five. This method is not suitable for cells that are sensitive to shear damage.

2.5 Methods based on measuring cell components.

These methods are usually employed where measuring cell mass, numbers or volumes are either difficult or impossible. Typical examples of these circumstances are when cells grow as filaments or form clumps. They are also applicable if the material contains contaminants such as non-cellular solids which would interfere with other estimations. They are also generally more applicable to multicellular material containing collections of highly differentiated cells such as those found in plant and animal systems.

The components that are selected for assay are those which are easy to assay and for which sensitive accurate assay procedures are available. Preferably ideal assays would be ones which are not sensitive to other non-cellular materials. Also important are, of course, the speed and cost of the assay procedure.

The success of measuring cell components as a means of estimating biomass depends on:

- the absence of related materials in the non-cellular solids;
- the correlation of the cell component with cell mass.

The latter point is particularly important as it is pointless using a cellular component that is only very loosely correlated with total biomass. For example, we might use a specific enzyme as a measure of the amount of biomass that is present. This may be satisfactory if the enzyme is constitutive. If, however, its production can be repressed, or it is inhibited by metabolites, the amount of this enzyme would better reflect the physiological status of the biomass, rather than the amount of biomass present.

In this section we will consider three of the most commonly measured cell components used for the estimation of biomass, namely protein, DNA and ATP.

2.5.1 Protein

protein

It is generally true that the protein content of cells does not vary greatly with different growth conditions. Protein therefore provides a reasonable estimate of cell mass. The major limitation of protein analysis is the presence of non-cellular and water insoluble protein in medium components.

Biuret, Lowry and Kjeldahl methods

There are several methods of protein analysis that are useful for biomass estimation. Each will give a different protein value because they measure different properties of the protein. However, the values are meaningful if the basis of the analysis is understood. Two of the most widely used methods of protein measurement are based on the development of colour by reaction with chemical reagents; these are the Biuret and the Lowry methods.

The Biuret method is easy, with good reproducibility since it measures peptide bonds, but it is not very sensitive. The Lowry method is more sensitive, but since it responds strongly to the aromatic amino acids, errors are introduced unless samples and reference proteins are similar in composition. Another approach is the measurement of nitrogen content by the Kjeldahl method. This involves chemical digestion of the protein to liberate ammonia which is then measured by titration. The nitrogen content of biomass can be estimated to within 1% by this method. The value when multiplied by 6.25 gives a value for crude protein, but is subject to error from non-protein nitrogen. Typically the Biuret method can estimate protein in the mg range. The Lowry method is about 10-100 times more sensitive (that is, it can be used to measure samples containing about 10-100 $\mu g.cm^{-3}$ protein).

In the Biuret method, the protein is usually precipitated by the addition of 10% (w/v) trichloroacetic acid. The pelleted protein is then redissolved in a known volume of alkaline copper sulphate solution. In these conditions, the Cu^{2+} ions chelate with the oxygen and nitrogen atoms attached to the peptide back-bone of the protein to form a purple coloured chelation complex.

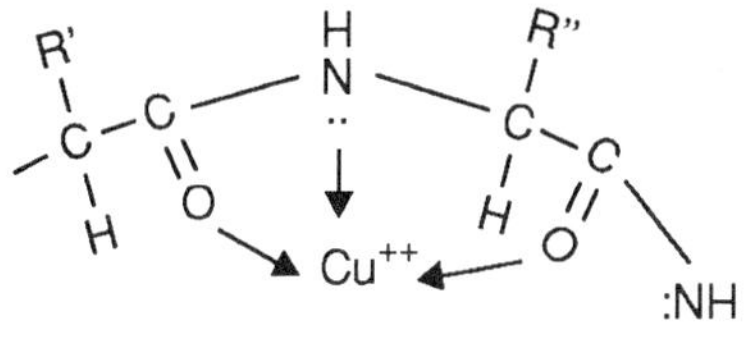

The absorbance of this coloured complex is measured spectrophotometrically at 550nm. A calibration curve of absorbance against protein concentration is prepared using a standard protein (usually bovine serum albumin fraction V) and this is used to calculate the protein content of the sample.

The Lowry method involves the use of Folin-Ciocalteau reagent. This reagent produces a coloured product by becoming reduced by phenolic (and other) groups. Thus protein rich in tyrosine produces much more coloured product than the same mass of protein which contains only a small amount of tyrosine. Again a standard protein is used to prepare calibration curve.

Π A suspension of bacteria was divided into two. Sample A was assayed for protein by the Biuret method and shown to contain 10mg cm^{-3}. Sample of B was assayed for protein by the Lowry method and shown to contain 15 mg cm^{-3}. In both cases bovine serum albumin fraction V was used as a standard. Can you work out a reason for this apparent discrepancy?

The reason is that neither method gives an absolute measure of the amount of protein present. When we determine protein by either of these two methods we should perhaps really report the values in the following way. For sample A, this contains as much protein ml^{-1} as to give the same amount of coloured product in the Biuret reaction as will 10mg of bovine serum album. Likewise, sample B contains as much protein cm^{-3} as to give the same amount of coloured product in the Lowry procedure as will 15mg of bovine serum album.

The discrepancy arises because the two reagents react with proteins in different ways. However, because the Biuret reagent reacts with the peptide backbone, a feature of all proteins, the differences in the amount of colour produced g^{-1} protein are usually small. In contrast, because the Folin-Ciocalteau reagent reacts with the side groups of particular amino acids and different proteins have different amino acid compositions, we should anticipate that the amount of colour produced g^{-1} protein will vary markedly from protein to protein.

We need to make one further point. In conducting assays depending on chemical interaction, it is important to understand the chemistry involved and to be aware of the possibilities of interference in the assay. For example, ammonia and urea can both chelate with Cu^{2+} ions in alkaline solution to produce purple complexes. The presence of either of these will interfere with the assay of protein by the Biuret method. Likewise, the presence of phenols or the phenolic amino acids will give misleading results in the Lowry method.

2.5.2 DNA

It is often assumed that the DNA per cell is constant for any particular cell line. This is erroneous especially with prokaryotic cells; faster growing cells have more DNA per cell than slower growing cells. The variation in DNA per unit of cell mass is, however, low and thus DNA content can be used as a measure of biomass. It has not, however, been extensively used because the method is slower than that used for protein determination.

The measurement of DNA in biomass is based on the estimation of deoxyribose using the Dische reaction. In this procedure, the biomass is first harvested and suspended in cold 0.25mol dm^{-3} perchloric acid at 0°C to remove RNA. After standing for 15 minutes or so, the biomass is reharvested and the extraction with cold perchloric acid is repeated. The biomass is then incubated in a suitable volume of 0.5 mol dm^{-3} perchloric acid at 70°C. This process hydrolysis the DNA and removes the purine residues from their deoxyribose moieties. The hydrolysate is collected and reacted with a diphenylamine reagent prepared by dissolving diphenylamine (0.3mg) into glacial acetic acid (20ml) containing sulphuric acid (0.3cm^3) and acetaldehyde (16mg cm^{-3}). Equal volumes of hydrolysate and diphenylamine reagent are allowed to react, in the dark, at about 30°C for 16 hours. During this time, the diphenylamine reacts with the deoxyribose released during DNA hydrolysis to form a blue product. The absorbance is measured at 600nm and compared with a standard curve relating D-deoxyribose concentration to absorbance at 600nm. To obtain an absolute measure of the DNA present in the biomass, allowance has to be made for the fact that only the purines are removed during the hydrolysis of the DNA and a small adjustment may be required to allow for the base composition of the DNA. There are many variations on this general reaction scheme.

Typically the method can be used to measure samples containing about 1mg DNA. It particularly finds use when the biomass is present in complex culture media containing non-cellular insoluble protein.

Π Why are measurements of RNA, lipids and carbohydrates not used for the estimation of biomass.

The main reason is that the concentrations of these cell components are influenced strongly by the growth conditions. Their correlation with cell mass is therefore poor.

2.5.3 Adenosine triphosphate (ATP)

firefly luciferase

All living cells contain ATP and this is present in fairly constant amounts in each cell type; it is rapidly lost after the cells die. It follows that its concentration should be proportional to the concentration of viable cell mass. Measurement of ATP involves measurement of light produced as a result of a reaction catalysed by the enzyme firefly luciferase. The amount of light produced is proportional to the amount of ATP taking part in the reaction. The light producing reaction is:

$$\text{ATP} + \text{luciferin} + O_2 \xrightarrow{\text{luciferase}} \text{oxyluciferin} + \text{AMP} + \text{PPi} + CO_2 + \text{light}$$

Commercial luciferin/luciferase reagents are available which give a constant light output with time. The light can be measured and recorded automatically using a luminometer. The procedure is very sensitive with light output being proportional to ATP concentration over the range 10^{-11} to 10^{-6} moles dm^{-3}. In a typical bacterium, the ATP pool has been estimated to be in the range 4.5-7.5 μ moles g^{-1} cell dry weight.

This means that the minimum amount of biomass required for biomass estimation using ATP is $10^{-11}/4.5 \times 10^{-6} = 2.2 \times 10^{-6}$ g.dm^{-3} cell dry weight.

Although very sensitive, there is one serious drawback with this method. It is assumed that the ATP content of the cell does not change during the sampling period and does not change with the growth environment. These assumptions, however, are difficult to justify. For example, the ATP pool in *E. coli* has been estimated to turn over 4-8 times per second. Therefore, if ATP synthesis is interrupted during sampling, this could rapidly upset the ATP pool size and introduce large errors in biomass estimation.

Nevertheless, the luciferase assay has been used to estimate yeast and other microbial biomass in beer. Indeed, the correlation between the amount of cellular ATP and yeast cell growth has been shown to be good, using the luciferase assay.

SAQ 2.5

For each of the methods of biomass estimation listed 1) to 5), select factors which could greatly reduce the accuracy of the determination.

Method of biomass estimation	Factors
1) Protein	Slow sampling speed.
2) ATP	Changes in biomass morphology.
3) Absorbance	A complex growth medium.
4) Wet weight	The presence of non-cellular particulate material.
5) DNA	Cell clumping.
	Cell lysis

SAQ 2.6

Below are listed some methods of biomass estimation and the minimum dry weight of bacteria required for an estimation with an error of 2%. Match these methods with the appropriate minimum dry weight of material required.

Method of biomass estimation	Minimum dry weight of material required (mg)
Biuret protein	50
cell count	1.0
Lowry protein	0.1
DNA	0.00001
dry weight	
absorbance	

2.6 Estimation of biomass by determining metabolic activity

There is a natural assumption that the more biomass present, the greater the total amount of metabolic activity. Providing the physiological status of the biomass remains constant and there is no limitation of substrate, then this assumption is a reasonable one. Thus in principle, we can use the rate of nutrient consumption and product formation as a measure of biomass. It is not however usually used for one-off measurements and it is more frequently applied to situations in which we wish to continually monitor the amount of biomass produced.

Nutrients which can be correlated with biomass are the carbon source, nitrogen source and oxygen. We may write the generalised reaction as

$$[\text{C - source}] + [\text{N - source}] + [O_2] \rightarrow \text{biomass} + [CO_2] + H_2O + [\text{products}] + [\text{heat}]$$

The rate of this forward reaction depends on the amount of biomass present. Monitoring the amount of biomass present using this relationship can be done by measuring any of the components indicated by brackets. The key to choosing one of these is the availability of suitable instrumentation. The assay of oxygen and heat production are described elsewhere in this text, so we will not examine the actual techniques here (see Chapters 10 and 11). However, we will show how they can be applied to measure biomass.

For the measurement of biomass using metabolic activity we must:-

- have an accurate measurement of the nutrients or products;
- be able to determine the ratio of biomass produced per unit of nutrient consumed or products formed (the so called growth yields).

Ideally there should be a direct proportionality, that is:

$$X = Y_{x,s} . S \text{ or } X = Y_{x,p} P$$

where:

X = biomass produced; $Y_{x,s}$ = growth yield coefficient for substrate (biomass produced per unit of substrate consumed); $Y_{x,p}$ = growth yield coefficient for product (biomass produced per unit of product produced); S = substrate consumed; P = product generated.

To use this approach to biomass estimation, the growth yield coefficients ($Y_{x,s}$ and $Y_{x,p}$) should be constant. This may not always be true if the culture conditions vary and sometimes the growth yield coefficients are dependent upon the growth rate.

Usually the consumption of carbon, nitrogen or oxygen sources are used for the determination. If products are used, then carbon dioxide or heat production are used. We will now examine each of these in turn.

2.6.1 Use of carbon source consumption for biomass estimation

carbon balance

Carbon is typically about 50% of the dry cell weight. This suggests that measurement of the disappearance of the source of carbon should be a good means of following an increase in biomass. But the carbon source is often the energy source and much of the carbon in converted to CO_2, and may also be converted to products. This balance can be described as follows:

$$C_{consumed} = C_{biomass} + C_{carbon\ dioxide} + C_{products}$$

It follows that to estimate biomass, carbon source consumption, CO_2 production and product formation must be measured independently. The accuracy of the determination then depends on the accuracy of each method of analysis. The method is further complicated if more than one major carbon source is present. For example, in complex media where glucose serves as the main energy source and amino acids, added as protein hydrolysates, are readily incorporated into biomass.

2.6.2 Use of ammonia consumption rates for biomass estimation

chemical analysis

ammonium selective electrode

When there are no nitrogen containing products formed, and ammonia is the major nitrogen source, its consumption is routinely and effectively used to estimate biomass in fermentation processes. Ammonia consumption can be measured by several methods, these include chemical analysis and the ammonium selective electrode. The main limitation of the chemical analysis is its difficulty to automate, whereas the ammonium electrode method suffers from interference by monovalent ions such as sodium and potassium. Cells growing on other nitrogen sources pose greater problems in estimation.

2.6.3 Oxygen consumption

On-line measurement of oxygen consumption is a useful method for estimating biomass in a reactor. The most common technique for this approach are mass spectrometry and paramagnetic oxygen.

mass spectrometry

We will discuss mass spectrometry in a later chapter. Essentially mass spectrometry is based on the separation of ionised molecules under vacuum. The separation, dependent on the mass to charge ratio, is achieved by magnetic instruments. Mass spectrometry has the advantage that it can be used to analyse for any vapour phase component simultaneously. Compared to other methods of O_2 analysis it has a shorter response time (in the order of seconds rather than minutes) and has greater accuracy. Use of mass spectrometry has the added benefit of being able to be interfaced with a computer and can be used to monitor several fermenters. This approach is finding increasing application. This is despite its main disadvantage - cost - which is typically ten times more than paramagnetic O_2 analyzer. This strategy is not, however, commonly used on laboratory scale experiments. An alternative method is to use paramagnetic oxygen analysers which are relatively inexpensive.

Estimation of biomass, by this approach, depends upon the determination of O_2 consumption rate by measurement of O_2 content in incoming and outgoing air. The exact oxygen transfer rate (OTR) is then determined by multiplying the aeration rate by the difference in O_2 concentration (inlet and outlet gases) and taking the absolute temperature and pressure into consideration. This is maybe expressed mathematically as follows.

$$OTR = \phi\ (C_{in} - C_{out})$$

where ϕ: = aeration rate; C_{in} = oxygen concentration of inlet gas (air); C_{out} = oxygen concentration of outlet gas; OTR = oxygen transfer rate.

To determine the biomass concentration (X) of a culture, the growth yield coefficient for oxygen substrate (Y_o = g biomass produced g^{-1} O_2 consumed) and the specific growth rate (μ) must also be known. The relationship between them is:

$$OTR = \mu.X\,(Y_o)$$

(Note this relationship is derived in the Biotol text 'In vitro Cultivation of Micro-organisms')

We can calculate X from this relationship providing μ, Y_o and OTR are known.

Let us examine a numerical example.

SAQ 2.7

Analyses of the inlet and outlet gases showed that 0.5mmoles O_2 dm^{-3} air were removed by a culture. The growth rate of the culture (μ) was $0.2h^{-1}$ and the volume of the culture was 1 dm^3 and this culture was being sparged at a rate of 1 dm^3 min^{-1}. Calculate how much biomass was present in the culture if the Y_o value = $0.5gg^{-1}$ and the molecular mass of O_2 is 32.

You should realise that Y_o can either be determined experimentally or determined theoretically by considering the oxidation state of the substrate and biomass.

Y_o is determined using the following equation.

$$\frac{1}{Y_o} = [\frac{2C + (H/_2) - O}{Y_s} M + \frac{O'}{1600} - \frac{C'}{600} + \frac{N'}{933} - \frac{H'}{200}]$$

where:

C, H, O = Number of carbon, hydrogen and oxygen atoms in substrate

C', H', O', N' = Percent of carbon, hydrogen, oxygen and nitrogen in biomass

M = Substrate molecular weight

Y_s = Growth yield coefficient for carbon substrate.

These relationships and similar relationships are derived in the BIOTOL text 'Bioprocess Technology, Modelling and Transport Phenomena' and will not be described further here.

2.6.4 The use of carbon dioxide formation

carbon dioxide

Carbon dioxide is generally the most useful product to measure because:

- it is the most common catabolite;
- it does not require sterile sampling since CO_2 is in the gas phase;

- it is easy to measure CO_2 accurately and rapidly.

CO_2 can be measured by a mass spectrometry. Indeed, determination of the CO_2 production rate also requires measurement of O_2 and gas flow rates.

The accuracy of biomass estimation by this approach relies on the correlation between CO_2 production rate and growth. The exact ratio of CO_2 produced per unit of cell mass synthesised depends on factors such as the catabolic pathway used and the efficiency of coupling of energy generation (catabolism) with energy consumption (anabolism). The correlation will break down when growth and catabolism are uncoupled as is the case for cells in the stationary (non-growing) phase of the culture cycle.

Other catabolic products that are useful for monitoring fermentations are the end products of catabolism using anaerobic pathways. These include ethanol, lactate, acetate, propionate and butanol.

2.6.5 Heat formation

Heat is generated during cell growth. The amount of heat generated is stoichiometrically related to rate of biomass formation (growth rate). It can therefore be used as a means of indirectly estimating microbial biomass. Measurement of heat is discussed in Chapter 10, so we will not elaborate on this aspect here.

2.7 Special techniques with eukaryotic cells and tissues

Our discussion of biomass estimations so far has centred on techniques that are best used for unicellular (especially bacterial) systems. Some of these methods can, however, be readily adapted for use with multicellular organisms (eg dry weight determination). A feature of multicellular systems is that they often show specialisation, they become differentiated. In such systems measuring some property such as protein content or metabolic activity may reflect more the state of differentiation rather than the amount of biomass. In such systems, we are often interested in the amount of growth of cells, and the proportion of cells that are growing. We need, therefore, methods which will enable us to measure growing and non-growing cells.

Multicellularity and cell differentiation are features particularly of eukaryotic cells. In this section, we will examine methods which enable us to measure cell proliferation in eukaryotic systems.

We remind you briefly of the cycle of events in the growth of eukaryotic cells. The cycle is made up of a series of phases. These are:-

- gap 1 phase (G_1);
- DNA synthesis period (S);
- Gap 2 phase (G_2);
- Mitosis (M phase) in which the chromosomes condense and separate into daughter nuclei;
- cytokineses in which the cells divide.

Cells that leave the growth cycle are said to enter G_o phase. In the G_o phase, cells may become irreversibly differentiated and may be prevented from re-entering the growth cycle. We can represent this by the figure:

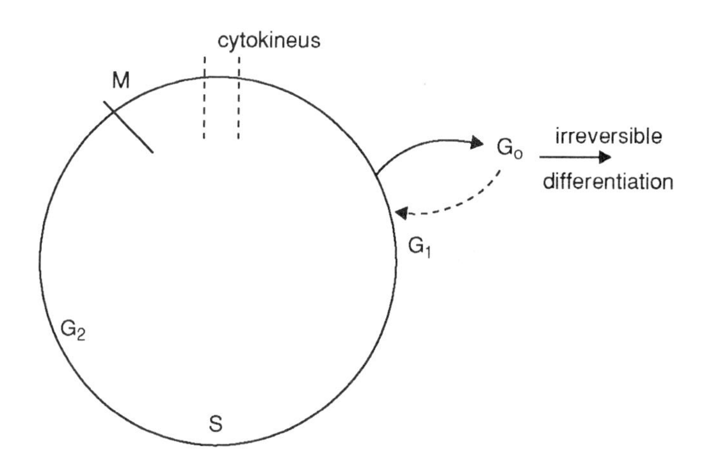

We do not intend to elaborate on this further here. Further details of eukaryotic cell cycles are given in the BIOTOL text 'Infrastructure and Activities of Cells'. The occurrence of a regular cycle of events during the cell cycle enables us to devise methods which enables us to distinguish cells which are in the active cell division cycle from those that are not. Here we summarise these methods. Further details are available in the BIOTOL texts '*In vitro* Cultivation of Plant Cells' and '*In vitro* cultivation of Animal Cells'.

2.7.1 Mitotic index

For many cases, determining the proportion of cells in mitosis provides a method of measuring cell proliferation rates. In principle the technique is straight forward. Examination of cells (either from cell suspension cultures or from tissue slices) by microscopy enables identification of cells that are in mitosis. One would anticipate that the greater the extent of cell proliferation, the greater the proportion of the cells would be in mitosis. The proportion of the cells which is in mitosis is usually described as the miotic index.

identification of cells in mitosis

This approach has some difficulties. First there has to be strict morphological criteria for recognising cells in mitosis. These include condensed chromosome, the absence of a nuclear membrane, the absence of a clear zone in the centre of the nucleus and the nuclear region being surrounded by cytoplasm that attracts basophilic stains. In practice, such strict morphological criteria is not often used and researchers simply look for 'condensed chromosomes' which perhaps more equates with metaphase rather then the whole of mitosis. Very frequently the criteria used to recognised mitotic figures is not listed and this means that the data are of limited value.

The second limitation of this technique is that the proportion of cells in mitosis at any one time is dependent on two main factors:

- the relative duration of the M phase compared to the full cell cycle time;
- the proportion of cells which are undergoing active cell division.

Π If all the cells in a culture were growing with a mean generation time of 10h and 10% of the cells were in mitosis, how long does mitosis last? (Assume that the cells are evenly distributed through the phases of the cell cycle).

The answer is 1hr (since 10h is represented by 100% of the cells; 1hr is represented by 10% of the cells).

Π If the duration of mitosis was reduced to 0.5h, what proportion of the cells would be in mitosis at any one time? (Assume that the cell cycle is still 10h long).

The answer is 5% (since 0.5h represents 5% of the cell cycle time).

Π If only half of the cells described above were in the division cycle (ie half were in G_o), what proportion of the cells would be in mitosis at any one time?

The answer is 2.5% (ie half of 5%).

Thus the proportion of cells in mitosis at anyone time is a reflection of the relative duration of the M phase and of the proportion of the cells in the division cycle. (Note that we made a major simplifaction in the calculations carried out above. Cells are not normally evenly distributed through the cell cycle. This is because for each cell that enters mitosis, two cells will emerge into the G_1 phase. This has to be allowed for in any calculation).

Because of the difficulties cited above, measuring the mitotic index of a cell population is not frequently used as an absolute method for measuring cell proliferation. But it is likely to remain an important method to compare proliferation rates in different systems. The method has the advantages of being 'low technology' and requires little specialised equipment. It can also be used with very small amounts of material.

In tissue slices, the method can be automated using a digital image analyser. In this the tissue is scanned microscopically and the image is analysed electronically. Providing suitable criteria are used, the instrument is able to distinguish and count mitotic and non-mitotic nuclei.

2.7.2 Metaphase arrest technique

In this technique, the tissue is exposed to an agent (often vincristine) which blocks the cell cycle in mitosis. The cells accumulate at the metaphase of the cell cycle and can be identified by the presence of condensed chromosomes on the metaphase plate. This technique involves incubation of tissues (cells) for various times in the presence of vincristine. The estimation of mitotic rate is then a rather complex function of vincristine concentration and duration of incubation. The drawback of the method is that metaphase - inhibited cells are often distorted and difficult to properly identify.

inhibition by vincristine

2.7.3 Detection of DNA synthesis and other proliferation - associated compounds.

The production of new cells inevitably means that DNA is synthesised. If large populations of cells are produced, we can use straightforward DNA estimations. However, with low cell densities (eg in a monolayer culture of animal cells) or in multicellular (tissue/organ) systems such chemical approaches may be not be applicable. Two other strategies have been developed. In one, readily detectable nucleotides (eg tritiated labelled thymidine) are incubated with the cells. Growth (and therefore DNA synthesis) is detected by uptake of the nucleotide. The amount of nucleotide taken up can best be measured by autoradiography (which provides evidence that nuclear DNA synthesis has taken place) or by scintillation counting. With tissues, care has to be taken because the proliferating cells may not be uniformly distributed through the tissue. Thus at best the method gives only a rough estimate of proliferation rates.

Certain proteins (eg histones, nucleotide biosynthetic enzymes) are also associated with cell proliferation. Another approach to measuring cell proliferation rates is to measure these. Two main approaches are used. One is to use monoclonal antibodies which react with the 'proliferation associated compound' to measure them (we will deal with immunochemical methods in a later chapter). The other method can be used with enzymes which are linked to proliferation. For example, the enzymes ribonucleotide reductase and DNA-directed DNA polymerase are both linked to DNA syntheses and to cell proliferation. The presence and amounts of such enzymes may be used to determine cell proliferation rates.

2.7.4 Flow cytometric analysis

This is an exciting technique which has one unfortunate limitation. It demands the use of expensive equipment. The cells are incubated with fluorescent dye which binds with DNA. The cells are then passed singly through a fine aperture where the fluorescence of each nucleus is measured. Cells that are in G_2 and M phases produce twice as much fluorescence as G_1 cells. Cells which are in the S phase give fluorescence values intermediate between those of G_1 and G_2 cells.

Typically several thousand cells are measured in this way and the results plotted as shown in Figure 2.6. From such a plot, the relative duration of G_1, S and G_2 +M phases can be determined. The analysis is however complicated by the presence of non-dividing cells.

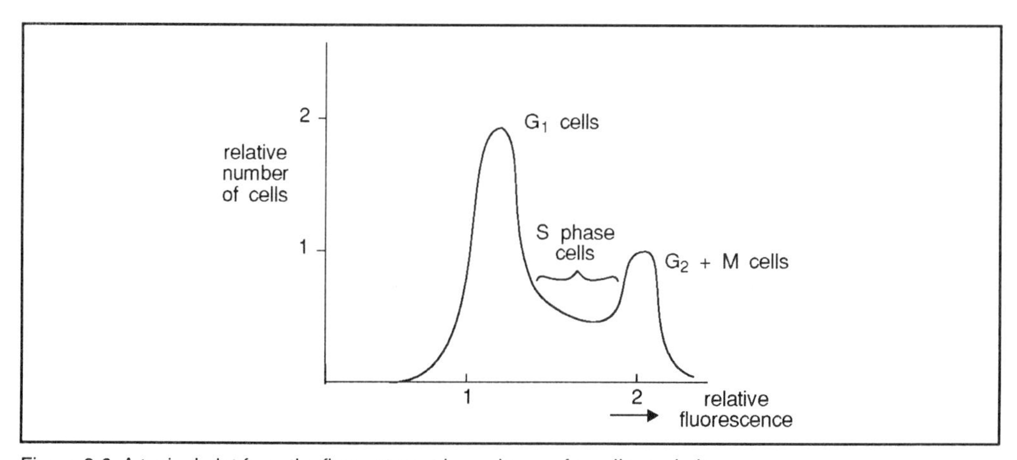

Figure 2.6 A typical plot from the flow cytometric analyses of a cell population.

Π Draw onto Figure 2.6 a plot of relative number of cells against relative fluorescence for:

a) a population of cells which had been incubated in vincristine for a long time;

b) a population of cells which had a very short G_1 phase.

For a), you should have plotted a graph in which all the cells had a relative fluorescence of 2. This is because the cells will all accumulate in metaphase (see 2.7.2).

For (b) you should have reduced the size of the peak for G_1 relative to that for G_2. The shorter the phase the fewer the cells will be present in that phase.

2.8 Concluding remarks

In Section 2.6, we examined a range of techniques used for determining biomass using some feature of metabolic activity. The use of these techniques are more-or-less confined to determination of biomass where on-line measurement is important. Their most widespread use is with monitoring the progress of cell cultures in bioreactors (fermenters) where rapid results are necessary. Such methods of course are not applicable to stored, inactive samples.

Π Now that you have completed this chapter, it might be helpful to produce yourself a revision table giving the advantages and disadvantages of the methods described, including their sensitivity and their general application. We would suggest you use the following headings.

Method of biomass estimation	Advantages	Disadvatages	Sensitivity	Uses

When you have completed this attempt the following SAQs and prove to yourself that you have understood the advantages and disadvantages of the techniques described and can choose appropriate techniques to meet particular circumstances.

SAQ 2.8

1) List as many of the factors which decrease the reliability of estimates of biomass based on metabolic activities as you can. You should be able to list six or more such factors.

2) Match the following methods with the appropriate merits and limitations.

Methods of biomass estimation

a) Dry weight measurement

b) Protein measurement

c) Measurement of O_2 consumption rate.

Merits and limitations

i) results obtained very quickly,

ii) distinguishes between viable and non-viable cells

iii) requires special equipment

iv) lacks sensitivity

v) suitable for giving on-line measurement of biomass.

SAQ 2.9

Select the most suitable method from the list provided below for determining biomass for each of the following circumstances.

1) To monitor the growth of cells at hourly intervals in a culture of yeast in order to harvest the culture when the biomass reaches a particular concentration. Cell densities are of the order of 0.2-20mg dry weight cm^{-3}.

2) To determine the amount of biomass present in the condensate collected from the cooling coils of an air conditioning unit.

3) Measure the biomass present in several cultures of unicellular micro-organisms being compared as potential sources of an enzyme.

4) To determine the amount of biomass in a bioreactor in order to control the rate of substrate addition.

5) To determine the survival of biomass in a culture of bacteria after exposure to toxic chemicals.

6) To determine the amount of biomass in $10cm^3$ samples of homogenised cultures of filamentous fungi from which an antibiotic is to be extracted. The aim of the study is to compare the amount of antibiotic produced by the various cultures.

Methods of biomass estimation

Total cell counts; dry weight measurement; viable cell count, protein measurement, measurement of O_2 consumption, most probable number measurement, DNA estimation, absorbance.

Summary and objectives

In this chapter we have examined the methods available for estimating biomass. We have predominantly, but not exclusively, focused our attention on biomass represented by unicellular systems. We began by discussing general issues of sampling, precision and accuracy before moving on to discuss the various methods available. The methods we described include the measurement of cell mass, total and viable cell numbers, cell protein, DNA and ATP and methods based on the metabolic activities of cells. For each of the methods we have examined we have included discussion of their sensitivity, accuracy and speed and the factors which may influence their reliability. Towards the end of the chapter we included a brief outline of approaches that have been used to determine biomass proliferation especially in tissues (eg tumours).

Now that you have completed this chapter you should be able to:

- describe a wide variety of techniques for estimating biomass;
- list the merits and limitations of the various techniques used for estimating biomass;
- calculate biomass concentration from supplied data obtained from a variety of techniques;
- select appropriate methods for estimating biomass in a wide variety of circumstances.

Methods of cell disruption

Methods of cell disruption

3.1 Introduction

This chapter considers cell and tissue homogenization as a single step within cell fractionation. The ultimate objective of the cell fractionation procedure and the nature of the starting material are important parameters in defining homogenization as an empirical method of cell disruption.

In order to isolate the complex of biological membranes and organelles of a cell, the plasma membrane must be disrupted. The ultrastructure and function of each individual separated component must be assessed while limiting as far as possible physical (eg heat) and biological (eg enzymatic degradation) damage.

'You can't make an omelette without cracking eggs' - sums up the rationale for devising effective methods for cell disruption. Much of biotechnology is concerned with exploiting the biosynthetic machinery of cells or subcellular systems and has advanced from discoveries made in biochemistry and microbiology. Both of these highly complementary disciplines are concerned fundamentally with correlating cell structure with function and make substantial use of *in vitro* methodology. This involves the removal of biologically -derived material such as subcellular fractions (mitochondria, plasma membrane), cells, tissues or organs from the natural or *in situ* state to an artificial environment. The term *in vitro* is literally defined as in glass because historically glass vessels were routinely used for the experimental incubation. Biochemical analysis of the resulting subcellular membrane fraction for example, has been shown to further our understanding of the role of each individual component within the intact and *in vivo* state. The criticism of much *in vitro* methodology is that it inevitably creates laboratory artifacts which do not represent the *in vivo* position. The challenge in much *in vitro* work is to choose methods of cell disruption and incubation of subcellular fractions, cells, tissues or organs that are appropriate to the particular investigation and which may or may not entail minimizing the presence of artifacts.

cell fractionation

Cell fractionation proceeds in two consecutive stages:

- homogenization which essentially disrupts the tissue and releases the cellular components into the resultant homogenate;
- centrifugation which separates the individual components within the homogenate in accordance with their density, size and shape.

This chapter will refer to preliminary stages of cell disruption in depth. Centrifugation is the subject of the following chapter.

The ideal homogenate would be comprised of all of the desired cellular components in an unaltered morphological and metabolic state and in an abundant yield. In practice, fractionation of selected material often produces components with at least one altered characteristic. The ultrastructure may be preserved to the detriment of metabolic integrity. If the ultimate objective of the cell fractionation is to isolate a particular compound, the retention of morphological and metabolic integrity throughout the preliminary stages of cell disruption may be unnecessary. However, a detailed analysis of the function of individual components within the cell would demand preservation of metabolic and ultrastructural integrity.

At present, homogenization is a highly empirical procedure. The preliminary stages of cell disruption are critically dependent upon the nature of the starting material, eg plant, animal or bacteria, and the objective of the fractionation. For example, if we wish to isolate an enriched plasma membrane fraction, we must ensure that the intact plasma membrane is not disrupted too greatly prior to differential centrifugation where particles are separated according to size and density. A high variation in the sizes of the plasma membrane fragments will yield an enriched plasma membrane fractions which is of low abundance and high variability.

3.2 Homogenization medium

3.2.1 Introduction

conditions of homogenization

In this section we shall consider the conditions of homogenization and attempt to rationalize the use of the various compounds incorporated in most isolation media.

The composition of the homogenization medium is dependent on the particular tissue and the objective of the particular investigation.

Π From your current knowledge of biological systems, make a list of two or three of the more important factors which govern the choice of the homogenization medium.

No doubt you included in your list the use of water as the most common solvent and the need for controlling the osmotic pressure and the pH. You probably did not think of including specific enzyme inhibitors.

We will consider each of these factors in more detail.

3.2.2 Osmotic pressure of the homogenization medium

Cell disruption is usually performed in an either slightly hypo-osmotic or iso-osmotic medium to preserve morphological integrity. Sucrose is commonly used as an osmoticum to prevent subcellular organelles or vesicles from adverse swelling or shrinkage. Mannitol and sorbitol have also been used when sucrose has been found to interfere with the biochemical analysis of the subcellular component. Sucrose is preferred to salt solutions because the latter tend to result in aggregation of subcellular organelles, for example in liver extracts . However, in some tissues such as spleen, 0.25 mol dm^{-3} sucrose promotes agglutination of subcellular organelles.

Iso-osmotic media such as those provided by 0.25-0.32 mol dm^{-3} sucrose are routinely used to preserve metabolic and morphological integrity in the isolation of subcellular organelles. Hypo-osmotic media are often used in the isolation of plasma membrane fractions because under these conditions, disrupted cells yield plasma membranes in the form of large ghosts (empty cells) or sheets. This significantly reduces the high degree of variation in size of these fragments prior to centrifugation.

Although the homogenization medium is usually aqueous in nature, non-aqueous media have been used in isolating subcellular organelles. Light or heavy organic solvents such as ether/chloroform are mixed to give a required density so that the required particles can be collected in the subsequent centrifugation step. Organic solvents have been used in the isolation of chloroplasts and leucocytes. Problems with the use of non-aqueous media include inactivation of some enzymes and loss of morphological integrity in some tissues.

3.2.3 Additives in the homogenization medium

The types of additives that might be included in the homogenization medium depend on the nature of the tissue and the objectives of the study.

Chelating agents (for example ethylene diaminetetra acetic acid disodium salt, EDTA,) may be added to the homogenization medium to remove divalent cations, such as Mg^{2+} or Ca^{2+} ions which are required by membrane proteases. It is important to note that protease inhibitors should still be included in the medium because EDTA and thiol reagents may activate some proteolytic enzymes. However, many membrane marker enzymes require cations for activity and may be inhibited after the cations have been removed from the extract. The addition of 1-2 mmol dm^{-3} Mg^{2+} or Ca^{2+} to the homogenization medium maintains nuclear integrity. The isolation of ribosomes requires Mg^{2+} at a concentration of at least 1 mmol dm^{-3} because at lower concentrations they tend to dissociate into their corresponding subunits.

Cell disruption of plant tissue often releases phenols which inactivate enzymes by forming hydrogen bonds with the carbonyl groups in peptide bonds of proteins. Polyvinylpyrrolidone in either a water soluble form or as the highly cross-linked insoluble 'Polyclar' may be added to the isolation medium in order to remove the phenols.

Π Can you suggest how polyvinylpyrrolidone prevents inactivation of enzymes by phenolic compounds released during homogenization?

Plants contain a high amount of phenolic compounds that are oxidized to quinones by phenol oxidase. Polymerisation of the quinones forms a dark pigment called melanin which attaches to proteins and inactivates many enzymes. Polyvinylpyrrolidone binds phenols effectively removing them from solution. Thus they are not available to bind to, and inactivate, proteins. It is important to note that sulphydryl reducing agents such as 2-mercaptoethanol can inhibit phenol oxidases.

proteases

Cell or tissue fractionation is invariably performed at 4 °C to reduce the activity of membrane proteases. There are three main classes of protease activity:

- metal-activated proteases, these may be inhibited by the addition of chelating agents such as EDTA or EGTA;
- serine proteases, these may be inhibited by diisopropylfluorophosphate and phenylmethysulphonylfluoride;
- sulphydryl proteases, these may be inhibited by N-ethylmaleimide or E-64 (L-trans-epoxysuccinyl-leucylamide-[4-guanidino]-butane).

Other compounds found within homogenization media include sulphydryl group reducing agents such as 2-mercaptoethanol, dithioerythritol, reduced glutathione or cysteine. Many enzymes contain an essential sulphydryl group in their active sites that must remain in a reduced state to ensure that the enzymes activity is maintained.

Each individual homogenization medium is different and may contain specific inhibitors for enzymes such as phospholipases or phosphatases. Glycerol for example, is often added to the isolation medium in order to inhibit phosphatidic acid phosphatase. Ethanolamine and choline chloride are commonly used to inhibit phospholipase D.

3.2.4 pH of the homogenization medium

Most isolation media are weakly buffered to the physiological pH 7.4 of animal cells. It appears, however, that a slightly alkaline pH 7.5-8.0 helps to minimise the fragmentation of the plasma membrane. The buffering capacity of the homogenization medium for plant cell fractionation must be high to offset the release of organic acids by the disrupted vacuole which constitutes a substantial volume of the cell. The pH of the isolation medium for plant cell disruption is usually 7.5-8.0 which is the pH of the cytoplasm.

SAQ 3.1

How are the components of the homogenization medium formulated?

3.3 Procedures of cell and tissue disruption

3.3.1 Introduction

In this section we shall discuss the principal techniques used to disrupt cells and tissues. Special emphasis is placed upon the more predominant physical methods that generate shearing forces to disrupt the cells. The use of chemical and enzymatic methods of cell disruption is critically evaluated.

temperature control

In order to minimize membrane protease activity and denaturation of protein by excessive heat, all stages of cell fractionation must be performed at 4 °C. All media and apparatus should be pre-cooled and maintained at this low temperature throughout the procedure.

Soft animal tissue usually requires mild shearing forces to disrupt the cells. However, the presence of the thick polysaccharide cell walls of plant and bacterial cells demands high shearing forces which may result in further damage to the cells by physical stress.

We will review several cell and tissue homogenizing procedures that impose different degrees of stress upon the biological material:

- physical methods: solid shear methods, liquid shear methods, pressure homogenization, tissue presses, sonication, osmotic shock, freezing/thawing;
- non-physical methods: use of organic solvents, detergents and enzyme digestion.

3.3.2 Solid shear methods for cell disruption

In these methods, disruption of the cells or tissues results from the shearing forces generated between the cells and a solid abrasive. These techniques are severe and tend to damage large subcellular organelles such as mitochondria, nuclei and chloroplasts within the cell.

mortar

In the simplest of all methods the cell or tissue is placed in a ceramic **mortar** (Figure 3.1) and the heavy round-ended tool referred to as the **pestle** is used to grind the material in the presence of a small amount of abrasive in the form of coarse sand or fine silica sand and alumina for more delicate tissue. This procedure is commonly used in disrupting plant tissue or bacterial cells because of the chemical nature of their cell walls and not for the soft animal tissue. The pestle and mortar are usually pre-cooled before use so that the material is maintained at a low temperature during disruption.

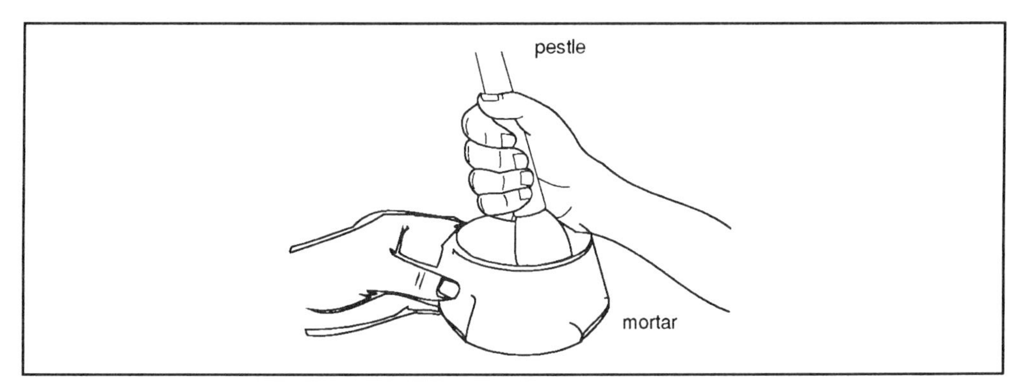

Figure 3.1 Pestle and mortar used to disrupt tissues and bacterial cell walls.

Mechanical shakers such as the Mickle shaker and ball mills can be used with vigorous shaking using glass beads of diameter 0.05-0.50 mm to disrupt bacterial cells. A cooling jacket is commonly used to remove the considerable amount of heat generated during the operation.

3.3.3 Liquid shear methods of cell disruption

These methods of cell and tissue disruption rely on the shearing forces generated between the tissue and a liquid medium.

blender

In a **blender**, which uses rotating cutting blades, a considerable shearing force can be generated to grind the tissue. The blades, orientated at different angles to each other to enhance the mixing of the homogenate, are driven at high speed only for a short period of time because considerable shearing forces can be generated after prolonged use. The material is placed in a pre-cooled capped mixing container with the homogenization medium and subjected to efficient blending for a short and defined period of time.

In **tissue homogenizers** (Figure 3.2) the tissue is ground by the relatively mild shearing forces generated by an upward and downward rotation of a plunger or pestle within a glass cylinder. The surface of the plunger and the inner surface of the cylinder are ground.

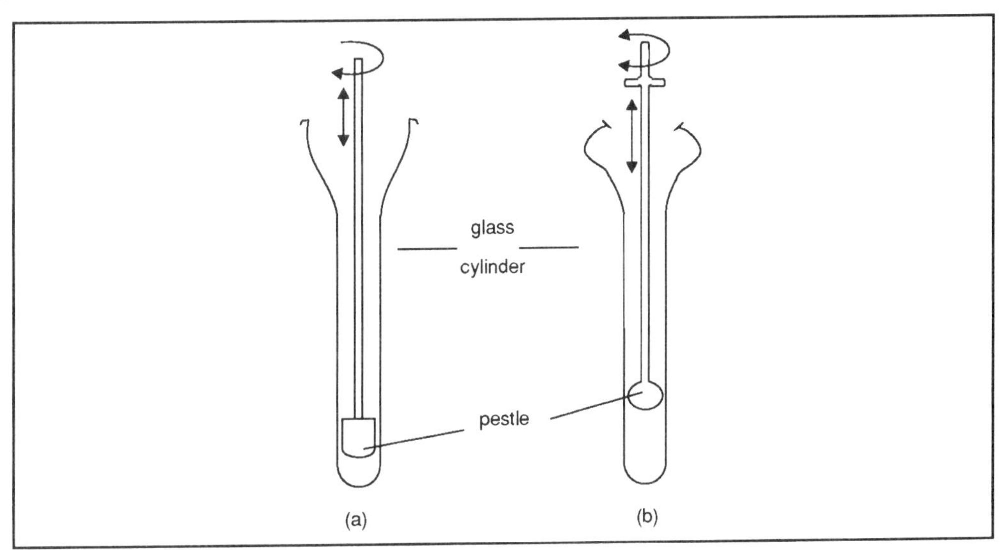

Figure 3.2 a) The motor-driven Potter-Elvehjem homogenizer and b) the Dounce homogenizer.

Potter-Elvehjem homogeniser

The **Potter-Elvehjem homogenizer** (Figure 3.2a) comprises a power-driven teflon, pyrex glass or lucite pestle and a glass cylinder that is of slightly larger diameter (0.05 to 0.5 mm clearance depending on the tissue) than the plunger. The speed of rotation of the pestle within the glass cylinder can be closely monitored with the use of a tachometer. The chopped tissue is forced between the walls of the glass cylinder and the rotating plunger as it passes up and down the vessel.

SAQ 3.2

What are the principal parameters that must be controlled in order to maximise the efficient mixing of the contents of the glass vessel with Potter-Elvehjem homogenizers?

The efficiency of the homogenizer also critically depends upon the amount of vascular and connective tissue present. This should be removed by filtration through muslin or cheese-cloth prior to homogenization. The homogenization conditions may vary considerably between different tissues. Plant material for example, often contains a high degree of vascular tissue and older animal tissue contains more collagen than tissue from younger animals.

In addition, in order to reduce the effect of excessive heat upon the biological material, the glass vessel may either be immersed in cold water within a water jacket or simply immersed in ice between each successive passage of the plunger.

Dounce homogenizer

The **Dounce homogenizer** (Figure 3.2b) operates with identical principles to that of the Potter-Elvenjem homogenizer. The difference is that the pestle is rotated and passed up and down the glass cylinder by hand and is not driven by a motor. This homogenizer generates a considerably more gentle and controllable disruption of tissue because the operator can adjust the ferocity of the shearing forces by hand to suit the tissue in question. In contrast to the Potter-Elvehjem homogenizer, high shearing forces that may generate excessive heat production and possible damage to the tissue are avoided. In addition, the smaller shearing forces generate large sheets of plasma membrane that sediment in a low centrifugal field.

micro-homogenizers

Microhomogenizers are designed to disrupt cells or tissues present in a small amount such as in biopsy material. They consist essentially of a loop of wire containing the material to be disrupted, connected to a motor, and immersed in the medium within a capillary tube. Razor blade fragments may be added to the medium to enhance the mixing of the material. The mixture can then be vortexed and the razor blade fragments subsequently removed with a magnet after microhomogenization.

Motorized homogenizers are commercially available as an effective replacement for the Potter-Elvenhjem or Dounce homogenizer. Examples include the Polytron or Ultraturrax. The cutting blades at the open end of the instrument are rotated at high speed to generate low shearing forces. These instruments are claimed to enhance the reproducibility of tissue disruption and provide greater ease in homogenizing large amounts of material. However, they are markedly more expensive than simple Dounce homogenizers and can prove expensive in maintaining a sharp rotor bit. There is also significant difficulty in standardizing the shearing conditions imposed upon a tissue preparation.

pressure homogenization

The technique of **pressure homogenization** is based upon the principle that cells equilibrated with an inert gas such as nitrogen or argon at high pressure imbibe large amounts of the gas. When the pressure inside the vessel is suddenly returned to atmospheric pressure, the newly formed gas bubbles inside the cytoplasm rupture the membranes of the cell and form vesicles. This procedure has been successfully used in disrupting cultured cells. In contrast to the shearing methods, there is no thermal damage to the tissue and the speed of disruption and the ease with which the cells are disrupted by only one decompression have proved advantageous. The problems associated with pressure homogenization are that there is a variable degree of subcellular organelle damage and that difficulties may arise in separating endoplasmic reticular and plasma membrane vesicles. As a result, this technique does not lend itself well to the isolation of an enriched plasma membrane preparation.

The French press and the Hughes press are typical examples of **tissue presses**. There are many different types of design for tissue presses.

French press

In a **French press** (Figure 3.3) the cell suspension is poured into a chamber which is closed at one end by a needle valve and at the other end by a piston. Pressures of up to about 10^7 Nm^{-2} (16000 $lbin^{-2}$)are applied by a hydraulic press against a closed needle valve. When the desired pressure is attained, the needle valve is fractionally opened to marginally relieve the pressure. The cells subsequently expand and rupture, thereby releasing the cellular components through the fractionally open valve. The French Press is often used to disrupt microbial cells.

Hughes press

The procedure in a **Hughes press** can be regarded as a solid shear method because the cells are disrupted with a solid abrasive. Moist cells are either placed in a chamber with an abrasive and then forced through a narrow annular space by pressure applied to a piston at -5 °C or the cells are ruptured by shearing forces generated by ice crystals acting as an abrasive at -25 °C and the frozen material is forced through a narrow orifice.

Sonication

Sonication techniques use ultrasonic energy to generate high transient pressures that are believed to disrupt the cells. The use of high frequency sonic or ultrasonic energy to disrupt cells is not commonly used because there are many problems associated with its use. There is a considerable amount of heat generated by sonic vibration that cannot be completely eliminated, and in addition, there is considerable subcellular organelle damage, DNA breakage, oxidation of sulphydryl groups and cleavage of electrostatic and peptide bonds.

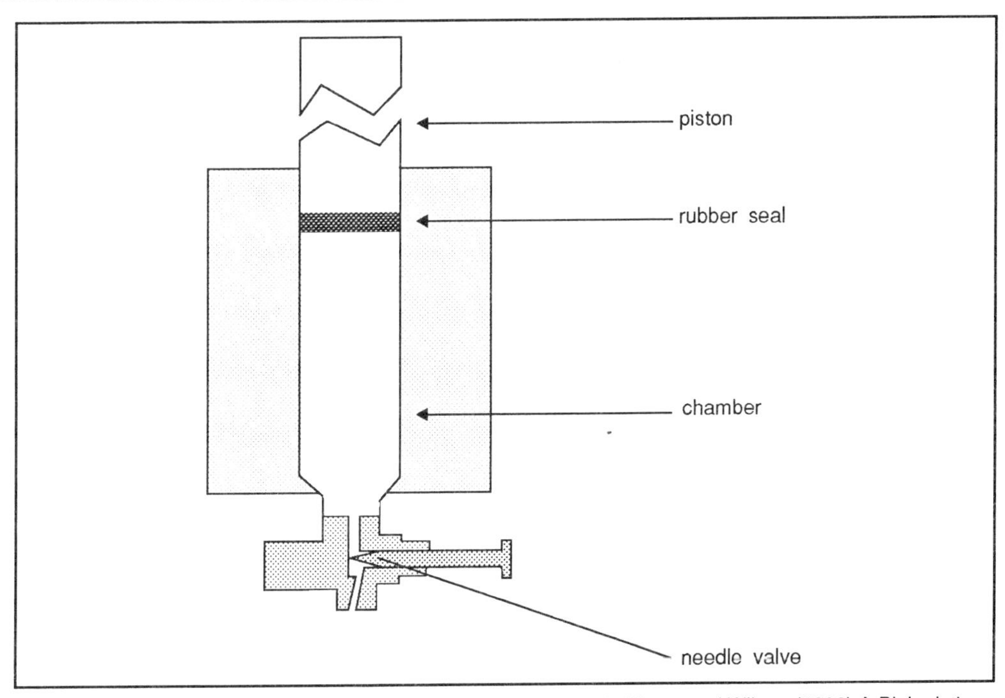

Figure 3.3 The pressure cell of the French press.Redrawn from Williams and Wilson (1986) A Biologist's Guide to Principles and Techniques of Practical Biochemistry pub.Edward Arnold, London.

osmotic shock

In the **osmotic shock** method, cells may be disrupted by placing them in a hypo-osmotic medium. The sudden change in water potential of the suspending solution results in cell lysis.

In the **freezing/thawing** technique, the cells may be broken open by successive periods of freezing and thawing. Ice crystals are generated during the freezing stage which disrupt the cells when they melt during thawing. This technique however, is slow and releases a limited amount of subcellular components.

The extraction of fermentation products from microbial cells needs a technique that has large scale application. Examples include intracellular enzyme extraction from micro-organisms that are protected by a thick cell wall. Many of the procedures described above have been used for the large scale extraction of the fermentation products from a fermentation broth; further details will be found in, the BIOTOL text 'Product Recovery in Bioprocess Technology'.

3.3.4 Non-physical methods of cell disruption

In the **chemical disruption** of cells, the phospholipid bilayer of the plasma membrane is disrupted by solubilisation of the integral membrane protein. This can be achieved primarily in three ways.

Organic solvents, such as chloroform/methanol mixtures are commonly used to dissolve membrane lipids and release the integral proteins and subcellular components within the cell. Organic solvent or mineral acid extraction does not however, preserve morphological or metabolic integrity.

Chaotropic anions that include compounds such as potassium thiocyanate, potassium bromide and lithium diiodosalicylate are believed to enhance the solubility of hydrophobic groups in an aqueous environment by influencing the order of the water molecules around the group. Disruption of the lipid membrane releases the subcellular components for further analysis.

Detergents are widely used for disrupting the membrane bilayer. Detergents solubilize the integral membrane proteins by interacting with the phospholipid bilayer. They form mixed micelles of the various components of the membrane and the detergent.

The detergents can be categorized into three classes. Sodium dodecyl sulphate (SDS) is an anionic detergent that disrupts the membrane. Non-denaturing detergents such as the bile salts, cholate and deoxycholate are also effective in disrupting the membrane and in contrast to denaturing anionic detergents, these permit observation of the integral membrane proteins in their native state. The third class of detergents are the non-ionic detergents, such as triton X-100 and octyl glucoside, that do not destroy enzyme activity of integral membrane proteins.

Enzymatic digestion of the cell wall of the plant, yeast and bacterial cells results in protoplasts (cell wall-deficient cells) which may then be disrupted by mild shearing forces generated by the Potter-Elvehjem or the Dounce homogenizer. A complex mixture of broad specificity enzymes may be used in hydrolysing the cell wall. These include the chitinases, pectinases, lipases, proteases and cellulases. Several commercial cellulytic enzyme preparations are isolated from wood-degrading fungi. *Cellulysin* (Calbiochem-Behring Corp.), a common cellulytic enzyme preparation is widely used in the isolation of plant protoplasts. Previous studies have revealed that the impure enzyme preparations often prove superior in protoplast isolation to the pure cellulase because of the diversity of constituent enzymes. However, batch quality is variable and this may adversely affect protoplast viability and purity within a heterogenous mixture of cells.

It is also important to appreciate that enzymic digestion of the connective tissue surrounding many animal cell types has been used to release cells from their organs. Collagenase and hyaluronidase are routinely used to isolate hepatocytes from the liver as they digest the surrounding matrix and generate individual viable cells.

Summary and objectives

We have learnt in this chapter that the plasma membrane must be disrupted in order to isolate subcellular components such as intracellular membrane fractions, organelles or a soluble compound. The conditions of homogenization are chosen to cater for isolation of these fractions.

We have discussed the components of the homogenization medium with reference to their function.

The bulk of the chapter described the physical and non-physical techniques of cell disruption are evaluated as an integral part of cell fractionation.

Now you have completed this chapter you should be able to:

- place cell and tissue homogenization as a necessary prerequisite for the subsequent separation of subcellular components;
- describe the complex nature of the isolation medium;
- list the main parameters that need to be controlled during homogenization;
- appreciate the difficulty in determining the components for the particular homogenization medium;
- understand the main methods of cell disruption;
- select suitable procedures for disrupting certain cells or tissues.

Centrifugation

Centrifugation

4.1 Introduction

Π In the previous chapter we considered the disruption of tissue and cells to give a homogenate. Before reading on write down a list of the various components which may be present in such a homogenate.

Depending on the original material disrupted your list should have included such components as debris from connective tissue, intact cells, sub-cellular fractions (mitochondria, plasma membranes, tonoplasts, lysosomes), macromolecules (proteins, nucleic acids, lipids), small molecules in addition to salts and other components of the original homogenization medium.

The most common method of separating such particles is centrifugation. This technique depends on the fact that particles of different density (buoyancy), size and shape sediment at different rates in a centrifugal field. Centrifugation can be used as a process for preparing angles of, for example a particular cell fraction. It can, however, also be used as an analytical tool to provide information about the size and shape of molecules. We deal with this aspect of centrifugation in the second part of the chapter.

4.2 Basic theory of centrifugation

The sedimentation rate of a particle is critically dependent upon the applied centrifugal force generated within a centrifuge. The applied centrifugal force is often expressed as:

$$\text{Applied centrifugal force} = \omega^2 x \qquad \text{(E - 4.1)}$$

where ω is the angular velocity of the rotor in radians per second (ie 2π times number of revolutions per second) and x is the radial distance of the particle from the centre of the rotor.

The applied centrifugal force is referred to as the relative centrifugal force (RCF) when the applied centrifugal force is expressed relative to the earth's gravitation.

$$\text{RCF} = \omega^2 x / g \qquad \text{(E - 4.2)}$$

where g is the acceleration due to gravity.

SAQ 4.1

For an ultracentrifuge operating at 60 000 rpm, calculate the applied centrifugal force and the RCF on a particle 6 cm from the axis of rotation ($g = 9.80 \text{ m s}^{-2}$).

The motion of a particle in a fluid medium under the influence of an applied gravitational field is determined by the balance between the gravitational force and the frictional resistance of the medium.

rate of sedimentation

In accordance with 'Stokes' law, the rate of sedimentation of a rigid, spherical particle under an applied force $\omega^2 x$ can be expressed in the following relationship:

$$v = 2\omega^2 xr^2 (\rho_p - \rho_m) / 9\eta \qquad \text{(E - 4.3a)}$$

where v is the sedimentation rate ($= \frac{dx}{dt}$,) r isthe radius of the particle, ρ_p and ρ_m are the densities of the particle and medium respectively, and η is the coefficient of the viscosity of the medium.

This equation is commonly written as:

$$dx / dt = 2\omega^2 xr^2 (1 - \vartheta \rho_m) / 9\eta \qquad \text{(E - 4.3b)}$$

where dx/dt is the sedimentation rate and ϑ the specific volume (ie volume per unit mass) of the particle.

The sedimentation rate of the spherical particle is therefore directly proportional to the square of the radius, the difference in density of the particle and suspending medium and the gravitational field. It is inversely proportional to the viscosity of the medium. It is important to bear in mind that sedimentation is not solely dependant upon size.

SAQ 4.2

Calculate the rate of sedimentation of a particle at a distance 5 cm from the axis of rotation in an ultracentrifuge operating at 50 400 rpm.

$\rho_p = 1.330$ g cm^{-3}, $\rho_m = 0.998$ g cm^{-3}, $r = 2.7 \times 10^{-3}$ µm, $\eta = 0.890 \times 10^{-3}$ kg cm^{-3}s^{-1}

SAQ 4.3

In what circumstances will a spherical particle suspended in a medium under an applied centrifugal field remain stationary?

This is as far as we will take the theory of centrifugation at this stage. This is as much as you need to know in order to apply centrifugation to the preparation of sub-cellular fractions and molecules. In section 4.7, we will, however extend our discussion of centrifuges to examine their use as tools for analysing the sizes and shapes of molecules. For this we will need to deal with a little more theory. This we will cover in sections 4.7.3 and 4.7.4.

4.3 Preparative centrifugation

4.3.1 Introduction

isolation

Preparative centrifugation is concerned with the isolation of a relatively large amount of biological material for further analysis.

There are principally three types of centrifugation that are used for this purpose:

- differential centrifugation;
- rate-zonal centrifugation;

- isopycnic centrifugation.

4.3.2 Differential centrifugation

A heterogenous suspension of particles can be separated into particular fractions when it is subjected to an increasing relative centrifugal field over an extended period of time. This technique separates particles as a function of size and density and is only suitable for materials that have markedly different sedimentation characteristics.

A particular centrifugal field is chosen over a period of time in order to sediment or pellet predominantly particles of larger mass and retain smaller or less dense particles within the supernatant. To improve the purity of the pellet, it is then subjected to repeated washing steps of resuspension in the suspension medium and centrifugation (Figure 4.1). As a consequence, the fraction will be of greater purity but the yield will be reduced.

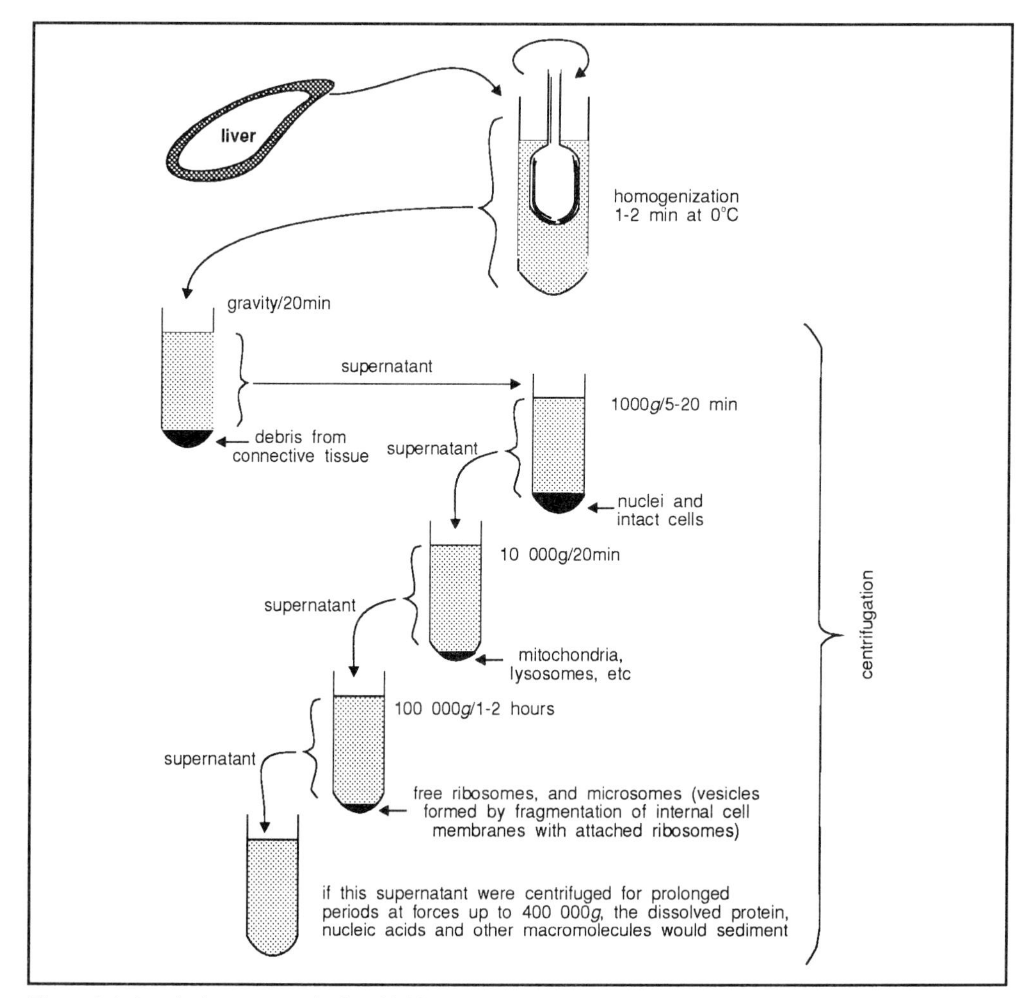

Figure 4.1 A typical program to isolate highly enriched organelle fractions from the liver homogenate.

Differential centrifugation of homogenized tissue is routinely conducted to remove contaminating cell components such as heavy nuclei, cell debris, cell walls, chloroplast and mitochondria and to form microsomes prior to density gradient centrifugation.

Π Examine Figure 4.1. At what centrifugal force do free ribosomes sediment?

You should have identified that subjecting a suspension of ribosomes to a centrifugal force of 100 000g for 1-2 hours will cause them to sediment. Note, however, the rate at which they sediment depends on the density and viscosity of the medium.

SAQ 4.4 Make a list of the shortcomings of differential centrifugation?

4.3.3 Rate-zonal centrifugation

In this technique, centrifugation proceeds in a medium of increasing density. The density of the particles must be greater than the maximum density of the gradient medium. The sample is carefully layered onto a pre-formed density gradient the density of which increases towards the bottom of the tube (see Section 4.5.3). A positive increment in density of the suspending medium prevents premature diffusion of the sample. Centrifugation subsequently proceeds for a fixed period of time so that the particles separate into a series of bands in accordance with parameters such as the centrifugal force, size and shape of the particles and difference in density between the particles and the suspending medium (Figure 4.2). However, the particles will sediment over a prolonged period of time.

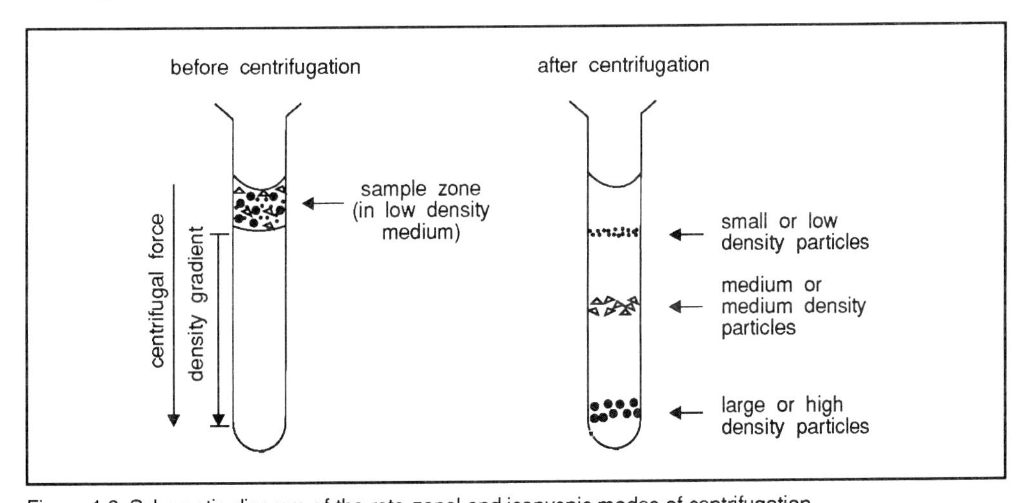

Figure 4.2 Schematic diagram of the rate-zonal and isopycnic modes of centrifugation.

Π Make a list of the purposes of the density gradient?

The sort of items you should have included are:

- to prevent the diffusion of the sample into the suspending medium prior to centrifugation;
- the increased density and viscosity of the medium prevents the mixing of the column by convection;

- to facilitate the separation of particles of varied sedimentation characteristics (size, density) from a heterogenous suspension.

Rate-zonal centrifugation is commonly used for the isolation of subcellular fragments such as plasma membranes, tonoplasts and golgi apparatus from plant microsomes.

4.3.4 Isopycnic (equal density) centrifugation

This technique is based upon the principle that the density of the particle never exceeds the maximum density of the gradient. In contrast to rate-zonal centrifugation, isopycnic centrifugation is based solely upon the buoyant density of the particles and is unaffected by the size or shape of the particles. Separation of the particles proceeds until their buoyant density is equal to the density of the gradient. No further sedimentation occurs towards the bottom of the centrifuge tube and as a result, this technique is often referred to as 'equilibrium isodensity centrifugation' and is independent of time. It is important to appreciate that the buoyant density of particles is critically dependent upon the extent of hydration of the particles by the density gradient medium. The choice of media is therefore extremely important and will be discussed in detail in Section 4.4.

There are mainly two types of isopycnic centrifugation: in one, the continuous density gradient is pre-formed prior to centrifugation; in the other, the gradient is formed during centrifugation.

pre-formed density gradient method

In the pre-formed density gradient method the sample is carefully layered onto a continuous density gradient which has a density in the same range as the particles of interest. Sedimentation will cease when the buoyant density of the particles is equal to the density of the gradient.

self-forming density gradient method

Salts of heavy metals have been used traditionally for separation of macromolecules by the self-forming density gradient method of isopycnic centrifugation. These salt solutions are ionic and have an associated low water activity. For example, concentrated solutions of caesium salts ($CsCl$ and Cs_2SO_4) are used in the separation of nucleic acids and solutions of rubidium salts are predominantly used in the separation of proteins. Detailed discussion of the properties of the heavy metal salts and the effects of an ionic environment on the buoyant densities of biomacro molecules is beyond the scope of this chapter. Note that the BIOTOL series contains a text dedicated to the isolation and characterisation of nucleic acids and proteins. ('Analysis of Amino Acids, Proteins and Nucleic Acids'). Buoyant density centrifugation of nucleic acids is described in detail in that text. Here we will summarise the procedure.

The sample, eg DNA, is mixed thoroughly with a concentrated solution of caesium chloride to form a uniform and homogenous suspension. Centrifugation generates a density gradient of caesium chloride and subsequent separation of the nucleic acids to their individual buoyant densities. This technique has been used to separate leucocytes and erythrocytes, and human lipoprotein complexes. It is however, most commonly used as a tool in molecular biology to separate nucleic acids.

4.4 Nature of density gradient material

4.4.1 Introduction

In this section we shall consider all of the main types of density gradient media used for centrifugation. Each individual type of medium is critically evaluated in terms of its chemical and physical properties. To see if you appreciate some of the problems involved attempt the following SAQ.

SAQ 4.5

A gradient medium should encompass many important properties to ensure that it is an ideal material for the separation of biological material by density gradient centrifugation. What would you suggest are these extremely important parameters?

4.4.2 Density gradient media

The variation of density, viscosity and osmotic pressure at different concentrations of a range of media is illustrated in Figure 4.3.

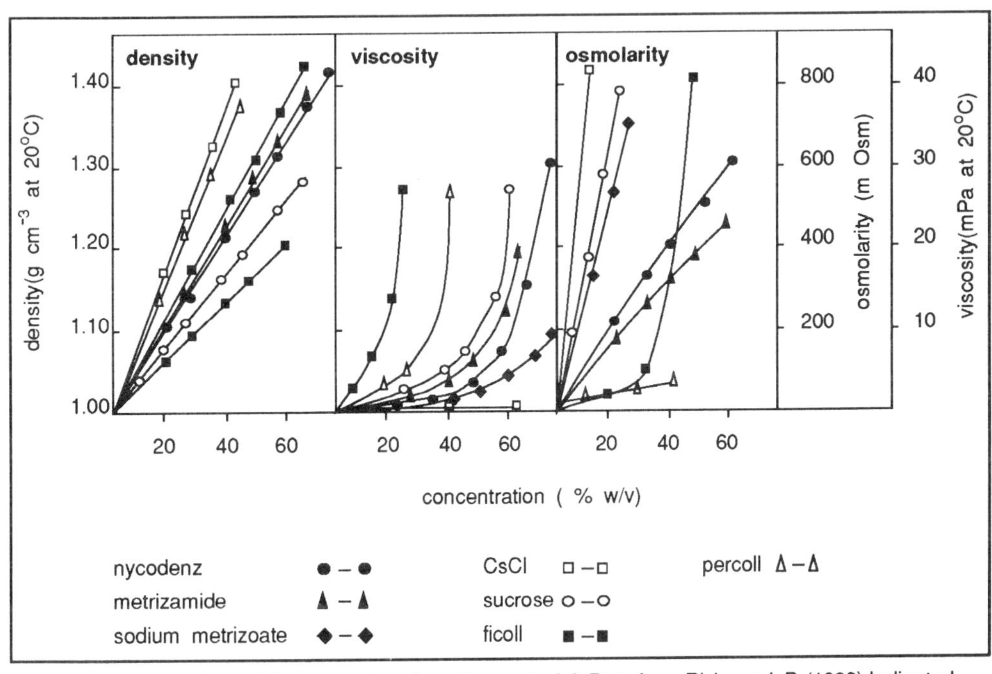

Figure 4.3 A comparison of the properties of gradient material. Data from Rickwood, D (1983) Iodinated Density Gradient Material. Information Retreval Ltd., Oxford University Press, Oxford.

sucrose

Sucrose is one of the most suitable gradient media for rate-zonal centrifugation and has been routinely used in the isopycnic centrifugation of subcellular organelles, membrane fractions and viruses. Sucrose is extremely stable in solution and inert towards biological material and it is also readily available and inexpensive. However, it exerts a high osmotic pressure at comparatively low concentrations and its high viscosity at concentrations exceeding 40% (w/w) precludes its capacity for use in isopycnic

osmotic pressure

centrifugation where certain particles would simply not be able to reach their buoyant densities.

glycerol

Glycerol has frequently been used as a gradient medium for rate-zonal centrifugation because it can preserve many enzyme activities. The main problem is that a glycerol solution of equal density to that of sucrose is considerably more viscous.

polysaccharides

Several different polysaccharides have been used to eliminate the problems of high osmotic pressure associated with sucrose. These include glycogen, dextrans and Ficoll (a copolymer of sucrose and epichlorohydrin). Ficoll is the most commonly used polysaccharide gradient material with a molecular mass of 400 000. Although Ficoll exerts a comparatively low osmotic pressure at low concentrations, it does have acute viscosity problems.

iodinated compounds

Most iodinated compounds used as gradient material are based upon the structure of tri-iodobenzoic acid. Examples of this class of gradient media include Metrizamide, Nycodenz, Triosil and Urografin. All iodinated compounds, except Metrizamide and Nycodenz have a free carboxyl group. Metrizamide and Nycodenz have blocked carboxyl group. As a result of the blocked carboxyl groups, these two compounds are soluble in many organic solvents, and are unaffected by either the ionic environment or the pH of the solution. Metrizamide and Nycodenz have been found to be extremely inert as gradient media and can be used to separate many particles such as nucleoproteins, nucleic acids and subcellular organelles. Metrizamide is commonly used to isolate intact vacuoles from plant protoplasts. Iodinated compounds exert a much lower osmotic pressure and viscosity than do sucrose and polysaccharides. Although caesium salts operate at high density and low viscosity in isopycnic centrifugation, their high osmotic pressure at low concentration precludes their use in the separation of osmotically-sensitive samples. However, Metrizamide absorbs fairly strongly in the ultraviolet region and tends to react weakly with some proteins, which makes it less useable for the separation of proteins.

colloidal silica

The most widely used colloidal silica solution is Percoll. This material has several advantages over other colloidal particles in that it is coated with polyvinylpyrrolidone to prevent adherence to biological material and to enhance the stability of the suspension. Colloidal suspensions exert very little osmotic pressure and can separate large particles such as viruses, large subcellular organelles and bacteria. Unfortunately, these particles absorb strongly within the ultraviolet region of the electromagnetic spectrum and therefore interfere with optical measurements of proteins and nucleic acids. In addition, only particles significantly larger than colloidal silica can be separated because silica will sediment before smaller sample particles separate and form bands in density gradient centrifugation.

As you will see from the above discussion there is no ideal medium for density centrifugation. Thus, it is necessary to compromise and choose the best for a particular situation.

4.5 Practical knowledge

4.5.1 Introduction

Having selected the most appropriate density medium (Section 4.4), it is now necessary to consider the practical aspects of forming a density gradient and then loading the sample onto the surface of the gradient. After the centrifugation, the separated fraction must be removed from the density gradient prior to biochemical analysis.

A density gradient can either be discontinuous or continuous.

4.5.2 Formation of discontinuous density gradient

For this type of gradient, solutions of decreasing density are carefully layered above each other with a pipette. The sample is then layered on top of the gradient prior to centrifugation. If the discontinuous gradient is left for a prolonged period of time, diffusion within the medium will occur and eventually result in a smooth continuous gradient.

4.5.3 Formation of a continuous density gradient

gradient former

To form a linear continuous density gradient, two cylindrical chambers of identical diameter, that are interconnected at their base with a plastic or glass tube, are required (Figure 4.4). The connecting channel must be clamped in order to control the mixing of the contents of the chambers. This apparatus is often referred to as the gradient former.

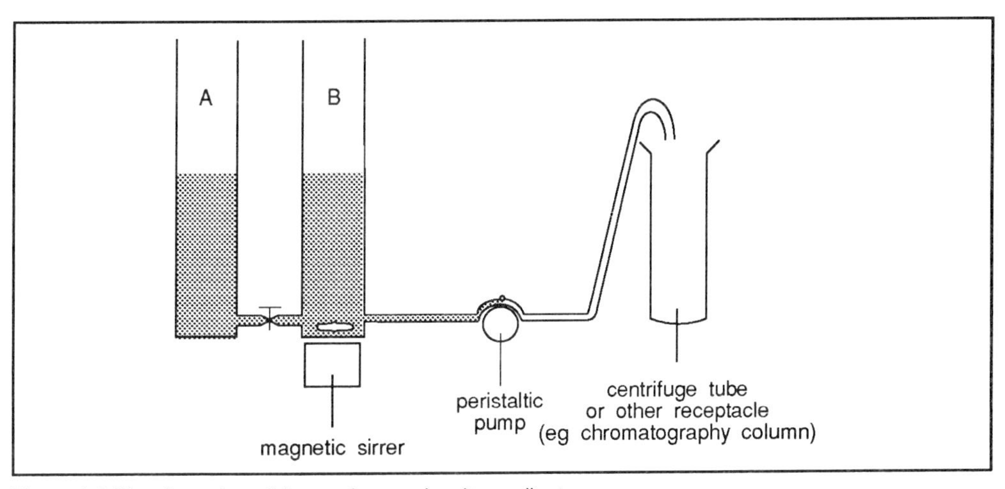

Figure 4.4 The formation of the continuous density gradient.

Reservoir A contains the less dense solution and the mixing chamber B contains an equal volume of the denser solution which is stirred by a magnetic stirrer and possesses an outlet to a peristaltic pump which in turn is connected to the centrifuge tube or other receptacle such as a chromatography column.

It is important to appreciate that the centrifuge tubes can be filled by either of two processes. Centrifuge tubes that are wettable in aqueous solutions such as polypropylene tubes can be filled by resting the flexible plastic tubing against the top of the centrifuge tube and allowing the solution to flow down and fill the tube by 'top loading'. As reservoir B is being stirred, the control valve is opened fractionally and reservoir A is raised slightly in order to remove any air bubbles and then opened fully to force the solution through the tubing at about 0.5 - 1.0 cm^3min^{-1}.

Alternatively, centrifuge tubes, such as polyallomer and polycarbonate tubes, that are not wettable in aqueous solution, must be filled by a process referred to a 'bottom loading'. In this case, the denser solution is placed in reservoir A and the less dense solution is poured into the mixing chamber B. The outlet tubing is fitted with a glass

Pasteur pipette which touches the bottom of the centrifuge tube and as soon as the connecting channel between the two reservoirs is opened, the less dense solution will fill the centrifuge tube first and then be replaced by a denser solution.

linear density gradient

When the diameters of the two cylindrical chambers are identical, a linear density gradient is formed. If however, the diameter of reservoir B is greater than the diameter of reservoir A, a concave exponential gradient is formed (with respect to concentration profile along the centrifuge tube). Alternatively, a convex density gradient is formed when the diameter of reservoir B is less than the diameter of reservoir A.

applications of density gradients

Discontinuous gradients are routinely used for separating subcellular organelles or membrane fractions. Linear continuous density gradients have been found useful in separating nucleic acids, subellular organelles and ribosomal nucleoproteins. Convex gradients, where the initial density gradient near the top of the tube is steep and the gradient near the bottom of the tube is shallow, may be useful in differentially separating particles. The smaller (light) particles near the top of the tube will separate further than the larger (heavy) particles at the bottom of the tube. For concave density gradients, the larger (heavy) particles near the bottom of the tube separate further than smaller (light) particles.

4.5.4 Loading the sample

Precise loading of the sample in rate-zonal centrifugation is critical because enhanced resolution of the bands is dependent upon the narrow layer of sample applied to the top of the gradient.

The sample is applied slowly and smoothly from a Pasteur pipette against the wall of the centrifuge tube and close to the meniscus of the gradient. The pipette tip is then withdrawn gently towards the top of the tube and the sample is allowed to run down towards the gradient to form a sharp and well-defined sample layer.

SAQ 4.6

Why is the precise loading of the sample not critical in isopycnic centrifugation?

4.5.5 Removal of the fractions after centrifugation

Following centrifugation, mechanical disruption of the gradient must be avoided and the centrifuge tubes must be handled with great care.

There are several prescribed methods for removing the bands from the gradient; only a few will be briefly mentioned:

- the bands may be removed carefully with a Pasteur pipette to prevent cross-contamination by another fraction;
- the bottom of the tube may be pierced and the fraction collected as the gradient passes through the hole. Some tubes however, such as polycarbonate tubes, cannot be easily pierced;
- the density gradient can be displaced upwards by a denser solution of the gradient medium introduced carefully at the bottom of the centrifuge tube.

Direct removal of the bands with a Pasteur pipette is only feasible if the fractions are visible. Upward displacement of the gradient is undoubtedly the most common technique for collection of undetectable fractions because there is no piercing of the centrifuge tube. The tubes can then be washed and re-used for another centrifugation step.

4.6 Preparative centrifuges and associated rotors

4.6.1 Introduction

In this section we will be discussing the application of different types of preparative centrifuges. The use of rotors in differential sedimentation, rate-zonal and isopycnic centrifugation is evaluated.

4.6.2 Preparative centrifuges

There are three main types of preparative centrifuge: general purpose, high speed and ultracentrifuges.

General purpose centrifuges are capable of a maximum relative centrifugal field of approximately 6000 *g*. They can utilize a variety of interchangeable fixed-angle and swing-out rotors that can hold large volumes of material. The centrifuges are used primarily to sediment particles rapidly such as plant protoplasts and red blood cells.

In **high speed centrifuges** a maximum relative centrifugal field of about 50 000 *g* can be generated. They are refrigerated to control the temperature of the rotor chamber and can utilize a variety of interchangeable fixed-angle and swing-out rotors. These centrifuges are routinely used to sediment cell debris, or large subcellular organelles such as mitochondria, nuclei and chloroplasts.

Ultracentrifuges can generate a maximum relative centrifugal field of approximately 600 000 *g*. The rotor chamber is refrigerated under vacuum to reduce heat production by the fast-spinning rotor. As a result, the rotors are made of high tensile aluminium alloy or titanium alloy to withstand the great forces generated within the chamber. These centrifuges can separate small subcellular organelles such as ribosomes and membrane fractions.

4.6.3 Rotors

These are five types of rotors available for preparative centrifugation. These are:

- fixed-angle rotors;
- vertical rotors;
- swing-out rotors;
- continuous action rotors;
- zonal rotors.

A detailed discussion of all rotors is beyond the scope of this chapter, but we will outline each.

fixed-angle rotors

Within fixed-angle rotors, the centrifuge tubes are maintained at an angle to the axis of rotation (15°-35°). The particles travel horizontally until they reach the walls of the centrifuge tube and then slide down the tube to form a pellet. Rapid sedimentation of the particles is enhanced by the strong convection current caused by this type of rotor. As a result, this type of rotor is primarily used in differential sedimentation of particles with significantly different sedimentation rates. The fixed-angle rotor is also used in isopycnic centrifugation because here convection currents have no effect upon the separation of particles.

vertical rotors

Vertical rotors consist of tubes that are permanently maintained in the vertical position. They are primarily used in isopycnic centrifugation and not in differential sedimentation because the pellet sediments along the whole length of outer wall of the centrifuge tube.

swing-out rotors

The centrifuge tubes in swing-out rotors are vertical prior to centrifugation but then swing into a horizontal position during the acceleration of the rotor. Swing-out rotors are routinely used in rate-zonal centrifugation and isopycnic centrifugation because particles travel over a longer path length and so this increases the resolution of the centrifugation step. They also encounter much reduced convective currents that will not disturb the resolution of the bands.

continuous action rotors

The continuous action rotors separate small amounts of material from large volumes of suspension. An example includes the isolation of cultured cells from a cell suspension. The zonal rotors were designed to reduce the wall effects encountered within swing-out, fixed-angle and vertical rotors. Their large capacity makes them ideal for large scale preparative rate-zonal centrifugation.

4.7 Analytical centrifugation

4.7.1 Introduction

In the previous sections we considered the use of the centrifuge and ultracentrifuge as a preparative device; ie the separation of a heterogeneous suspension of cells, debris macromolecules etc. It is possible using an ultracentrifuge to obtain values for the molecular mass of molecules and, in conjunction with diffusion data, an idea of the shape of the molecule. To use an ultracentrifuge in the analytical mode we must be able to determine the concentration or concentration gradient of the component of interest in the ultracentrifuge cell.

It is not possible to remove samples and alalyse them while the rotor is in motion so we have to find a way of solving this problem.

In the next sections we will discuss the principles of monitoring concentrations and concentration gradients in analytical centrifuges and we will then go on to discuss the theory and practice of determining sedimentation velocity and the sedimentation equilibrium methods for determining molecular mass and shape.

4.7.2 Monitoring the concentration gradient in an ultracentrifuge

Figure 4.5 depicts an ultracentrifuge rotor head and details of the sample cell. The cell, fitted with thick quartz windows, has a sector shape when viewed at right angles to the plane of rotation since sedimentation occurs radially. Initially the sample is uniform throughout, but as a result of rotation the 'top boundary' of the solute (macromolecule) moves outwards as sedimentation proceeds.

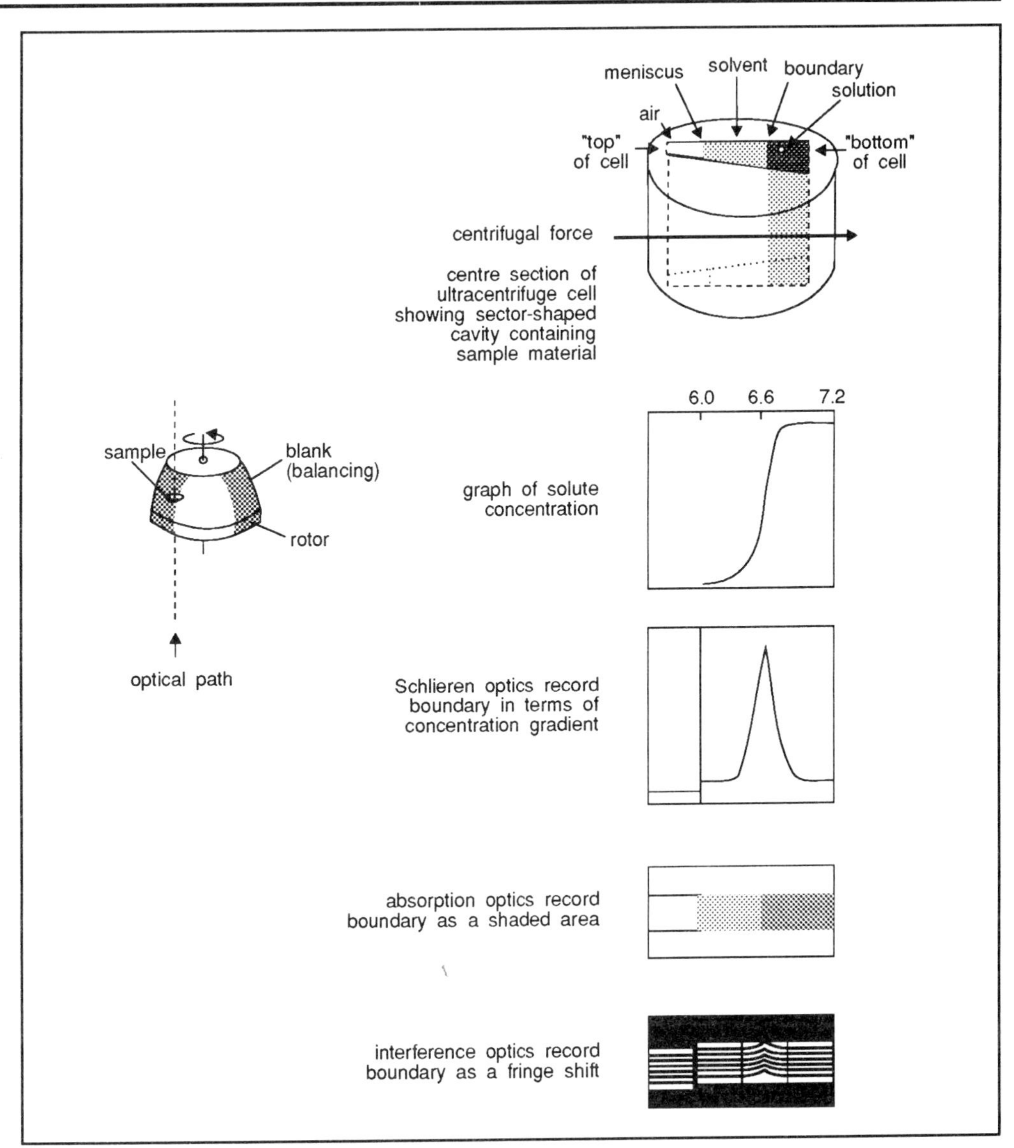

Figure 4.5 Ultracentrifuge rotor and summary of methods use for the study of macromolecules in an analytical ultracentifuge. Note the graph of solute concentration down the centrifuge tube. Note also the 'break' in the recorded parameters (Schlieren optics record, absorption optics and interference optics) at the menisus at the 'top' of the tube. (See text for further details).

The concentration profile of the sample in the sector cell can be monitored spectroscopically, (Figure 4.6), although it is now more common to detect it by making use of its effects on the refractive index of the sample.

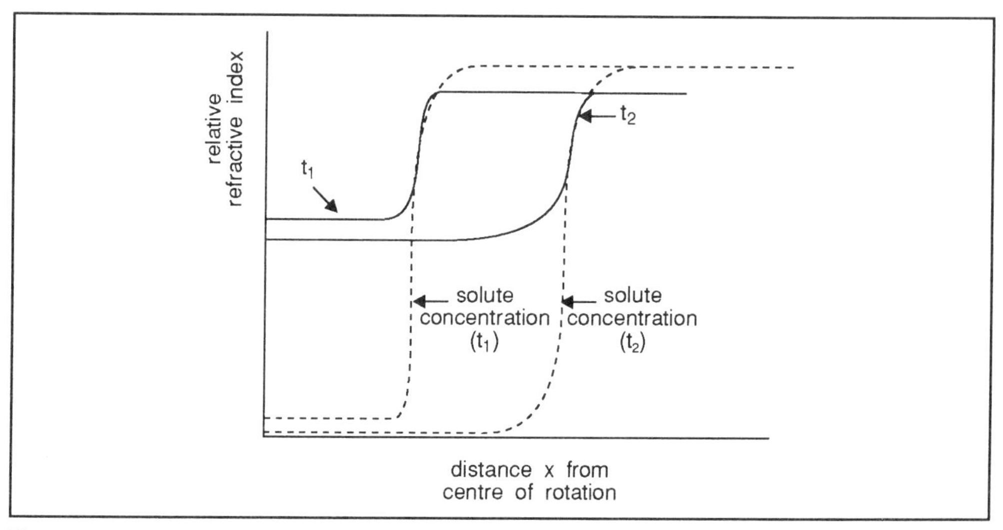

Figure 4.6 Plot of refractive index against distance from the centre of rotation (continuous line), at two times (t_1 and t_2, where $t_2 = t_1 + t\text{min}$). The dotted line gives the solute concentration along the centrifuge tube.

The boundary between the top surface of the solution and the remaining solvent as sedimentation occurs gives rise to an abrupt change in the refractive index between the two parts of the sample. The Schlieren optical system turns this refractive index gradient into a trace of rate of change of refractive index with distance down the cell (Figure 4.7). An alternative interference system monitors the refractive index gradient through its effect on the interference of two beams of light, one coming through the sample, the other through the reference blank cell.

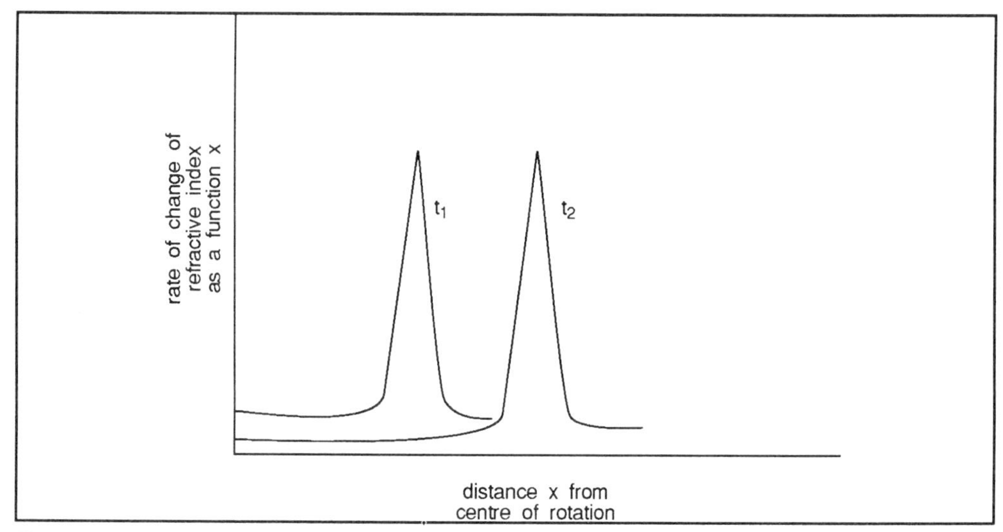

Figure 4.7 Schlieren optics record at two times (t_1 and t_2). In this, the rate of change in refractive index is plotted as a function of x, the distance from the centre of rotation.

Whichever method is adopted the position of the boundary can be determined at regular intervals.

Π What do you think will happen if there is more than one type of macromolecule in the sample?

Quite simply there will be several boundaries formed and these will be evident from the Schlieren pattern - each moving at its own rate down the cell. In this instance, absorption spectroscopy cannot be used to detect the various boundaries.

In the next two sections we will examine the details of how we can determine important parameters such as molecular mass using analytical centrifugation. It is the nature of this approach that it is rather mathematical. Try to work through the mathematics so that you understand, at least in principle, why it is we can determine such parameters molecular mass from analyzing how molecules behave under a centrifugal force. We will build on the basic theory of centrifugation we dealt with in section 4.2. You should recall that sedimentation rates depend on the distance from the centre of rotation (x), the square of the angular velocity (ω^2), the difference in the density of the particle and the suspending medium and the radius of the particle. (If you cannot recall this, re-read Section 4.2 before reading on).

4.7.3 Sedimentation velocity method

In a centrifugal field a solute molecule is accelerated until the frictional force resisting the motion is equal to the acceleration of the centrifugal force multiplied by the effective mass, [$m\ (1 - \vartheta\rho_m)$ where ϑ is the specific volume of the solvent m is the mass of the particle and ρm is the density of the solvent].

The effective mass of a molecule is of course related to the difference between the mass of that molecules minus the effects of buoyancy of the suspending medium, ie $m\ (1 - \vartheta\rho_m)$

The frictional force is the product of the velocity, dx/dt and the frictional coefficient f. When the steady state velocity is attained it can be shown that:

$$f\frac{dx}{dt} = m\ (1 - \vartheta\rho_m)\,\omega^2 x = M\frac{(1 - \vartheta\rho_m)}{N_A}\,\omega^2 x \qquad \text{(E 4 - 4)}$$

where:

ω is the angular velocity of the rotor (ie revolutions per second x 2π)

M is the molor mass and N_A the Avogadro constant.

The sedimentation coefficient (constant) s, defined as:

$$s = \frac{1}{\omega^2 x}\frac{dx}{dt} \qquad \text{(E - 4.5)}$$

is constant for a given solute molecule in the medium used. If the boundary moves from a position x_1, at time t_1, to a position x_2 at time t_2.

$$s = \frac{1}{\omega^2 (t_2 - t_1)} \ln x_2/x_1 \qquad \text{(E - 4.6)}$$

This is the important relationship to remember. Since ω has the units of reciprocal seconds, it follows that *s* has the units of seconds. The sedimentation constants of proteins fall in the range 10^{-13} to 200×10^{-13}s , the unit 10^{-13}s is called a Svedberg (the IUPAC recommended symbol is Sv, most books use the symbol S), named after the worker The Svedberg of Uppsala, who initiated the development of the ultracentrifuge in 1923. Typical values of sedimentation coefficients of biomolecules are shown in Figure 4.8

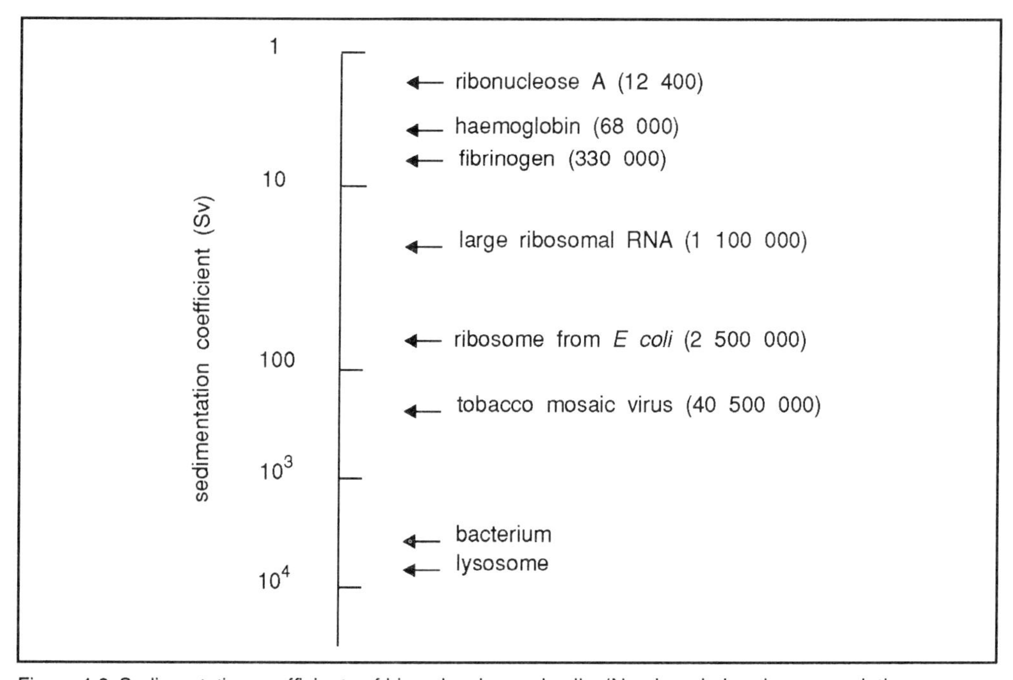

Figure 4.8 Sedimentation coefficients of biomolecules and cells. (Numbers in brackets are relative molecular masses).

Proteins and nucleic acids are often classified in terms of their sedimentation constants. The sedimentation constant is determined experimentally by determining the position of the boundary after fixed time intervals.

SAQ 4.7

In an ultracentrifuge the distance of the boundary of a protein from the axis of rotation was 5.949 cm after 60 min and 6.731 cm after 130 min. The speed of the centrifuge was 50 400 rpm. Calculate the value of the sedimentation constant.

If the value of *s* for successive intervals of time is constant, then the system is monodisperse, ie it contains particles of uniform size.

For a spherical particle, substituting Equation 4.5 into Equation 4.3a gives:

$$s = 2r^2 (\rho_p - \rho_m) / 9\eta \qquad \text{(E - 4.7)}$$

from which r and hence the molecular mass $M \left(= \frac{4}{3} \pi r^3 N_A \rho_p \right)$ can be calculated.

SAQ 4.8

Using the value of s from SAQ 4.7, calculate the effective radius and hence the molecular mass of the protein if

$$\rho_p = 1.333 \text{g cm}^{-3}, \ \rho_m = 0.998 \text{g cm}^{-3}, \eta = 0.890 \times 10^{-3} \text{kg m}^{-1} \text{s}^{-1}$$

State any assumptions which you make.

If the particle is not spherical it is necessary to know the value of the frictional coefficient f. Substituting Equations 3.5 into Equation 4.4 gives.

$$s = M(1 - \vartheta \rho_m) / N_A f \qquad \text{(E - 4.8)}$$

Thus the sedimentation constant itself cannot be used to determine the molar mass of non-spherical particles.

diffusion coefficient

Fortunately, there is a standard relationship between f and the diffusion coefficient, D. D is a measure of the rate at which molecules spread across a concentration gradient, it can be measured by observing the rate at which a boundary spreads. The crucial equation is:

$$f = RT / N_A D \qquad \text{(E - 4.9)}$$

This relationship is independent of the shape of the species.

Combining Equations 4.8 and 4.9 yields.

$$M = \frac{RTs}{D(1 - \vartheta \rho_m)} \qquad \text{(E - 4.10)}$$

Values of f obtained from Equation 4.9 are normally compared to the value f_o ($= 6\pi\eta r$) for a spherical molecule. The ratio f/f_o gives an indication of the molecular geometry of the molecule. For a rod-like (cigar shaped) molecule the axial ratios (length/diameter) for different values of f/f_o is given in Table 4.1

f/f_o	1.04	1.25	1.54	2.95
axial ratios	2	5	10	50

Table 4.1 Values of the axial ratio for different f/f_o values for a prolate ellipsoid.

SAQ 4.9

For bovine serum albumin at 25°C, $s = 5.04 \times 10^{-13}$s, $D = 6.97 \times 10^{-7}$ cm^2 s^{-1}, $\rho_m = 1.0024$ g cm^{-3}, $\vartheta = 0.734$ cm^{3}g^{-1}. $\eta = 0.89 \times 10^{-3}$ kg m^{-1} s^{-1}, $R = 8.314$J K^{-1} mol^{-1}.

Calculate the value of the molar mass and the frictional coefficient. Hence deduce a value for the axial ratio.

4.7.4 Sedimentation equilibrium method

The difficulty with the sedimentation velocity method to determine molar masses lies in the inaccuracies adherent in the determination of the diffusion coefficient. The problem of knowing D can be overcome by allowing the solution to settle into thermal equilibrium. When a quantity of material driven outwards by a centrifugal force is exactly balanced by the amount diffusing in the opposite direction, a state of equilibrium is reached and the boundary remains unchanged with time.

The sedimentation rate is: (compare Equation 4.4)

$$\frac{\mathrm{d}n}{\mathrm{d}t} = c\,\frac{\mathrm{d}x}{\mathrm{d}t} = c\omega^2 x\,\frac{M}{N_A}\left(1-\vartheta\rho_m\right)\left(\frac{1}{f}\right) \qquad \text{(E - 4.11)}$$

where $\frac{\mathrm{d}n}{\mathrm{d}t}$ = the amount of solute flowing per unit time. The diffusion rate by Fick's law:

$$\frac{\mathrm{d}n}{\mathrm{d}t} = -\left(\frac{kT}{f}\right)\frac{\mathrm{d}c}{\mathrm{d}x} \qquad \text{(E - 4.12)}$$

where k, the Boltzmann constant is R/N_A

At equilibrium, the sum of these rates is zero and we obtain:

$$\frac{\mathrm{d}c}{c} = \frac{M\,(1-\vartheta\rho_m)\,\omega^2 x\,\mathrm{d}x}{RT} \qquad \text{(E - 4.13)}$$

Integration between concentration c_1 at x_1 and c_2 and x_2 gives:

$$M = \frac{2RT\,\ln c_2/c_1}{(1-\vartheta\rho_m)\,\omega^2\,(x_2^2-x_1^2)} \qquad \text{(E - 4.14)}$$

To use this technique the centrifuge is run more slowly than for the sedimentation velocity method since it is of little use to have all the solute molecules pressed in a film at the bottom of the cell. Optical methods are used to determine the concentrations (or ratio of concentrations) are different positions in the cell. This thermodynamic method does not require an independent measurement of D to fix the molecular mass, but the time required to establish equilibrium is so long that the method is not much used with substances of molar mass greater than 5000 g mol^{-1}

SAQ 4.10

Calculate the molar mass of hemoglobin from the fact that in an equilibrium ultracentrifuge experiment at 20°C, $c_2/c_1 = 9.40$ where $x_1 = 5.5$ cm and $x_2 = 6.5$ cm. The ultracentrifuge operated at 120 rps, $\vartheta = 0.749 \times 10^{-3}\ m^3kg^{-1}$ and $\rho_m = 0.9982 \times 10^3\ kg\ m^{-3}$.

Summary and objectives

In this chapter we have shown that particles can be separated according to their size, shape and density by centrifugation. However, different types of centrifugation are necessary to separate particles that have either similar or markedly different sedimentation characteristics. In analytical centrifugation, a small amount of material is needed and there is a need to continuously monitor the sedimentation characteristics of the material.

We also described that there is no ideal medium for density gradient centrifugation. Several types of media have been developed and each was discussed with particular reference to application in rate-zonal or ispycnic centrifugation.

Various types of centrifuges and associated rotors were described and their applications discussed. Each type of rotor has specific advantages and limitations.

The theory of the use of analytical ultracentrifuges was also discussed and methods for the determination of molar masses of macromolecules described.

Now that you have completed this chapter you should be able to:

- explain the simple theory of centrifugation;
- calculate values for the angular velocity, the applied centrifugal force and sedimentation rates;
- explain the principles of preparative and analytical centrifugation;
- given an account of the theory of differential, rate-zonal and isopycnic centrifugation;
- appreciate the subtle differences between all of the main types of gradient media;
- describe the important steps in density gradient centrifugation;
- understand the criteria in choosing a particular rotor for separation of certain molecules;
- explain the theory and practice of the determination of molar masses by the sedimentation velocity and sedimentation equilibrium methods;
- calculate sedimentation coefficients and molecular masses from data relating to analytical centrifugation.

Separation methods

Separation methods

5.1 Introduction

In order to analyse and study biological compounds it is often necessary at some stage, to purify them, and much of biochemistry is concerned with separating compounds or groups of compounds, from all the others present in the biological material.

∏ Why might we want to purify a compound?

We might want to purify a compound because we want to:

- use it;
- determine its structure;
- identify the active component from a biologically active mixture;
- study its properties free from the constraints and ambiguities imposed by working with complex mixtures;
- assay it;
- ascertain whether it is present.

Mixtures of biomolecules in essence are complex mixtures of rather fragile molecules in aqueous solution in dispersion. For their purification, the usual methods of purification of organic chemistry, such as distillation, are inadequate. Instead we have to use physical techniques which separate molecules on the basis of only small differences in their properties. Often, extremes of pH and temperature, the use of organic solvents, oxidizing and reducing agents have to be avoided when dealing with molecules separated from biological material, otherwise there is a risk that biological activity will be lost or the molecules denatured.

This chapter serves to explain the general principles involved in separation techniques which generally employ mild conditions and utilize differences in basic physical properties of the molecules such as their size, mass and charge as well as their solubility and adsorption properties.

We begin the chapter by describing the preliminary steps taken in purifying compounds from biological sources to produce clarified solutions or suspensions of the compounds of interests. The bulk of the chapter, however, is concerned with describing the procedures used to separate the various compounds present in the clarified solution. We begin this part of the chapter by briefly reminding you of the types of interactions which can occur between molecules. Knowledge of these interactions are essential if you are to properly understand the principles involved in purification procedures. We then discuss purification procedures which depend upon differences in the solubility of molecules before examining the important area of chromatography. This we extend into a discussion of separation methods which depend upon the electrical charge on molecules. Towards the end of the chapter, we describe separation methods which depend on the differences in the sizes of molecules. Finally, we examine separation methods which rely upon biological specificity, especially the important and potent technique of affinity chromatography.

Here we are primarily concerned with the general principles involved in these separation procedures. For more detailed descriptions of experimental procedures you are recommended to consult the appropriate books in the list given in 'Suggestions for further reading' at the end of this tet. This is a long chapter, so we advise you not to try to study it all at one sitting.

5.2 Preliminary purification

When faced with a complex mixture, such as a cell homogenate, fermentation broth or plant extract, the molecule of interest, perhaps a certain enzyme or an antibiotic or particular form of RNA, is likely to represent only a minute proportion of the total material. What is likely to be required first then, is a number of simple preliminary steps that remove the bulk of the unwanted material. Such steps might be simple physical techniques such as centrifugation or filtration. If for example, a cell culture has been homogenized and the compound of interest released into free solution, the remaining cell debris may be removed by centrifugation of the homogenate in a standard bench centrifuge for a short time, or by filtration of the homogenate through a fairly coarse filter such as paper, muslin, Miracloth, cotton wool, glass fibre filters. This filtration process may be accelerated, if necessary by applying a vacuum to force the filtrate through the filter. (Figure 5.1)

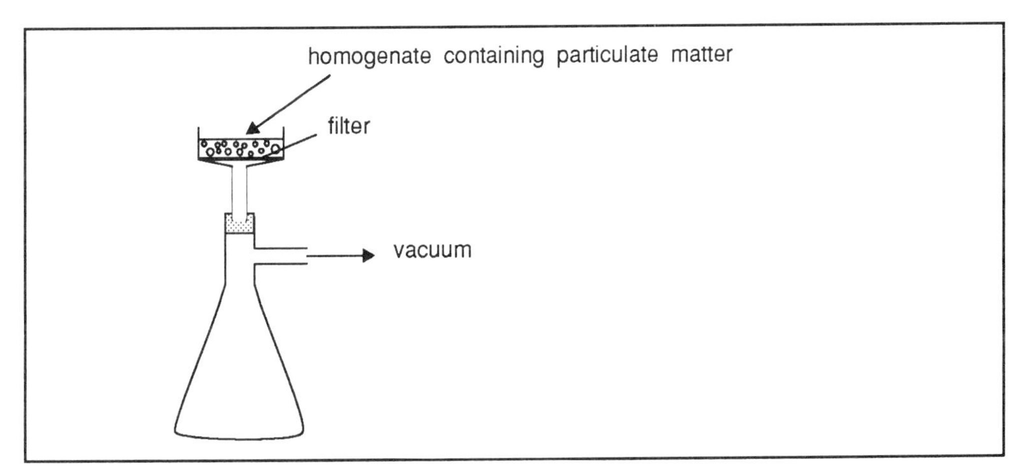

Figure 5.1 Side-arm flask used for vacuum filtration

Physical techniques such as these are suitable for removing particulate matter that may be present in the starting material. We should then be left with a particle-free solution with which it is much more convenient to work. This is where the real purification begins.

Π In attempting to purify an antibiotic from a microbial culture, you mechanically disrupt the cells and remove the cell debris by centrifugation. On testing the supernatant however, no antibiotic activity is detected. What does this tell you about the antibiotic and what additional/alternative preliminary step might you carry out in order to purify it?

Presuming that it has not been destroyed by the centrifugation process, the antibiotic has been removed with the cell debris and is therefore still associated with the biomass. It might be necessary to extract the antibiotic from the biomass using an organic solvent.

structural features

Obviously, different molecules possess different structural features and different physical properties. It is these differences in chemical and physical properties that provide the basis for the separation of compounds and we will look at these properties more closely later.

5.3 Intermolecular forces

Ultimately, the behaviour and properties of a molecule are determined by its structure. Structure determines the mechanism by which molecules interact with other molecules and it is these interactions upon which all purification methods depend.

Let us summarize the basic intra- and intermolecular forces which are so important in determining the properties and behaviour of a molecule in all the various purification systems encountered in biochemistry.

5.3.1 Ionic (Coulombic) forces

electrostatic forces

These are the strong electrostatic forces between oppositely charged ions. These are for example the major forces involved in the structure of inorganic salts such as sodium chloride, and they account for the high melting point of such compounds.

5.3.2 Dipole-dipole forces

Most molecules are not ionic but because their bonding electrons are distributed unevenly over the molecule, they have permanent dipoles. Different atoms have different electronegativities, ie they have different electron-attracting power, and so the molecule is partly polarised: eg HC1.

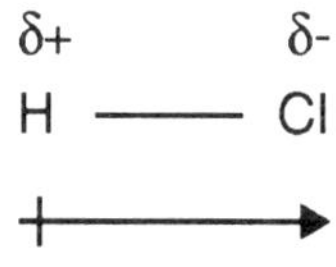

dipole moment

The dipole moment, p, is the product of one of the charges, q, in the dipole and the distance of separation, r of the charges ($p = qr$). The dipole moment of a molecule as a whole is a vector sum of the group moments. Dipole moments are usually defined in terms of the Debye, D (D = 3.336×10^{-30} C m).

The positive end of a molecule with a permanent dipole will be attracted to and align itself at, the negative end of a similar molecule; dipole-dipole interaction takes place (Figure 5.2).

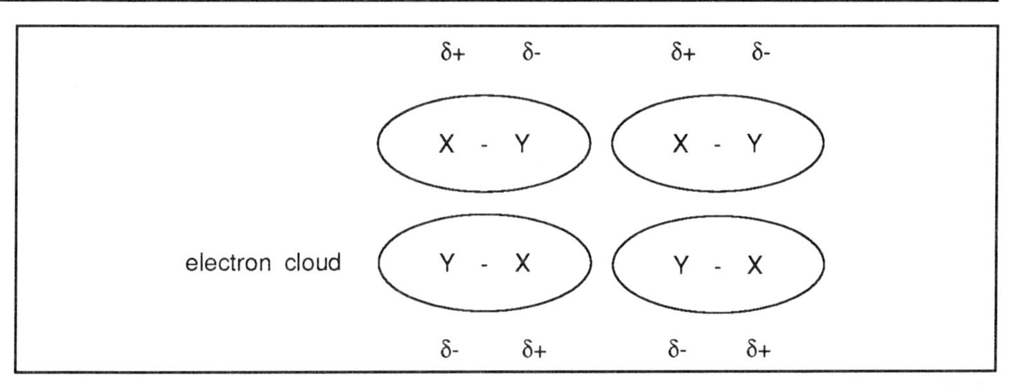

Figure 5.2 Dipole-dipole interactions. Note that molecules may be aligned in various orientations provided opposite charges come together.

A molecule will not have a permanent dipole, however, if it is completely symmetrical, even if it consists of atoms with different electronegativities, eg CCl_4.

5.3.3 Van der Waals forces and London - dispersion interaction

A symmetrical molecule (eg H_2, CCl_4) is non-polar and has no permanent dipole. The distribution of charge over such a molecule therefore is uniform over a given period of time. At any one time however, there might be a small temporary dipole due to the random fluctuation of the electron cloud causing a transient non-uniform distribution of charge. The negative or positive charge of one part of the molecule can distort the electron cloud of an adjacent molecule to induce in that molecule a small opposite charge in the region of contact between the two molecules. These temporary dipoles continuously induce temporary dipoles in adjacent molecules thus giving rise to the attractive forces known as Van der Waals forces (or dispersion forces); these are the reason that non-polar compounds can be liquified or solidified.

Van der Waals forces

London-dispersion interaction

London - dispersion interaction is caused by the attraction of electrons of one molecule by the nucleus of another molecule. This type of interaction occurs both in non-polar and polar molecules. Generally London-dispersion interaction increases with molecular mass because the number of electrons that can be attracted increases with molecular mass. Because of this H_2, F_2 and Cl_2 are gasses at room termperature while Br_2 is a liquid and I_2 a solid under these same conditions. The dipole-dipole interaction, the dipole-induced dipole interaction and the London-dispersion interaction are often included under the general heading of Van der Waals interactions.

The important facts to remember are that the size of the force is determined by the relative polarizability of the electrons of the constituent atoms. This in turn depends on the size of the atoms and the degree to which the electrons are held by the nucleus. Generally, in a given family of elements, the larger atoms are more polarizable as their outer electrons are more shielded from and are therefore less tightly held by, the nucleus.

5.3.4 Hydrogen bonds

Between hydrogen atoms and electronegative atoms or lone pairs of electrons on other elements, there is a very strong dipole-dipole interaction known as a hydrogen bond. These bonds are stronger than normal dipole-dipole interactions and the strength of the bond again depends upon the electronegativities of the atoms. Hydrogen bonding between molecules of water (Figure 5.3a) accounts for many of the special properties of that liquid, including its relatively high boiling point. Hydrogen bonds can also be formed intramolecularly (Figure 5.3b) as well as intermolecularly.

The fact that some non-polar molecules such as hydrocarbons are not water-soluble is due to the fact that they contain no groups capable of hydrogen-bonding with the water molecules and they cannot therefore break the relatively strong bonds between the water molecules as is required for solvation.

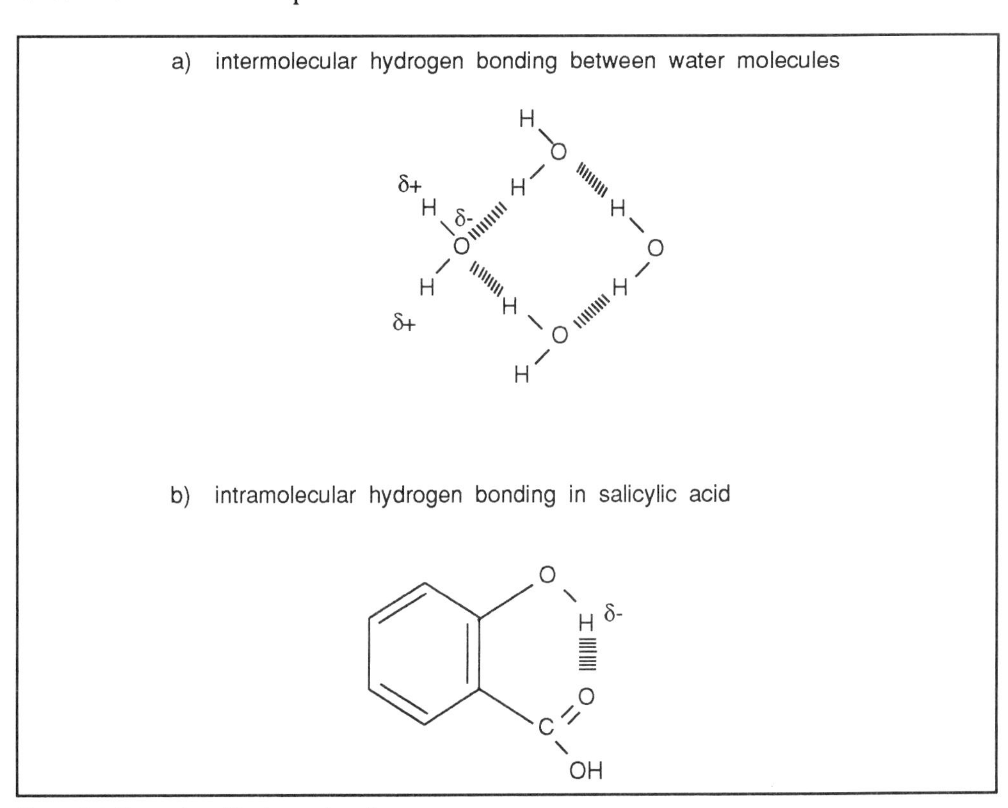

Figure 5.3 Examples of hydrogen bonding

SAQ 5.1

Explain how you would expect each of the following molecules to interact with like molecules. Suggest with reasons, their likely order of boiling point.

1) CH_3CH_2OH

2) CH_3CH_2Cl

3) $CH_3CH_2CH_2CH_2CH_2CH_3$

4) CH_3COONa

5) CH_3CH_3

5.4 Separation methods dependent on solubility

5.4.1 Introduction

Let us now look at the solubility properties of molecules which we can use as a 'handle' in order to isolate them from other molecules.

Π Can you describe in general terms, what sort of organic molecules are most likely to dissolve in water?

Compounds which can disrupt the intermolecular forces between water molecules and establish stable interactions between themselves and the water molecules are likely to be soluble and stable in water. This means compounds that are similar to water, in that they can hydrogen bond in the same way, are likely to be water-soluble. Such compounds are termed hydrophilic - 'water-liking'. (Note: More accurately we can describe that the likelyhood of a compound to dissolve is greatest if it is thermodynamically favoured (that is ΔG is negative). Compounds which give large negative ΔG values when dissolved in water are very soluble and are hydrophilic. Compounds which when mixed with water give positive ΔG values will not dissolve and are hydrophobic-water hating).

Non-polar compounds on the other hand, are less likely to be water soluble as they are less likely to be capable of breaking the strong hydrogen bonds. Many non-polar liquids, such as hexane, are immiscible with water. They interact with other hexane molecules by van der Waals forces and are incapable of forming a stable association with water or most other hydrophilic liquids. Such liquids are termed hydrophobic.

So in general, compounds are soluble only in liquids of similar polarity. These principles of solubility form the basis of a range of separation procedures. The first step in many purification strategies is simple extraction of a substance or a liquid by a range of different solvents, each of which will extract a different range of molecules. Care must be taken in the choice of solvents for the extraction of molecules of biological origin, to avoid loss of activity or denaturation. You should also remember that, as well as the techniques described below, solubility also forms the basis of partition chromatography described in a subsequent section.

5.4.2 Liquid-liquid extraction

solvent extraction

Liquid-liquid extraction (or solvent extraction) is the simplest separation technique, based solely on the solubility properties described above. The principle of separation is the way in which different molecules distribute themselves between a mixture of two immiscible liquids (or phases) - usually an aqueous solvent and an organic solvent. At equilibrium a molecular species will be distributed between the two phases in a characteristic way as described by the distribution coefficient:

distribution coefficient

$$K_D = \frac{\text{concentration in phase A}}{\text{concentration in phase B}} \qquad \text{(E - 5.1)}$$

For distribution between an organic and an aqueous phase K_D is expressed as:

$$K_D = C_o / C_{aq} \qquad \text{(E - 5.2)}$$

where C_o is the concentration of species M in the organic phase and C_{aq} the concentration of M in the aqueous phase. Equation 5.1 and 5.2 are only valid if the species is the same in both phases, ie it is not associated or dissociated in one or other of the phases.

K_D is therefore a measure of the way in which a solute partitions between the two phases. It has no units as it is a measure of relative concentrations. A non-polar molecule will therefore have a relatively high K_D because it is likely to be more readily soluble in the organic phase whereas polar or charged molecules are likely to partition mostly into the aqueous phase. In this way the compound can be selectively extracted into one of the two phases, for example, from an aqueous into an organic solvent 'leaving behind' material which is soluble only in the aqueous phase. If most of the compound is extracted into the organic phase, the aqueous phase can be re-extracted with fresh organic solvent to extract most of the remainder and so on, until almost all of the compound is dissolved in the organic solvent extracts which can be pooled.

SAQ 5.2

The distribution coefficient of a compound A between ether and water is 3. 100 cm^3 of an aqueous solution containing 10 g of A is shaken with:

a) 100 cm^3 of ether

b) Four successive quantities of 25 cm^3 of ether

Calculate the mass of A remaining in the aqueous phase in the two experiments.

Π Can you see any reason for extracting an aqueous solution with an organic solvent if the molecule of interest is insoluble in the organic phase?

Extraction with an organic solvent might remove from the solution, organic-soluble components, eg lipids, fatty acids thus serving to purify the mixture.

In practical terms liquid-liquid extraction is very simply carried out by mixing the two phases in a separating funnel and then allowing them to settle out, after which point the denser solvent can be drained off.

Π Can you think of a way in which you could alter an aqueous solution in order to influence the distribution coefficients of the solutes?

By changing the pH of the solution, solutes would become more or less charged, thus affecting their relative solubilities in the two phases.

So far we have only considered the distribution of a single compound between two immiscible phases. But what happens when a mixture of two or more components is shaken with two immiscible liquid phases?

For any given pair of phases each compound will have its own K_D and separations are based on utilising these differences in distribution coefficients of different compounds. For example, between water and chloroform, compound X may have a K_D of 0.5 whilst compound Y may have a K_D of 4.0. This means that, after equilibrium has been established, in the chloroform phase there will be half the concentration of X but four times the concentration of Y, compared to the water phase. It is this difference then, in the way compounds distribute themselves between two phases, of whatever nature, that enables them to be separated. If these two phases are separated and each again shaken with a fresh sample of the second phase further enrichment of the two layers would occur and the components X and Y would be further separated. This is the basis of countercurrent distribution. (5.3) and chromatography where the components distribute or partition themselves between two phases. These phases may be immiscible liquids, as we have discussed, or more commonly a liquid and a solid or a liquid and a gas.

5.4.3 Counter current chromatography

Counter current chromatography (CCC) is essentially a more elaborate method of liquid-liquid extraction in which a fresh sample of one or both phases is added after each distribution. At its simplest CCC basically consists of a series of interconnected separating funnels. More often a CCC apparatus consists of a helical coil in which one of the liquid phases is held in place by centrifugation whilst the other phase is then pumped through the coil enabling partition to take place without mixing. Another version involves passing droplets of mobile phase up through a column of liquid stationary phase.

Under these conditions each component in a mixture passes through the system at a rate dependent on its K_D. If the components have different K_D values they will travel at different rates and thus can be separated. The advantages of this method are that it can handle relatively large quantities of material (eg about 20 g), individual zones of the separate components can be collected as they emerge, and since it is a multistep process, closely related compounds (K_D values very little different) can be separated.

5.4.4 Precipitation

salting out

A crude purification method based on solubility properties is precipitation whereby the dissolution of relatively large amounts of a compound such as an inorganic salt, causes less soluble components of a mixture to precipitate, or 'salt out'. These can then be separated by centrifugation or filtration.

Precipitation can be brought about either by addition of organic solvents which lower the dielectric constant of the medium; this in turn causes charged molecules to interact more strongly with each other so that their solubility decreases, or by the addition of salts which compete with the compound for interaction with the water molecules.

This technique is particularly suited to biomolecules which are relatively insoluble, such as some proteins and polysaccharides.

5.5 Chromatography

5.5.1 Introduction

In this section we will consider the basic principles of separation of components in a mixture by various techniques which are dependent on the distribution of the components between two phases.

5.5.2 Basic principles of chromatography

All forms of chromatography have two features in common, namely they all consist of a solid (or liquid) which is immobile - the **stationary phase**, over which is passed the **mobile phase** - a liquid (or gas) containing the sample. Separation is based on the way the sample molecules interact with the two phases.

Chromatography was first employed to separate biological molecules in the 19th century when the Russian scientist Tswett, passed a solution of a plant extract over a column of powdered chalk. This resulted in the separation of the individual pigments as distinct bands on the chalk.

Π From Tswett's experiments, explain what constitutes the stationary phase and what constitutes the mobile phase.

The mobile phase is the solution of plant extract containing the compounds to be separated. This passes over the stationary phase - in this case the powdered chalk.

In practice, almost all chromatography is carried out in essentially this form. The stationary phase in the form of fine particles is packed into a column and is saturated with the mobile phase. Then the sample to be separated is loaded onto the top of the column, and as the mobile phase carries the sample through the column, sample components are separated and eventually elute from the end of the column. (Figure 5.4).

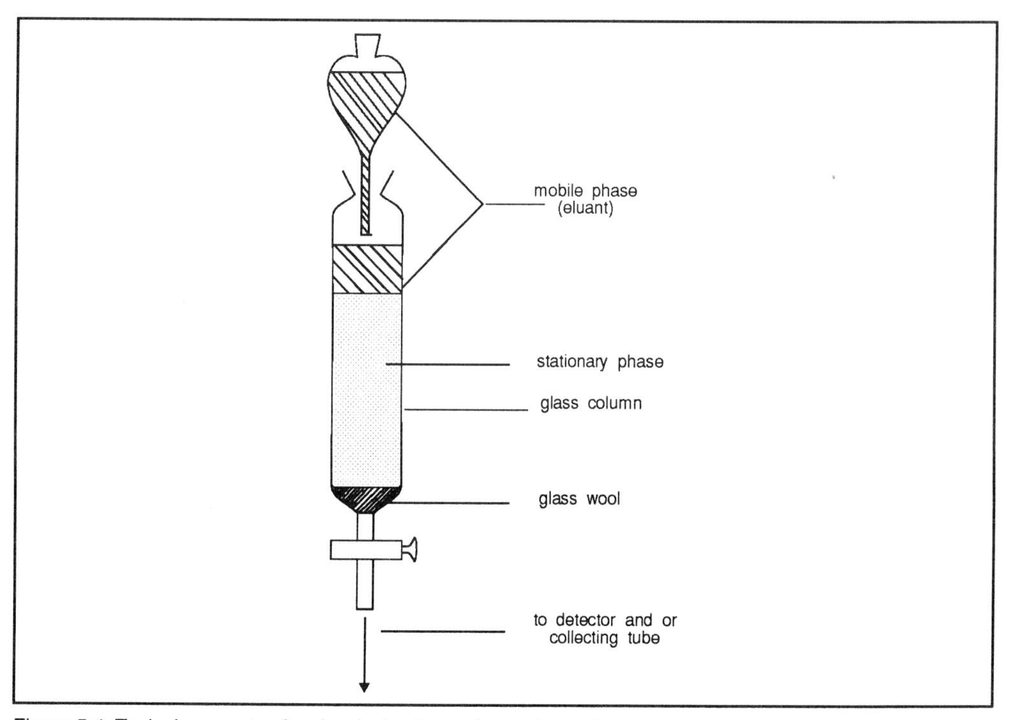

Figure 5.4 Typical apparatus for classical column chromatography.

The chromatography is followed by monitoring the solution as it elutes from the column. This is done by various means, most commonly by measurement of absorbance in the UV region (Figure 5.5).

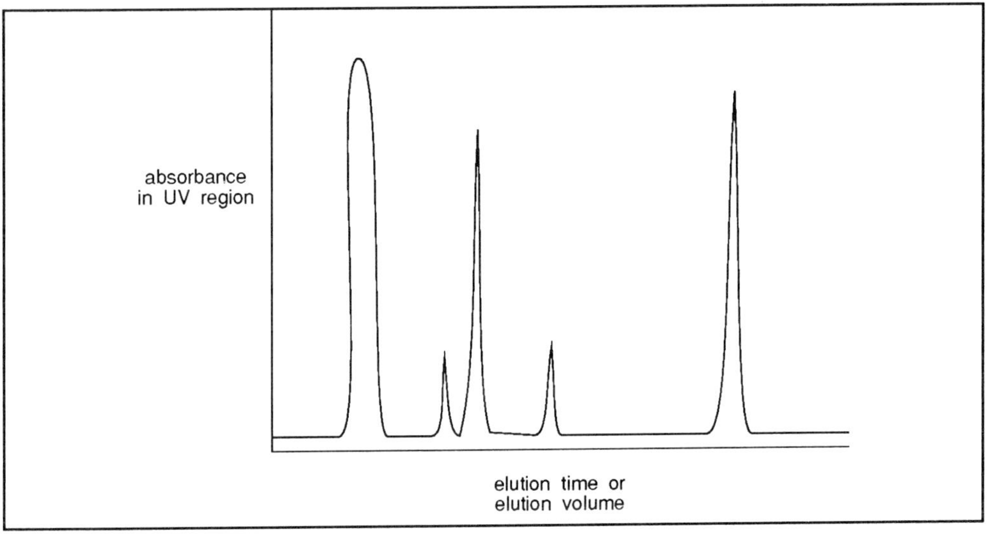

Figure 5.5 A typical chromatogram. Each peak represents a single component.

Π The sample is generally applied to a column in a relatively small volume of solution. Why do you think this is so?

To achieve the best separation the sample components should move as sharp bands of material so that the separated bands do not overlap with one another. This is facilitated by loading the starting material as a small volume.

As the mixture passes down the column, the solutes are in constant dynamic equilibrium between the stationary phase and the mobile phase (Figure 5.6).

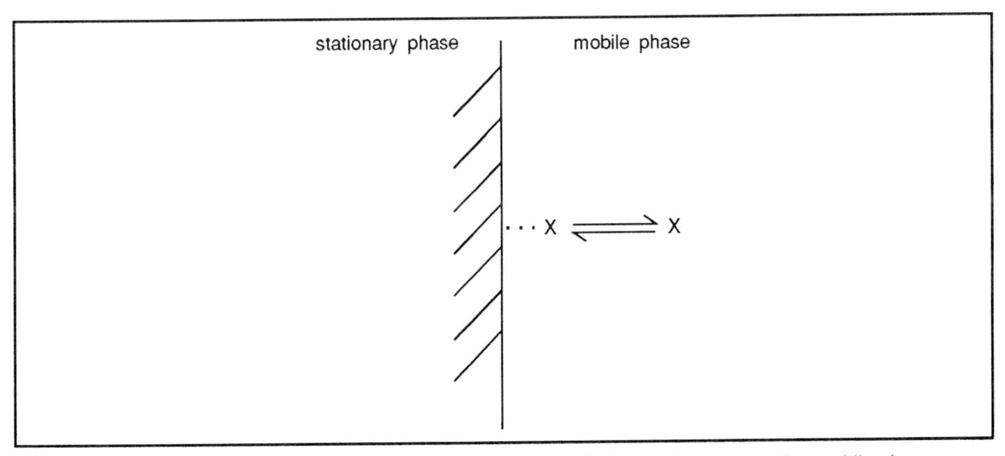

Figure 5.6 Molecule X is in constant equilibrium, interacting with the stationary, and the mobile phases.

The separation is based on the differences in distribution of the compounds of the mixture between the stationary and mobile phases. Thus, although molecules are constantly transferring between phases, certain molecules will tend to 'spend longer' in one of the phases, than other molecules. For example, molecule A may have less affinity for the stationary phase than for the mobile phase - so that the equilibrium in Figure 5.6 will be shifted to the right, whereas for molecule B the equilibrium may lie to the left as it has a greater 'affinity' for the stationary phase.

The position of equilibrium is determined by the strength of interaction of the sample compounds with the stationary phase and with the mobile phase (as well as by the interaction of the mobile phase with the stationary phase). As the mobile phase moves over the stationary phase at a constant rate, A will travel down the length of the column more quickly than B. A is said to have a shorter retention time (t_R), as it is retained on the column for less time than B. This separation of compounds is illustrated in Figure 5.7.

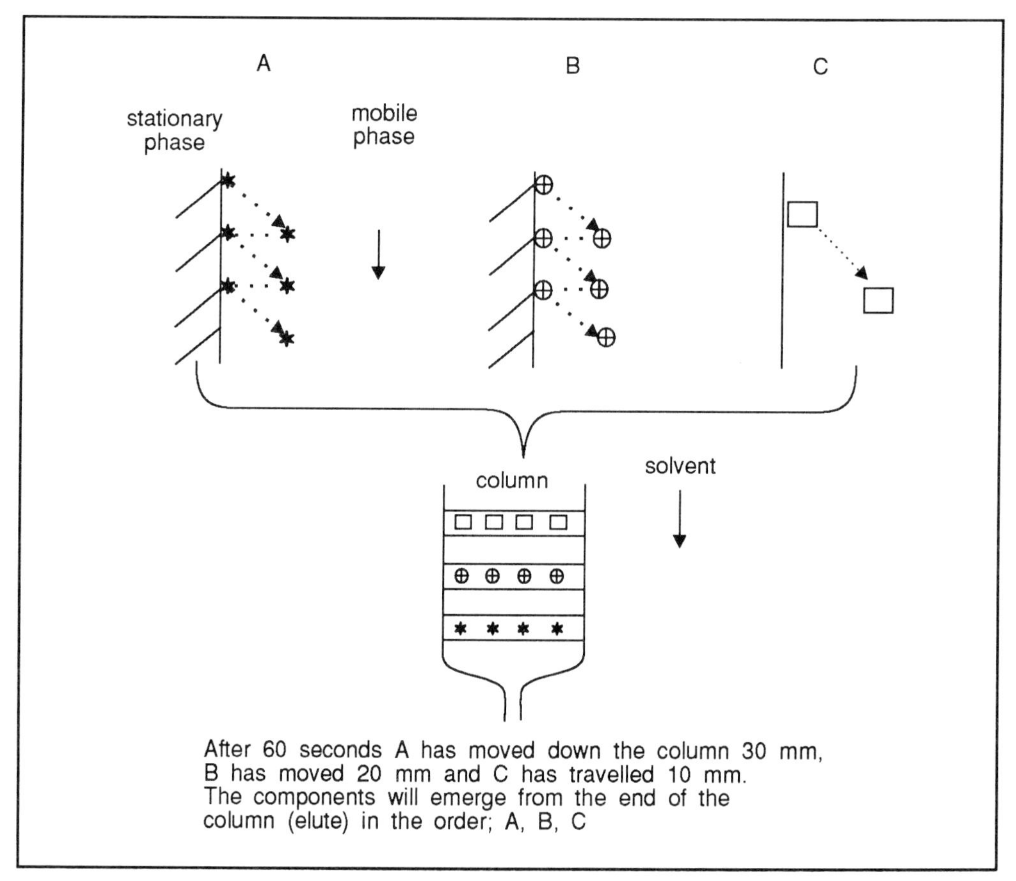

Figure 5.7 The way in which molecules separate chromatographically can be shown diagrammatically with molecules A, B and C. (See text for explanation).

Π Which of the compounds (A,B or C) illustrated in Figure 5.7 spends least time in the mobile phase?

You should have concluded that it was C as it is the slowest moving of the three compounds illustrated.

SAQ 5.3

Which of the following interactions are important in determining the behaviour of a molecule on a chromatography column?

1) interaction between the molecule and the column packing (stationary phase);

2) interaction between the molecule and its solvent (mobile phase);

3) interaction between the solvent and the column packing.

Π If X interacts with the stationary phase more strongly than does Y on the same column, which is likely to be eluted first and which will have longer retention time?

Clearly Y will elute first whilst X will be retained longer by the stationary phase and will therefore have a longer retention time.

particle size vs flow rate and resolution

Stationary Phases: A wide range of different materials involving different forms of molecular interaction are employed as stationary phases in chromatography. The stationary phase usually takes the form of fine particles which provide a large surface area for separation to take place. By using smaller particles a larger area for exchange between mobile and stationary phases in available per unit length of column, and therefore the smaller the particles used, the greater the separating (resolving) power of the column. However, this advantage is countered by the corresponding decrease in flow rate which allows the molecules time to diffuse which counteracts good resolution. An optimal particle size balances good resolution with reasonable flow rate.

Mobile Phases: The choice of the mobile phase is very important as it needs to be appropriate for the sample, (ie the sample must be both soluble and stable in the mobile phase) and compatible with the particular stationary phase being used. It is the one factor that can be most readily altered when establishing a chromatographic system and much of the skill of chromatography lies in selecting a suitable mobile phase.

This is particularly so in adsorption and partition chromatography (see 5.5.3 and 5.5.5) where the choice of an unsuitable mobile phase may for example, cause all the sample components to interact very weakly with the stationary phase giving rise to uniformly short retention times and consequent poor separation. Another unlucky choice of the mobile phase, however, may lead to the solutes being retained too strongly by the column and not being eluted. The trick is to find a happy medium whereby components have different retention times and are thus separated but are also eluted within a reasonable time. The separated components elute ideally as narrow bands of material, in small volumes of the mobile phase, (represented on a chromatogram as sharp peaks), before they have had time to diffuse to a significant extent which leads to less efficient separation.

isocratic elution/gradient elution

If development (ie the process by which the sample is carried through the column by the mobile phase) is carried out with an unchanged mobile phase, this is termed isocratic elution. Gradient elution involves altering the mobile phase during the column run, changing the composition of the mobile phase. This can be done continuously, or in a stepwise manner in which the mobile phase is changed at a number of particular points during the chromatography. In this way, the 'eluting power' of the mobile phase can be changed during the chromatography. This can be useful for several reasons; using an unchanged mobile phase may mean that some sample components elute from the column satisfactorily whilst others elute only after an unreasonably long time. Having eluted the faster migrating components, it may then be possible to increase the eluting strength of the mobile phase and elute the rest of the sample components. Gradient elution can therefore be used to separate compounds with greatly different chromatographic properties, a process which might be impractical with isocratic elution. Gradient elution can also sometimes lead to better resolution of compounds which have very similar retention times in an isocratic system.

Monitoring chromatography: To discover if and how a mixture is separated we need to follow the chromatography in some way. This can be done by online monitoring of the solution as it elutes from the chromatography column, commonly by measurement of absorbency in the UV region (Figure 5.5), or by collecting the eluent in a number of relatively small, discrete fractions which can then be individually tested, perhaps chemically, for the presence of different components.

5.5.3 Adsorption chromatography

This is perhaps the simplest of all chromatographic techniques where there is partitioning of molecules between a solid stationary phase and a liquid mobile phase. In adsorption chromatography the solvent and solute molecules compete for sites on the stationary phase and separation occurs because different molecules adsorb and are displaced differently. The adsorption is a purely physical process - no chemical bonds are formed as the molecules are continuously becoming bound and unbound from the solid surface of the adsorbent.

SAQ 5.4

Which of the following types of interaction do you think might be involved in the process of adsorption.

a) ionic bonding;

b) dipole-dipole interactions;

c) Van der Waals forces;

d) hydrogen bonds.

As adsorption is a physical process not involving a chemical reaction, a wide range of different materials can be used as adsorbents.

Π Make a list of the essential properties of a material to be used as an adsorbent.

The adsorbent of course, must not react with either the sample or the mobile phase and neither must it dissolve in the mobile phase.

adsorbents

The materials most commonly used for adsorption are silica, alumina, carbon and cellulose, all in a fine particle form, to provide a large surface area. Other materials that are sometimes used include magnesium silicate, magnesium oxide and calcium carbonate. The surface consists of discrete adsorption sites; in silica for example, these adsorption sites are the hydroxyl groups attached to the silica atoms (silanol groups). The number and geometric arrangement of the adsorption sites determines the level of activity of the adsorbent and the arrangement of the adsorption sites can be such that it is even possible to separate geometrical isomers.

Most adsorption chromatography carried out is normal phase chromatography, ie the stationary phase is polar and the mobile phase is non-polar. This is generally the case with chromatography employing silica and alumina. Polar molecules and those capable of hydrogen bonding will adsorb strongly to such materials whereas non-polar compounds will be less strongly adsorbed. Non-polar adsorbents include carbon (eg graphitized charcoal) and these can adsorb non-polar molecules which bind by dipole-dipole interactions and Van der Waals forces.

Π What type of compound would you expect to elute first from a silica column?

As described above, adsorption to the silica takes place by the formation of hydrogen bonds. Polar molecules therefore will bind to the silica most strongly and non-polar molecules will elute from the column first.

solvent strength parameter

Retention of solutes on a silica column will therefore decrease with increasing solvent polarity. Solvents can be arranged in the order in which they displace solutes from an adsorbent. As the solvents displace solutes by competing with them for adsorption sites, this order depends upon the strength of interaction between the solvent and adsorbent. This strength is termed the solvent strength parameter (E°) and represents the adsorption energy per unit area of standard adsorbent. In general, the higher the solvent strength parameter, the more rapidly will the solvent displace the solute from the adsorbent surface.

Table 5.1 shows the order - (or elutropic series), of a range of solvents on alumina (Al_2O_3). The elutropic series for solvents on alumina and silica are almost the same but both are very different from that on carbon.

Solvent	E^{o}
pentane	0
cyclohexane	0.04
tetrachloromethane (carbon tetrachloride)	0.18
l-chlorobutane	0.26
diisopropyl ether	0.28
benzene	0.32
diethyl ether	0.38
chloroform	0.40
dichloromethane	0.42
tetrahydrofuran	0.45
butanone	0.51
propanone (acetone)	0.56
ethyl ethanoate (ethylacetate)	0.58
l-pentanol	0.61
ethanenitriile (acetonitrile)	0.65
pyridine	0.71
propanol (-1-ol and -2-ol)	0.82
ethanol	0.88
methanol	0.95
1,2 - dihydroxyethane (ethylene glycol)	1.11
ethanoic (acetic) acid	Large
water	Large

Table 5.1 Eluotropic series for solvents on alumina

Thus if a solute will not elute from a column it may be necessary to alter the mobile phase by using a different, more polar solvent or more likely, by introducing or increasing a more polar component so that the overall E° of the mobile phase is raised. By selecting proportions of solvents with known E°, it is possible to construct a mobile phase with a suitable solvent strength.

uses of adsorption chromatography

Adsorption chromatography can be used for a wide range of substances but is used particularly for separating non-polar compounds. Because the nature of the solute-adsorbent interaction is largely dependent on the functional groups of the solutes, samples tend to be separated according to class type rather than any other single parameter and the technique is useful therefore for separating a class of molecules from other classes. As mentioned above, the regular geometrical position of adsorption sites means that various positional isomers such as 1,2-, 1,3- and 1,4-dihydroxybenzenes can be separated by adsorption chromatography. The technique can be used to separate very hydrophilic substances such as carbohydrates, that are not retained on reverse phase columns which have a non-polar stationary phase. (We will explain reverse phase chromatography in section 5.5.5).

5.5.4 High performance liquid chromatography

uniform small particles used

High performance liquid chromatography (HPLC) differs from classical column chromatography in that the stationary phase is in the form of very small, regularly sized particles (generally about 5 μm in diameter). This means that there exists a relatively large surface area on which separation can take place - the column is said to have a high capacity and thus gives rise to more efficient separation than can be achieved by classical column chromatography. The size of the particles means that they are very densely packed and high pressure is therefore required to force the mobile phase through the column. This entails the use of specialized apparatus such as metal columns, and pumps to pump through the mobile phase. This in turn allows relatively high flow rates to be achieved and enables the use of narrow-bore columns. The combination of these factors confers upon HPLC enormous advantages over gravity-fed columns in terms of speed, reproducibility and separative capacity.

Thus, the principles of separation and the mechanisms of interaction with the stationary phase (sorption mechanisms) are exactly the same but the technique offers the advantages of more efficient separation and the ability to work with much smaller sample volumes. HPLC cannot be used, except at great expense, on an industrial scale but is extremely widely used as an analytical or semi-preparative technique.

5.5.5 Partition chromatography

Normal phase chromatography

In partition chromatography, the stationary phase consists of adsorbed liquid. The principle of partition chromatography is basically the same as that for liquid-liquid extraction except that one phase is supported by a matrix and constitutes the stationary phase. Partition chromatography is also rather similar to adsorption chromatography, except that molecules are retained to different extents on the basis of their solubility in the stationary phase rather than on the basis of their absorptive properties. (The way in which solubility can aid separation was discussed at greater length in Section 5.4). In **normal phase chromatography** the stationary phase is polar whilst the mobile phase is non polar or less polar. In the case of classical column chromatography, the stationary phase is water adsorbed on a matrix such as cellulose, starch or silicic acid. The mobile phase consists entirely or partly of an organic solvent(s) and may even be miscible with water, as long as it is still able to behave as an effectively immiscible mobile phase. This demonstrates that the aqueous phase is best regarded not simply as water but as an insoluble complex with the matrix.

These days most partition chromatography is carried out as a form of HPLC. In this case, a liquid stationary phase can be coated onto an inert support material such as silica. The method suffers from the disadvantage that the liquid stationary phase may 'leach' of the column.

The liquid stationary phase can be chemically bonded to the support matrix (bonded phase chromatography). This, of course, prevents loss of the stationary phase from the column.

Reverse phase chromatography

reverse phase chromatography

When the mobile phase is more polar than the stationary phase this is termed **reverse phase chromatography**. Reverse phase chromatography is widely used, usually as a form of HPLC. Columns generally consist of silica particles to which are chemically bonded alkylsilyl groups of either 1,2,4,6,8,18 or 22 carbons and these are often employed in conjunction with mobile phases consisting of water and/or ethanenitrile (acetonitrile) or methanol.

bonded phase chromatography

Thus bonded phase chromatography can be described as a form of modified partition chromatography; although solutes are retained on the column in part according to the way that they partition between the two phases, (ie according to their solubility), the sorption mechanism is in fact more complex than this. The solutes are retained mainly by hydrophobic interactions and therefore, solutes generally elute in order of increasing hydrophobicity. Thus the most hydrophobic molecules are eluted last. Solutes generally have longer retention times on columns with stationary phases of increasing hydrocarbon chain length.

The fact that there is effectively more than one form of interaction at work and the fact that almost all organic molecules interact by hydrophobic forces to at least some extent, are reasons why reverse phase HPLC is such a valuable and widely used technique.

Π What kinds of molecules (polar or non-polar) would you expect to have the longest retention times on reverse phase HPLC columns?

Non-polar molecules are likely to be retained for longer by the non-polar stationary phase and will therefore have longer retention times than polar molecules. As with solubility generally, the rule of thumb is 'like dissolves like'.

5.5.6 Paper partition chromatography

Cellulose in the form of paper sheets makes an ideal support medium where water, adsorbed between the cellulose fibres forms the stationary phase. In this method the mixture is spotted onto the paper, dried and subsequently the chromatogram is developed by allowing the solvent to flow along the sheet. The solvent front is marked and, after drying the paper, the positions of the components present in the mixture are visualized by a suitable staining reagent (eg ninhydrin for amino acids). The ratio of the distance moved by the compound to that moved by the solvent is known as the R_f value.

$$R_f = \frac{\text{distance moved by compound}}{\text{distance moved by solvent front}} \qquad \text{(E - 5.3)}$$

The R_f value of a compound is more or less constant for a particular solvent system used under carefully controlled conditions of temperature and pH. For complete identification of a compound it is advisable to include pure known compounds as markers and run these on the same sheet and under the same conditions as the mixture.

Whatman No. 1 or similar grade is the most frequently used paper for analytical purposes. The paper can be impregnated with buffer solution before use, or chemically modified. For separation of lipids and similar hydrophobic molecules silica-impregnated papers are available commercially. The choice of the solvent, like that of the paper, is largely empirical and will depend on the mixture investigated. Ascending chromatography, with the bottom of the paper dipping in the solvent, is frequently employed.

5.5.7 Gas-liquid chromatography (GLC)

Partition chromatography is the most widely used form of gas-liquid chromatography, although the partitioning occurs between a liquid and a gas rather than between two liquids. The liquid stationary phase is coated onto an inert support in a narrow coiled column, 1-3 m long and 1-4 mm internal diameter. Commonly used stationary phases are polyethylene glycols and silicone gums; through this the gaseous mobile phase containing the sample is passed. The mobile phase is an inert carrier gas, usually nitrogen, helium or argon. The column is maintained at a very high temperature to volatilise the sample components.

suitable for volatile compounds

GLC is used mainly to separate non-polar, volatile compounds and has to some extent been replaced by reverse phase HPLC. It is used almost exclusively as an analytical tool and has the advantage that it is capable of dealing with extremely small amounts of sample.

5.5.8 Thin layer chromatography (TLC)

As well as being carried out in columns, separation of compounds by chromatography can also be carried out on a layer of supporting medium spread in a thin film over a sheet of inert backing material, such as glass or plastic. This is similar to paper chromatography but has the advantage that a variety of supporting media can be used. Separation can be by adsorption, partition, ion-exchange (5.6.2) or gel filtration (5.7.2) depending on the nature of the medium employed.

In this simple and widespread technique, the sample is applied as a spot on the film of say, silica (stationary phase) and then one end of the plate is placed in solvent so that solvent diffuses upwards acting as the mobile phase, carrying the sample with it. The sample components, as in column chromatography, are in dynamic equilibrium between the solvent and the adsorbent. By the time the solvent reaches the top of the chromatography plate the solutes are separated. Those retained most strongly by the adsorbent remain near their original starting point and those least retained by the adsorbent migrate near the solvent front. The solutes thus separated can be visualized by chemical or physical means and their positions noted as an R_f value (Equation 5.3).

two dimensional chromatography

Separated components can also sometimes be recovered from the TLC plate but the nature of the technique means that it is primarily analytical. Like paper chromatography, TLC has the advantage that it can be carried out in two dimensions. The mixture is separated in the first solvent and then, after drying, the paper or plate is turned through 90° and separation carried out in a second direction using a different solvent. This results in a map of the various components which can be identified by comparing their positions with a map of known compounds developed under the same conditions.

Π Can you see any advantages in carrying out two-dimensional chromatography?

You should recall that each compound has a specific R_f value in a given solvent system. This means that if there are two compounds which happen to have the same or similar R_f values in one solvent system they are unlikely to have the same R_f values in the second system and so will be separated.

Also, if a mixture is very complex it will probably result in a single dimensional chromatogram which is correspondingly complex and may be difficult to interpret because it consists of a large number of spots which are very close to each other. Developing this chromatogram in a second dimension with a different solvent system can help to further resolve the spots.

5.6 Separation methods dependent on charge

5.6.1 Introduction

Most biological molecules carry an electrical charge, the magnitude of which depends on the particular molecule and also on the pH and composition of the medium. This property can be exploited as the lever for separating a mixture by such methods as ion exchange chromatography, electrophoresis and isoelectric focusing.

5.6.2 Ion exchange chromatography

Ion exchange chromatography can be broadly defined as the separation of compounds on an insoluble matrix containing dissociable groups capable of exchanging with ions in the surrounding medium.

Mixtures are passed through a column of material to which are attached charged groups. If these groups are negatively charged they will attract positively charged species in the mixture and are therefore termed cation exchangers. Anion exchangers on the other hand possess positively charged groups and attract negatively charged species.

The ion exchange process itself takes place by substitution of ionic species. The counter ions of the exchanger are displaced by the sample ions which are then bound to the exchanger (Figure 5.8).

Π In what ways could you change the mobile phase so that you could then elute the bound ions?

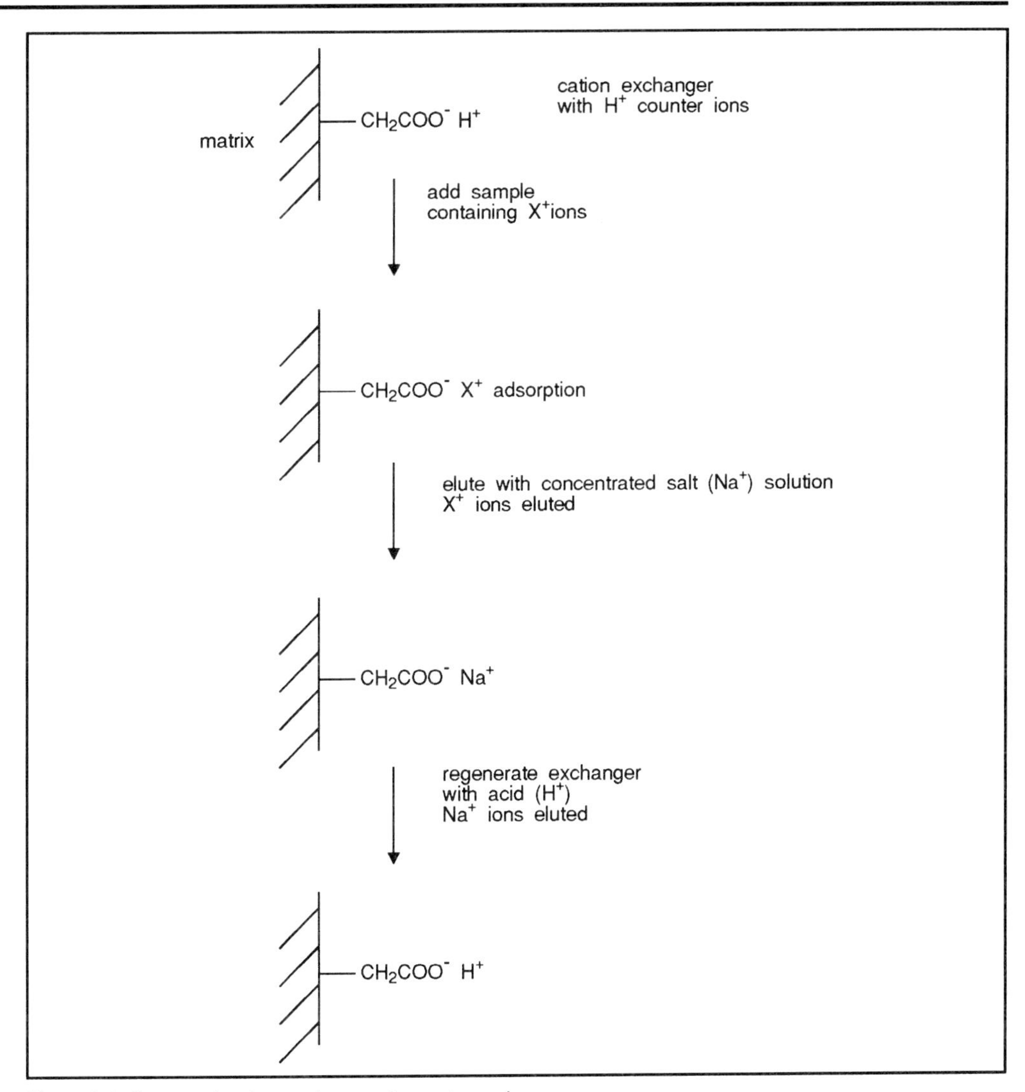

Figure 5.8 Stages of cation exchange chromatography

The sample ions bind to the exchanger by displacing the counter ions; in the same way it is possible to displace the sample ions by other ions. This can be done by increasing the ionic concentration of the mobile phase so that the sample ions are displaced by mass action, or by introducing an ion with a greater affinity for the exchanger.

The bound ions can also be eluted by changing the pH of the mobile phase. This will affect the degree of ionisation of the sample as described by the Henderson-Hasselbach equation:

for a weak acid:

$$pH = pK_a + \log [\text{ (ionised form)/(unionised form) }] \qquad \text{(E - 5.4)}$$

and for a weak base:

$$pH = pK_a + \log [\text{(unionised form)}/\text{(ionised form)}] \quad \text{(E - 5.5)}$$

Therefore by changing the pH or the ionic concentration of the mobile phase either isocratically or more usually in the form of a gradient, the bound ions can be selectively desorbed from the ion exchanger. A mixture of bound proteins for example, will elute sequentially as the pH reaches their individual isoelectric points.

As well as dividing into cation and anion exchangers, ion exchangers can be further sub-divided into strong and weak exchangers. Strong ion exchange groups are completely ionised and exist as charged forms at all except extreme pH values, whereas weak exchangers possess groups which are ionised only over a narrow pH range. Some of the functional groups to be found on the range of commercially available ion exchange resins are listed in Table 5.2. These groups are covalently bound to a insoluble matrix such as cellulose, silica, or often a form of the polymer, styrene divinylbenzene.

Strong anion exchangers	quaternary aminoethyl (QAE) $-CH_2-CH_2-N^+(C_2H_5)_2-CH_2-CH(OH)CH_3$
Weak anion exchangers	diethyl aminoethyl (DEAE) $-CH_2CH_2-NH^+(C_2H_5)_2$
	aminoethyl (AE) $-CH_2CH_2NH^+_3$
Weak cation exchanger	carboxymethyl (CM) $-CH_2COO^-$
Strong cation exchanger	sulphopropyl (SP) $CH_2CH_2CH_2SO_3^-$

Table 5.2 Functional groups of some ion exchangers

capacity of an ion exchange resin

The capacity of an ion exchange resin is a quantitative measure of the number of charged groups available for interaction per gramme of material. This may vary for different sample ions as for example, large molecules may not be able to penetrate the pores of the matrix, in which case the available capacity will be considerably reduced. The matrix is often made of polymer molecules linked together (that is cross-linked) to form a net-work. The degree of cross-linking can in some cases be varied by, for example, varying the proportions of divinylbenzene and styrene during polymerisation. Thus we may produce a loose or tight matrix.

As in most such techniques, in practice there is usually more than one process at work during separation. The pores of the matrix may give rise to some molecular sieving and there may be a significant degree of adsorption of the solutes onto the matrix.

Ion exchange chromatography is generally fairly simple to carry out and can be utilised in the study of almost all forms of biological molecules. It is a very powerful technique capable of separating closely related molecules, such as proteins differing in only one or two amino acids. It is also used widely on a technical scale, in water purification and in the purification of enzymes and antibiotics.

5.6.3 Ion-suppression chromatography and ion-pair chromatography

As mentioned earlier (5.5.5), reverse phase chromatography is one of the most useful and efficient methods of separating mixtures of organic molecules. (You will remember that in reverse phase chromatography the stationary phase generally consists of a non-polar liquid chemically bonded to a solid support and that separation occurs by a form of modified partition chromatography). This is not always ideal however, for weak acids and bases as these may be partially ionised at neutral pH values and so resolve poorly on reverse phase columns. This problem can be overcome by suppressing the ionisation of the molecule so that they exists solely in the unionised form and are consequently better resolved on a reverse phase column.

Π How might you suppress the ionisation of a weak acid?

By carrying out chromatography in the presence of a strong acid, the molecule will exist only in the protonated form, rather than alternating between its ionised and unionised form, as a consequence the separation will be correspondingly improved. Such acids might be sulphuric, perchloric or phosphoric acids.

Similarly, for chromatography of weak bases, the molecule can be retained solely in the unionised form by the presence of a stronger base, such as ammonia. This technique is known as ion-suppression chromatography.

ion-pairs

Ion-pair chromatography (IPC) is an increasingly used method which operates by essentially the same principle. Ionic species may be separated by reverse phase chromatography by conversion to a neutral molecule upon addition of a counter-ion. This method is particularly useful because it allows for simultaneous separation of a mixture of acidic and basic molecules which would not be readily separated by a single exchange column. Inorganic ions may also be separated by their conversion to ion pairs which can be retained by hydrophobic stationary phases.

Commonly used IPC counter-ions include detergents such as sodium dodecyl sulphate (sodium lauryl sulphate). These have large hydrophobic groups which give rise to long retention times on reverse phase systems.

$$M^+ + B^- \rightleftharpoons MB$$

sample ion — ion-pair reagent — neutral ion pair

So although the charge of the molecule is employed to make it amenable to chromatography, the actual mode of separation is by reverse phase, ie modified partition chromatography; you would have a case then, if you are now arguing that these techniques should not really be included in this section at all. However, in ion-pair chromatography the situation is slightly more complex than this. In many cases, what probably happens instead of, or as well as, the formation of a neutral ion pair is that the hydrophobic portion of the counter-ion binds to the non-polar stationary phase by hydrophobic interactions thereby exposing its charged site at which ion-exchange chromatography of the sample molecules can take place.

5.6.4 Paper or zone electrophoresis

electrophoretic mobility

The charge carried by a biological molecule such as a protein, is due largely to carboxyl and amino groups built into the matrix. The relative numbers of these groups on the protein will determine the electrophoretic mobility of the molecule, that is the rate of movement of the protein in a unit applied electric field. The mobility is directly related to the charge of the molecule. In solutions of low pH the carboxyl groups will exist in the undissociated form, while the amino groups will be positively charged. At higher pH values the carboxyl groups are fully dissociated and therefore negatively charged whilst the amino groups are uncharged (Figure 5.9).

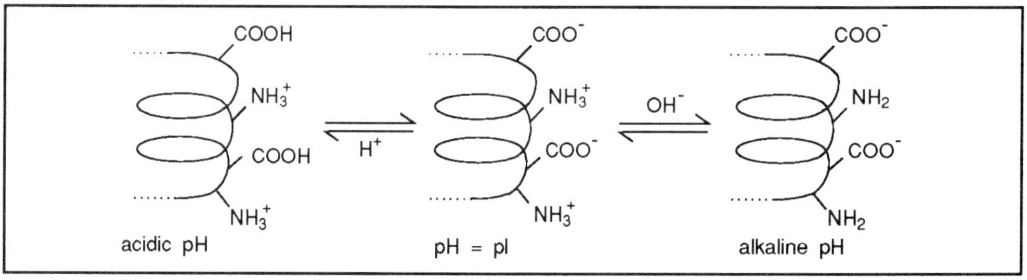

COOH
NH_3^+
COOH
NH_3^+
acidic pH
H^+
COO^-
NH_3^+
COO^-
NH_3^+
pH = pI
OH^-
COO^-
NH_2
COO^-
NH_2
alkaline pH

Figure 5.9 The change state of a protein molecule in solution at different pH values.

Thus in passing from a solution of low pH to one of high pH the charge carried by the molecule will change from positive to negative (Figure 5.10).

isoelectric point

The pH value at which the molecule has no net charge (ie the numbers of positive and negative charges are equal) is known as the isoelectric point, p*I* of the molecule. The isoelectric point is a characteristic of a molecule but it depends on the concentration and nature of the buffer solution.

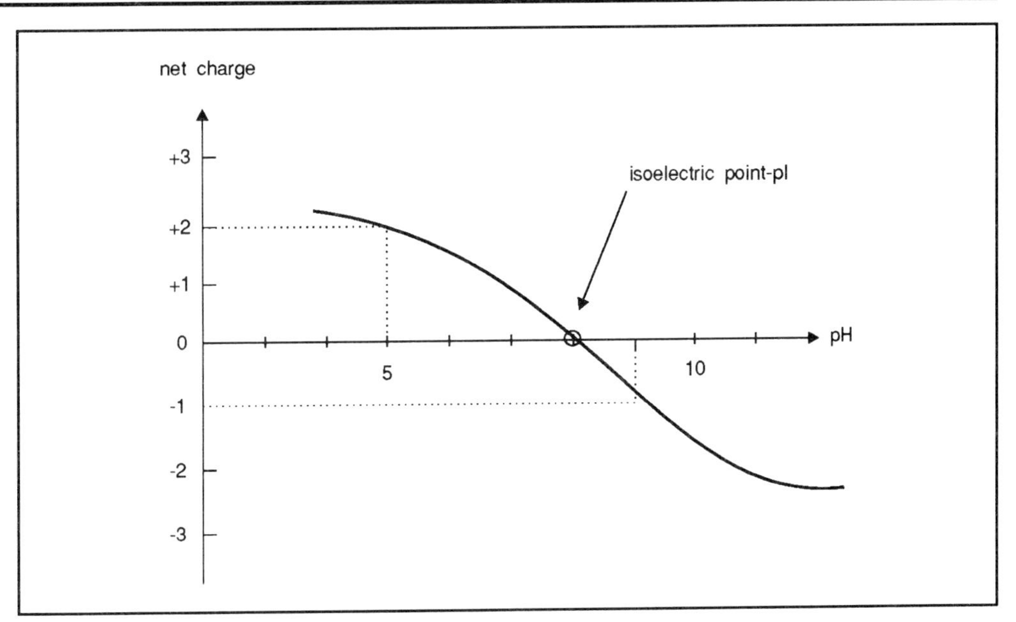

Figure 5.10 Variation of the net charge of a protein with pH

When a potential is applied across a solution containing proteins, each species will move according to its charge. This technique was first introduced by the Swedish biochemist Tiselius about 60 years ago. Complete separation of proteins in a mixture in solution by this means is very difficult to achieve in practice because gravity affects the boundaries and mixing of the zones occurs by convection. However, by carrying out electrophoresis on a supporting medium impregnated with buffer solution the effects of convection can be minimized and a mixture can be completely separated into distinct zones. Supporting media include filter paper and cellulose acetate; agar and polyacrylamide gels are also used (see 5.7.3). A typical apparatus is shown in Figure 5.11. The sample is applied as a spot or band across one end of the separation medium and a stable electric field (2 to 8 V cm^{-1}) is applied. High voltage electrophoresis apparatus (field strengths up to 100 V cm^{-1}) available commercially, is used for the separation of low molecular mass compounds. When electrophoresis is complete the supporting medium is dried and the zones visualized by an appropriate method (eg staining, UV radiation). Figure 5.12 shows the principles of electrophoretic separation of two proteins.

In Figure 5.12, the pH of the medium is 9. At this pH, compound A is well above the pI value (pH 5) and will carry a large net negative charge. Compound B is only just above its pI value (pH 8) and will have only a small net negative charge. Thus, when these two compounds are placed in an electric field, compound A will have a higher electrophoretic mobility than compound B and will therefore migrate faster. The electrophoretic mobilition of A and B at different pHs are shown on the right hand side of Figure 5.12.

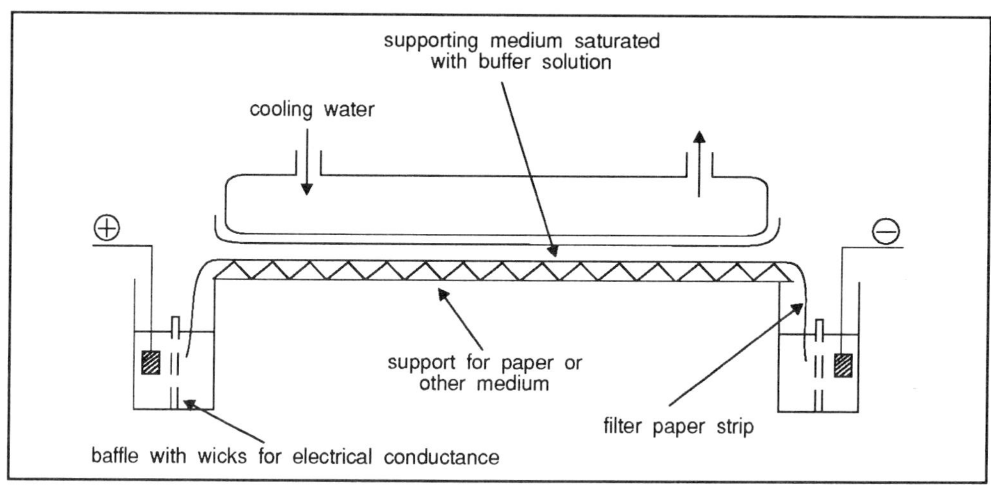

Figure 5.11 Typical apparatus for paper or zone electrophoresis

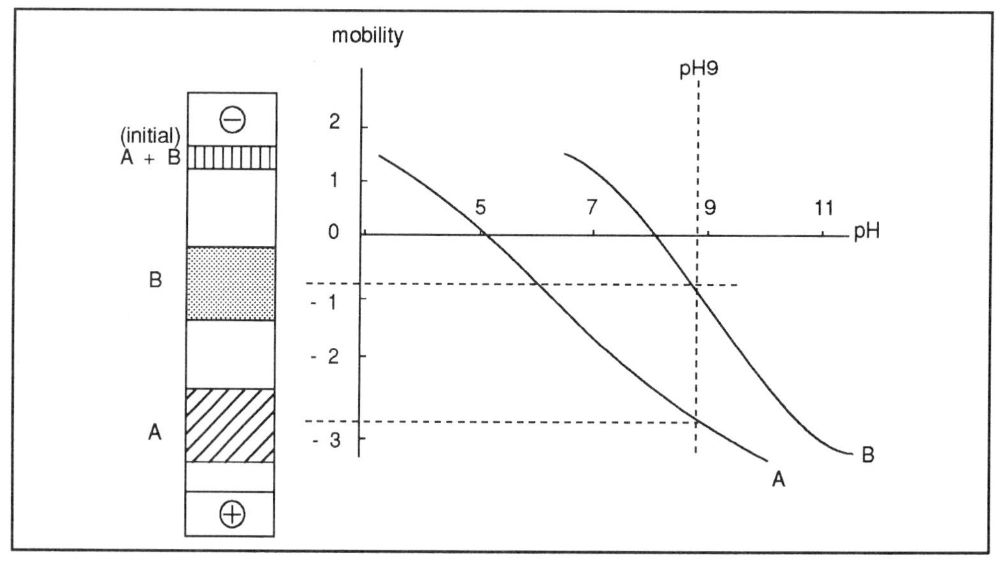

Figure 5.12 Principles of electrophoretic separation of two proteins A and B. Note that pI values for A = pH 5 and for B = pH 8. (see text for description).

Π In the diagram of the apparatus (Figure 5.11) can you explain why:

1) it is necessary to pass cooling water across the medium;

2) the electrode compartments are separated from the filter paper strips by a baffle and wicks.

The passage of an electric current through a thin layer of buffer solution will produce considerable amounts of heat, a phenomenon especially serious in high voltage electrophoresis. Heat causes evaporation of water from the medium, this in turn gives rise to convection currents and also causes changes in the ionic strength of the buffer solution. Both of these factors cause irregular and irreproducible movement of the zones.

The electrodes and electrode compartments are separated from the filter paper strips by wicks to minimise any products of electrolysis (eg H^+ or OH^-) reaching the buffer solution thereby causing a change in pH.

5.6.5 Capillary zone electrophoresis

electroosmosis

Capillary zone electrophoresis (CZE) (also known as capillary electrophoresis, CE) or, high performance capillary electrophoresis (HPCE) is a relatively recent and powerful form of electrophoresis in which separation is carried out in solution in a silica microbore capillary tube. When a high voltage (up to 20 kV) is applied across the tube the sample components separate according to their electrophoretic mobility whilst all being carried towards the negative electrode by electroosmosis (electroendosmosis). Electroosmosis is caused by the negatively charged silanol groups on the capillary tube at pH values greater than 3. These ionised silanol groups result in the formation of an electrical double layer, creating a region of charge separation at the wall-electrolyte interface. When a voltage is applied, there is therefore movement of anions and liquid towards the negative electrode; this is the electroosmotic flow towards the negative electrode. The electroosmotic velocity is greater than the electrophoretic migration velocity of the molecules so all the components migrate in the same direction although separation occurs essentially on the basis of their charge. As separated substances approach the cathode they are monitored by a UV detector. Uncharged species will not normally separate in these conditions so micellar additives such as sodium dodecyl sulphate (SDS) are included in the solution. Neutral molecules have different affinities for the charged additives and this forms the basis of their differential electrophoretic migration.

Gel electrophoretic migration (5.7.3) is limited by the loss of resolution due to zone broadening; CZE is such a powerful technique because it avoids many of the factors which give rise to this loss of resolution for example, by using a narrow bore capillary, heat is more easily dissipated during electrophoresis so that there is no temperature gradient across the medium, a factor which contributes to zone broadening. Typically, very small sample volumes are loaded (5-30 nl) and this too helps to improve resolution.

CZE can be used in the separation of peptides, proteins, oligonucleotides, nucleic acids and almost all forms of small organic molecule and can even be used for chiral (stereo chemical) separations.

5.6.6 Isotachophoresis

leading and trailing ions

In this technique charged substances are separated on the basis of their electrophoretic mobilities. It is characterised by the fact that, once equilibrium is established, all the ions in the mixture will ultimately travel through the medium (solution or gel) at the same speed. For the separation of a mixture of anions the sample ions are loaded with a leading ion (L) in the electrophoretic medium which has a higher mobility (μ_L) than the sample ions and a trailing ion (T) in the reservoir buffer solution which has a lower mobility (μ_T) than the sample ions. (All the ions must have a common counter ion). When the potential is initially applied, the faster migrating ions will move in front of the slower moving ions (see Figure 5.13b). Eventually they will form descrete bonds (Figure 5.13c). Once they have separated into descrete bonds they then each migrate at the same rate otherwise there would be a break in the electrical circut. The ions of the mixture therefore move through the medium at the same speed but are separated as discrete bands which can be monitored (eg by conductance, UV absorption) as they approach the end of the capillary tube in which the process is carried out, (Figure 5.13).

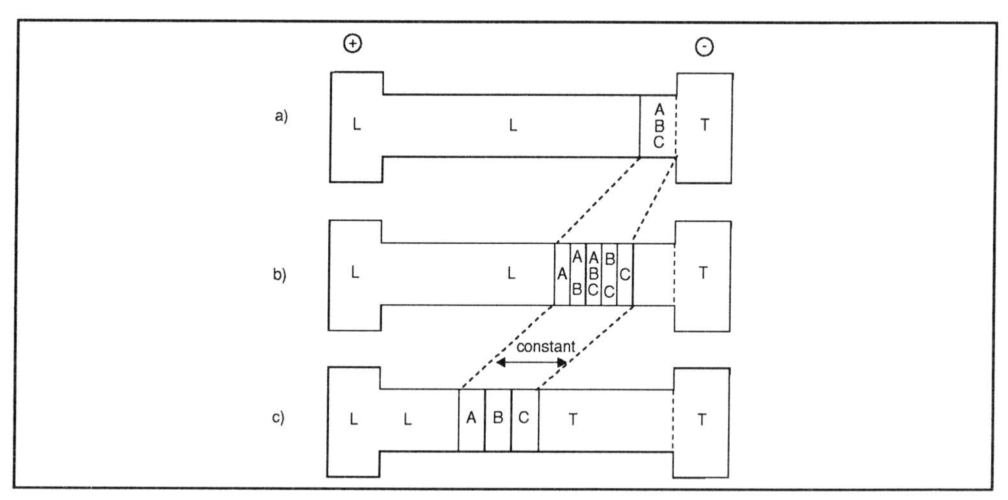

Figure 5.13 Isotachophoretic principle for separation of a mixture of anions ($\mu_A > \mu_A > \mu_B > \mu_T$ where μ = electrophoretic mobility). See text for explanation.

This techniques can be used to separate any charged substance, including inorganic ions, and is used to some extent in quality control in the food, brewing and pharmaceutical industries and in pollution control, to detect inorganic ions and pollutants in effluent water.

5.6.7 Isoelectric focusing

This technique, also called electofocusing, separates substances in an electric field across which there is both an electrical potential and a pH gradient. The anode region is acidic and the cathode region alkaline and an immobilised pH gradient is maintained between the electrodes.

Π If a protein has a pI value of 5 and it is in a solution at pH 6, will it carry a net positive or negative charge?

Because it is above its pI value it will have a net negative charge. At pH values below its pI value, it will be protonated and carry a net positive charge.

Isoelectric focusing is used for amphoteric substances, (molecules possessing both acidic and basic groups). Molecules applied to the gradient at a pH lower than their isoelectric point will be positively charged and will migrate towards the negative electrode but as they do so the surrounding pH steadily increases so that the molecules become steadily less positively charged and more negatively charged. Eventually the molecules reach their isoelectric point, (the pH where they have no net charge) and migration ceases (Figure 5.14). In the same way, molecules initially at a pH higher than their isoelectric point will be negatively charged and will move towards the positive electrode until they reach their isoelectric points and become stationary. The substance will be focused as a very sharp band at this point because if it tends to diffuse away it becomes charged and immediately migrates back to the region of its isoelectric point.

This focusing effect means that very high resolution can be achieved making the technique particularly suitable for separating molecules with very small differences in charge, notably isoenzymes. Such proteins may differ by only one or a few amino acids which may represent a difference in isoelectric points of as little as 0.01 pH units but even this is sufficient for separation by isoelectric focusing.

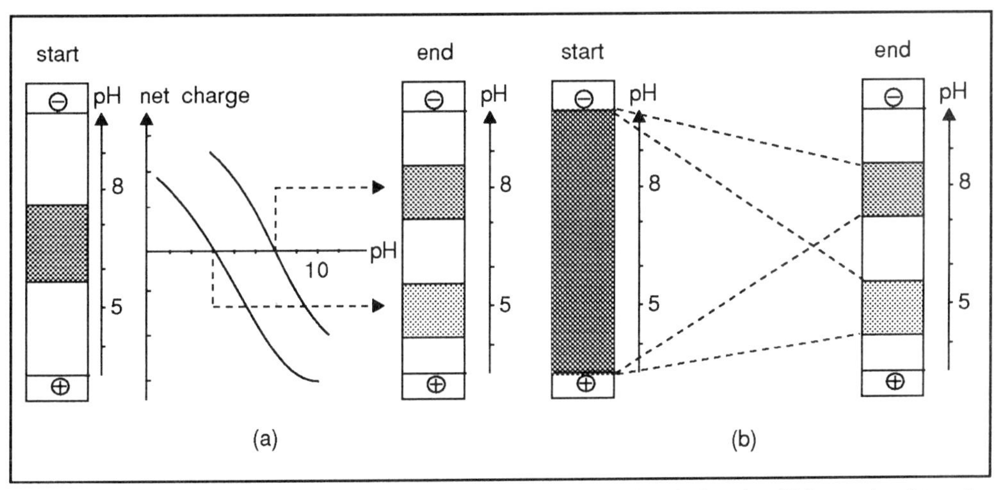

Figure 5.14 Principles of isoelectric focusing of two proteins A & B a) technique in which protein mixture is added to a prepared pH gradient; b) technique in which the protein mixture is distributed throughout a solution of mixed ampholytes, the ampholytes under an applied potential move to their isoelectric points so creating a pH gradient in which the proteins then take up their correct positions.

The separation is carried out in a liquid filled column or more usually, in a horizontal polyacrylamide or agarose gel. The pH gradient is formed by adding to the gel a mixture of low molecular mass carrier ampholytes. These are usually a mixture of synthetic aliphatic polyamino-polycarboxylic acids whose individual isoelectric points cover a certain pH range. These are prepared commercially to provide pH gradients over a wide range of pH values (eg pH 3-10) or over narrow ranges (eg pH 4-5). After being separated by isoelectric focusing, the samples can be visualized by staining the gel. The isoelectric points of the samples can be determined by measuring the pH of the gel at various points on the gradient using a surface electrode or by running molecules with known isoelectric points.

5.6.8 Chromatofocusing

Isoelectric focusing is essentially an analytical tool; for preparative work the closely related technique of chromatofocusing can be employed. A linear pH gradient is generated in a column and compounds are focused and eluted from the column in the order of their isoelectric points.

5.7 Separation methods dependent on size

Separation of molecules on the basis of their size is particularly appropriate for biological molecules as these encompass an enormous range of different sizes. Separation methods based on centrifugation and especially ultracentrifugation have already been described in Chapter 4.

5.7.1 Membrane filtration

dialysis

The simplest form of size separation is probably membrane filtration whereby mixtures are filtered through membranes which allow through only molecules below a certain relative molecular mass (RMM). Such filters may be constructed of nitrocellulose with a controlled degree of cross-linking to give RMM cut-off points of for example, 500 and 1000. This is also the basis of dialysis in which a relatively small volume of a mixture is sealed in a strip of dialysis tubing and placed in a much larger volume of water or buffer. The dialysis sac acts as a semi-permeable membrane and small molecules such as inorganic ions, diffuse freely down the concentration gradient into the water, whereas the larger molecules are physically prevented from leaving the dialysis sac. This is one of the main techniques classically used for 'desalting', eg ridding a solution of purified nucleic acids of their buffer ions.

The major drawback of separation methods involving membranes and/or filtration is that they lead to only one level of separation rather than varying degrees of fractionation. That is, the molecules in the mixture are simply divided into those above and those below, the cut-off point.

5.7.2 Size exclusion chromatography

gel filtration/gel permeation chromatography

To fractionate a mixture by utilising the difference in size of the components, it is possible to subject the mixture to size exclusion chromatography, also known as gel filtration or gel permeation chromatography.

The column packing material consists of beads formed of a cross-linked gel of a polymer such as polyacrylamide, agarose or silica. The degree of cross-linking is controlled and uniform and is such that large sample molecules cannot enter the pores in the beads and will simply pass rapidly through the column, migrating through the spaces between the beads. Smaller compounds are able to diffuse into the pores in the beads - in fact, the smaller the molecule, the more space will be available to it and hence the slower it will migrate through the column. Therefore, large molecules elute from the column early whilst smaller molecules spend more time in the pores of the stationary phase and hence take longer to elute from the column (Figure 5.15).

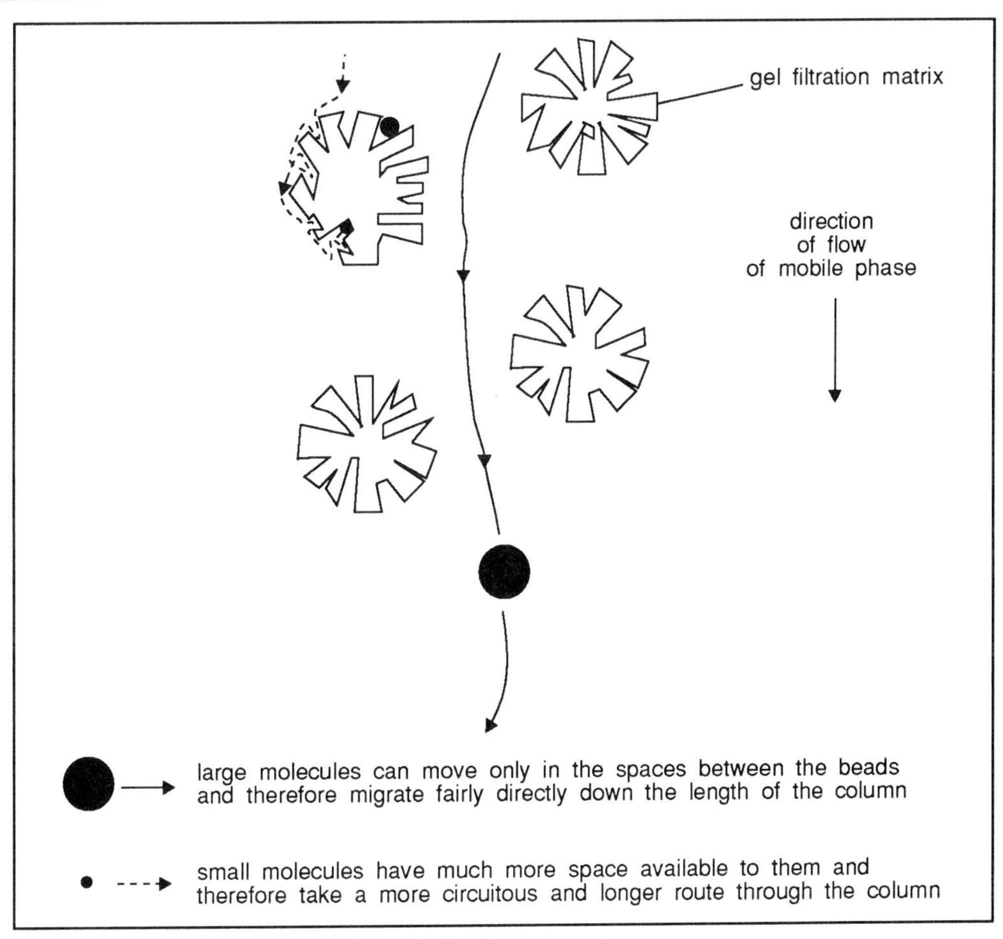

Figure 5.15 Behaviour of solutes in size exclusion chromatography

Size exclusion chromatography therefore differs from other forms of chromatography in that there is no physical interaction between the sample molecules and the stationary phase except in as much as the stationary phase is represented by the interstitial spaces of the gel polymer beads. The stationary phase is therefore (theoretically) entirely inert. However, sometimes other forms of interactions such as hydrophobic and ion-exchange processes do in fact play a part. The separation occurs because smaller molecules effectively travel a greater distance whilst passing through the column and therefore elute from the column more slowly than do larger molecules. By plotting a calibration curve of molecular mass of known molecules against their elution volume - ie the volume of mobile phase eluting from the column prior to the compound's elution, it is possible to estimate the molecular mass of unknown samples based on their elution volumes. Of course, it should be remembered that strictly speaking, the behaviour of the solutes is determined not by their molecular mass but more specifically by their three-dimensional size and shape.

Theory: As in other forms of chromatography it is possible to describe the behaviour of a solute on the column in terms of a distribution coefficient (K_D) with diffusion into the gel pores as the sorption process. (Sorption is a general term which describes the interaction of a solute with the stationary phase).

$$K_D = (V_e - V_o)/V_i \qquad \text{(E - 5.6)}$$

where V_e is the elution volume of the solute, V_o is the void volume - the elution volume of compounds completely excluded from gel pores and V_i is the volume of liquid inside the gel pores available to the very smallest solutes.

K_D therefore represents the proportion of the volume inside the beads, available to the solute.

SAQ 5.5

What can you say about solutes which have distribution coefficients, K_D on a gel filtration column, of:

a) 0;

b) 1;

c) greater than 1.

Π Protein A has a relative molecular mass of 10 000 and is a tightly folded spherical molecule. Protein B also has a relative molecular mass of 10 000 but is a long rod-shaped molecule. Would you expect these two molecules to co-elute on a size exclusion chromatography column?

influence of shape

Although the two proteins have the same molecular mass, protein A due to its compact, spherical shape, has available to it a greater volume into which it may diffuse and is therefore effectively a smaller protein than protein B. Molecules are constantly and randomly rotating during diffusion and the size of the spaces that they may enter is restricted by their length of their longitudinal axis. Protein B therefore effectively occupies a greater three-dimensional volume and therefore migrates through the column more quickly (Figure 5.16).

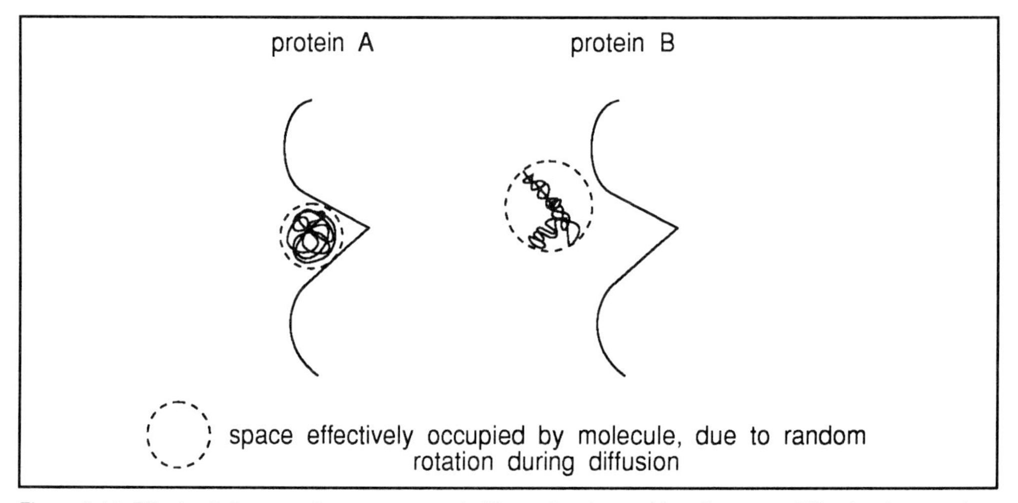

Figure 5.16 Effects of shape on the space occupied by molecules and its effect on gel filtration (see text).

Different gel filtration media are available, each designed to fractionate a specific molecular mass range. Like ion exchange media these are constructed of beads of cross-linked polymers such as dextran, agarose and polyacrylamide. The degree of cross-linking is carefully controlled and it is this which determines the pore size and fractionation range. It is also possible to obtain hydrophobic packing material for use with organic solvents.

desalting

Gel filtration is a very simple technique and very gentle, involving no high pressures (unless carried out as a form of HPLC) or chemical reactions and is therefore suitable for molecules such as large proteins. It is used commonly for: desalting solutions; buffer exchange; separating of enzymes from their inhibitors - in fact in almost any process in which one wishes to separate molecules of markedly different sizes.

SAQ 5.6

To determine the relative molecular mass of protein X, the unknown protein was run on a gel filtration column (50 cm x 2.5 cm) along with a range of proteins of known RMM. Listed below are the elution volumes and the RMM of each of the proteins. Use this information to plot the elution volume against log (RMM) and then estimate the relative molecular mass of X.

Protein	RMM	elution volume(cm^3)
cytochrome c	12 400	205
chymotrypsinogen	25 000	188
ovalbumin	45 000	170
transferrin	66 000	150
lactate dehydrogenase	135 000	130
catalase	240 000	105
β-galactosidase	520 000	90
protein X		140

5.7.3 Gel electrophoresis

As explained previously, many biological molecules can exist as charged species and in an electrical field these will migrate according to their charge towards the appropriate electrode. Electrophoresis is usually carried out in a polyacrylamide or agarose gel. When carried out in a gel, migration of solutes takes place due to their charge but separation occurs as a consequence of their size or a combination of their size and charge. Buffer solutions used ensure that all the components are charged with the same sign and so migrate in the same direction. This movement is opposed by the cross-linked mesh of the gel which acts as a molecular sieve. Smaller molecules are able to move relatively fast through the gel whereas larger molecules are hindered more by the matrix and migrate more slowly.

Proteins for example, are commonly separated by gel electrophoresis in polyacrylamide gels. If we do not wish to retain the activity of the protein the proteins are denatured with an anionic detergent - usually sodium dodecyl sulphate (SDS), which denatures all but the primary structure of the protein and binds to it in a fairly constant ratio of SDS molecules to amino acids. The large negative charge of the SDS molecules swamps the native charge of the protein which means that, all other factors being equal, the proteins will migrate towards the positive electrode at essentially the same rate. The cross-linking of the polyacrylamide however, restricts the migration of the proteins so that large proteins move slowly through the gel while small proteins are less hindered and move more rapidly. This is carried out vertically in thin sheets of gel held between two glass plates. After some time the compounds in the gel can be visualized by staining. The distance moved by a protein in the gel is more or less directly proportional to log (RMM) so that by including marker proteins of known RMM in the sample it is possible to calculate the RMM of the components of the mixture.

The nature of the technique means that it is used mainly for biological macromolecules, particularly proteins and nucleic acids and is discussed in greater detail in the BIOTOL text 'Analysis of Amino Acids, Proteins and Nucleic Acids'.

5.8 Separation methods dependent on biological specificity

5.8.1 Affinity chromatography

One of the most obvious and elegant means of separation in biotechnology is to utilise the biological specificity of the molecule to be purified as a handle for picking it out from a mixture. This method can be used for molecules that bind specifically and reversibly to a suitable ligand. This process is termed affinity chromatography.

Π Can you give any examples of types of molecules that bind specifically in such a way as to suggest that they might be used in affinity chromatography?

Some of the most commonly used types of interaction are those between antibodies and antigens, enzymes and inhibitors (or substrates), even viruses and their membrane receptor proteins. These interactions may be very specific or, more commonly, the ligand may select for particular groups of biological molecules. Some examples of commonly used ligands (components we attach to the stationary phase) and their bound molecules are given in Table 5.3.

Ligand	Bound molecules
DNA	DNA-binding proteins
poly (U)	mRNA possessing poly (A) tail
lysine	plasminogen, rRNA
tryptophan	α-chymotrypsin
concanavalin A	glycoproteins, glycolipids
protein A	antibodies of the IgG class, (molecules containing Fc region of IgG)
antibody	interferon
heparin	coagulation factor, various lipoproteins and lipases
adenosine	adenosine deaminase
sulphanilamide	carbonic anhydrase
apoflavodoxin	FMN (flavin mononucleotide)
cibacron blue F3GA (anthraquinone-type dye)	nucleotide-requiring enzymes

Table 5.3 Commonly used ligands and their bound molecules

Affinity chromatography is carried out in a column in which the ligand is bound covalently to the gel matrix. The ligand can of course, be either half of the interactive pair, eg the antibody or the antigen. It is also usually necessary to introduce a spacer arm which holds the ligand away from the surface of the supporting matrix. This prevents steric hindrance thus allowing access for large molecules to bind to the ligand. This forms the stationary phase. The mixture is then passed down the column and those molecules with a specific binding affinity for the ligand are retained by this stationary phase, whilst the rest of the mixture remains unbound and is eluted from the column (Figure 5.17).

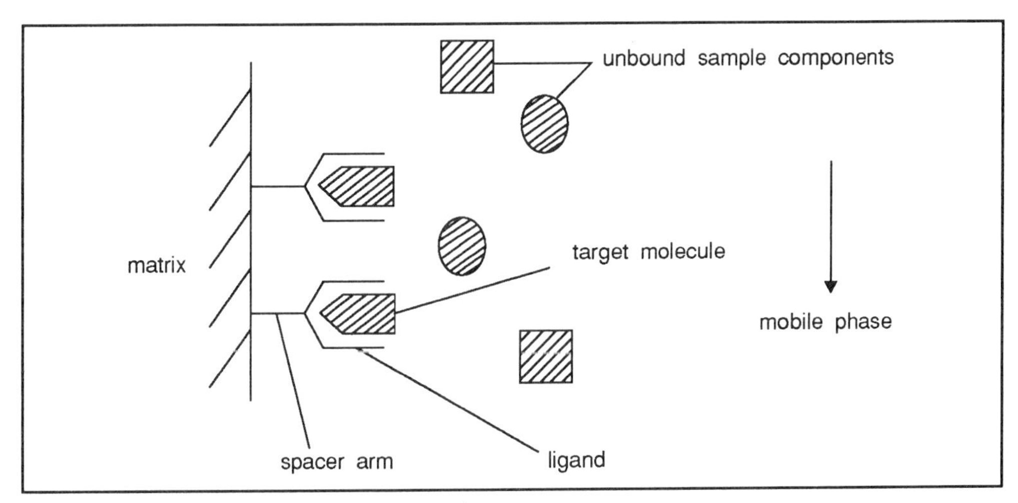

Figure 5.17 Mechanism of affinity chromatography

By altering the conditions of the mobile phase, eg changing pH, ionic concentration, it is possible to break the reversible bonds between target molecules and stationary phase and elute the bound molecules.

Alternatively, the purified compound may be recovered by specific elution whereby a compound with greater affinity for the ligand is added to the mobile phase, thus displacing the bound material

The ligand therefore, selects from the mixture, the one molecule or group of molecules for which it has a binding affinity. The ligand can be seen to act as a spatial template, or as a 'lock' for which the 'key' is the specifically binding target molecule.

5.9 Separation methods dependent on stereochemistry

Stereoisomer are molecules whose constituent atoms are joined in the same order but which differ in the arrangement of these atoms in space. They can be grouped into two main types.

optical isomers/chiral

Optical isomers are molecules which have a three-dimensional structure which is not superimposable upon its mirror image. An analogy is a pair of hands which are identical in so far as they are mirror images of each other but which are distinct by virtue of the fact that (even were it physically possible) they are not interchangeable - they are not superimposable on each other. Such molecules are said to be chiral and are distinguishable by the fact that they do not possess a plane or centre, of symmetry. The two different mirror image forms are termed enantiomers.

All amino acids (except glycine) for example, can exist as optical isomers (Figure 5.18).

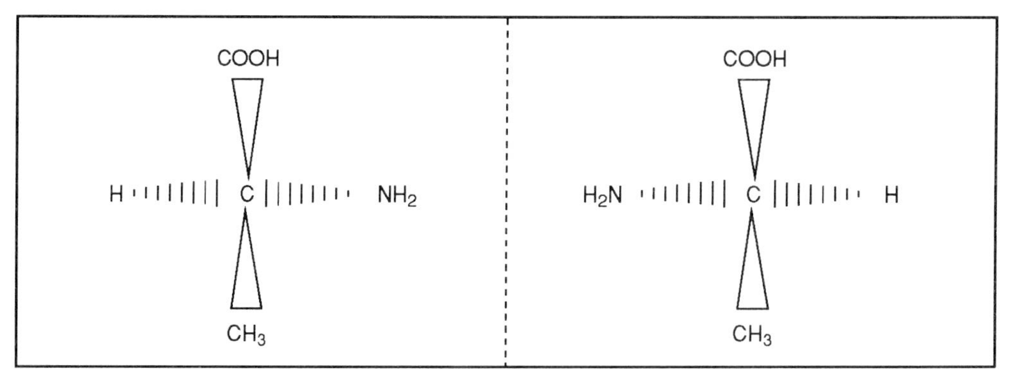

Figure 5.18 Optical isomers of alanine

Such isomers exhibit identical physical properties except with respect to their optical activity; enantiomers rotate a plane of polarised light to an equal extent but in opposite directions.

diastereo-isomers

Diastereoisomers are isomers which have their atoms arranged in the same order but differ in the configuration of one or more groups and are not therefore mirror images. An example is the pair of compounds, erythrose and threose (Figure 5.19). It should be noted that both erythrose and threose can also exist as optical isomers. All **cis/trans** isomers are diastereoisomers.

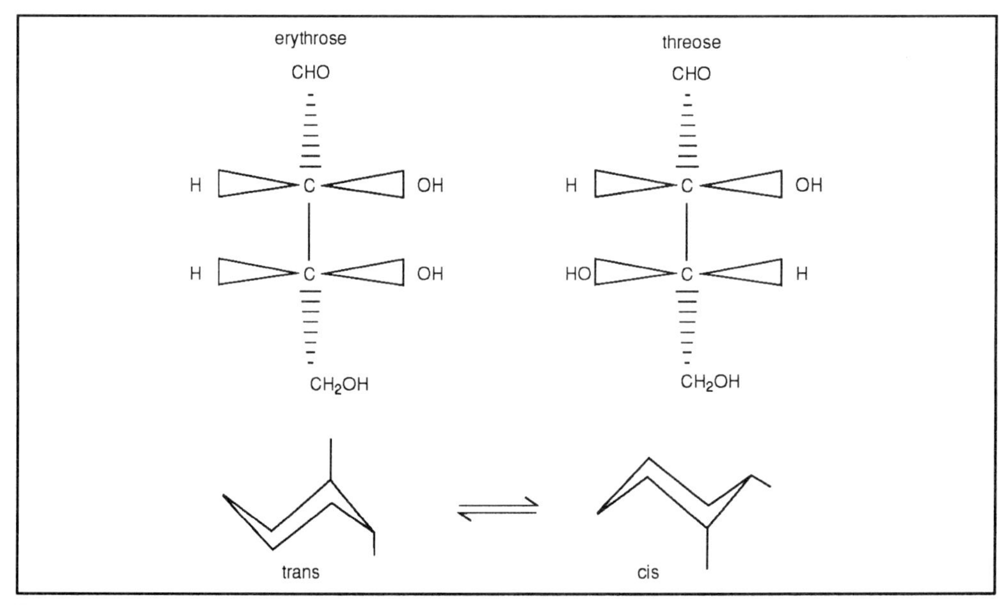

Figure 5.19 Examples of diastereoisomers

Most biological molecules are chiral and many exist in biological systems only in one of the enantiomeric forms. With a few exceptions, all amino acids present in living organisms consist only of one enantiomeric form, indeed, often only one of a pair of enantiomers will exhibit biological activity. Therefore all biological systems have evolved to selectively incorporate and deal with specific stereoisomers; this is important in biotechnology where chemical synthesis of a product may lead, for example, to the formation of a racemic mixture (equal mixture of a pair of enantiomers) which need to be resolved for biological use or study.

racemic mixtures

In the case of diastereoisomers separation can be done by utilising their difference in physical properties and separation between isomers can often be achieved by means of one of the techniques already mentioned such as adsorption chromatography or reverse phase chromatography. Optical isomers on the other hand, are more difficult to resolve as they have identical physical properties. There are two main ways of resolving enantiomers.

1) **Formation of diastereoisomers**: By derivatizing enantiomers by reaction with an optically pure reagent, it may be possible to produce diastereoisomers. This conversion of the enantiomers into diastereoisomers means that two enantiomers effectively possess a fundamentally different spatial configuration enabling them to be much more easily separated by conventional chromatography techniques - usually reverse phase HPLC. This derivatization can also be carried out during the process of reverse phase HPLC by the use of chiral mobile phase additives such as chiral metal chelates, these give rise to diastereoisomeric forms of optical isomers which can then be resolved by a standard reverse phase system. Similarly, enantiomers may be converted to diastereoisomer ion-pairs by adding a chiral counter ion to an HPLC mobile phase. The ion-pairs can then be separated by HPLC.

2) **Use of a chiral stationary phase:** A relatively recent but increasingly useful method for racemic resolution involves the use of a chiral stationary phase in HPLC. As optical isomers can differ in their interaction with other chiral molecules this allows enantiomers to be separated directly, without converting them to a diastereoisomeric form. Such stationary phases include several forms of cyclodextrin, α-acid glycoprotein immobilised on silica and bovine serum albumin covalently bonded to silica.

Π How might you monitor the separation of enantiomers on HPLC using a chiral stationary phase?

As mentioned above, enantiomers are detectable by their optical properties. Enantiomers rotate a plane of polarised light to an equal extent but in opposite directions. This optical activity is extremely useful and is the most important means of keeping track of enantiomers. Each molecule exhibits its own 'specific rotation' of the plane of polarised light which can be measured in a polarimeter. By measuring the optical activity of a column eluent therefore, it is possible to determine if and to what extent a racemic mixture has been resolved.

5.10 General aspects of purification

It is therefore apparent that there are many features of a molecule which can be used as handles for pulling it out of a mixture. As well as the properties mentioned, density is a physical characteristic used to fractionate some biomolecules notably nucleic acids.

There is no 'correct' way of purifying a particular molecule but some knowledge of the chemical and physical characteristics of the molecules to be separated can provide extremely useful clues as to the type of purification technique which might be most appropriate. In practice, the behaviour of compounds in various purification systems is rarely entirely predictable. In many forms of chromatography there may well be more than one type of interaction at work. For example, the support matrix of an ion exchange column may well interact with the solutes by adsorption giving rise to a more complex separation, and in reverse phase HPLC, one of the most widely used forms of chromatography, different types of interaction influence retention times to different extents depending on the exact nature of the sample, the stationary phase, the mobile phase and the conditions of chromatography. Therefore, while it is often possible to predict the general appropriate conditions for the separation of certain materials, the exact conditions are generally best determined empirically.

Π You have a mixture possessing interesting biological activity and are attempting to isolate the active component. What sort of preliminary information about the component might it be useful to obtain before beginning purification?

Perhaps the most important feature of the component to be separated is its solubility and hence a measure of its hydro- or lipo-philicity. If hydrophillic, the molecule may be charged in which case ion exchange chromatography or isoelectric focusing might be employed. If insoluble in aqueous systems, it may be possible to use partition papers. If volatile on the other hand, the sample might be separated by GLC. If a specific biological receptor for the component of interest is known, this might form the basis for a separation by affinity chromatography.

It is not often that you can take a complex biological mixture and isolate the desired component in a single step - usually a series of purification techniques is required to separate a mixture. Some techniques such as filtration, liquid-liquid extraction and some forms of liquid column chromatography are on the whole, more suitable for the cruder, early stages of purification. These tend to be processes which are fairly simple, have a fairly high capacity, do not use large amounts of expensive material and which generally, remove a large proportion of the unwanted material with the minimum of fuss. It is at the later stages that the more refined, subtler separation methods, such as HPLC tend to come into play.

During each stage of a purification procedure, it is necessary to monitor the compound(s) that is the target of the purification. It is important to have a marker by which the progress of the compound can be followed and by which its level of purity at various stages can be determined. In chromatography systems this is often done by measuring the absorbance of the eluate in the UV region as it comes off the column; the retention times of the various components can be compared with those of samples of known compounds separated under identical conditions.

Fractions from a purification system can be collected and tested individually for the presence of material by absorbance, by TLC, chemical assay or testing for biological activity, amongst other means. These methods can also give a measure of the level of purity of the compound of interest and can also often supply a quantitative measure of the recovery yield from various stages of the purification.

Purity is generally established for example, by the presence of a single peak on a chromatogram, a single spot on a TLC plate or a single band on a gel after electrophoresis. It is usually important to be able to account for all or most of the material throughout a purification procedure. If, for example, a large proportion of a valuable material was lost in one of the purfication steps it might be prudent to find an alternative procedure. If on the other hand, time was the major consideration, this procedure might be satisfactory.

Large Scale Purification: This chapter deals mainly with purification of material for the purposes of identification and analysis. Large scale purification brings its own problems and is dealt with in other books in the series (see for example the BIOTOL text 'Product Recovery in Bioprocess Technology'). The major problems involve the fact that many of the techniques described are not applicable to a larger scale (though in many cases this is changing as demonstrated by the development of industrial plant scale HPLC).

SAQ 5.7

Suggest with reasons, which techniques you might use to separate the following:

1) a mixture of amino acids;
2) a mixture of D- and L-valine;
3) an antibody from a serum solution;
4) a mixture of fat-soluble vitamins;
5) a mixture of long-chain polymers;
6) a mixture of organic acids;
7) a mixture of biliverdin methyl ester isomers.

SAQ 5.8

A solution of a compound was believed to be pure by the fact that when analyzed in a chromatography system the chromatogram revealed a single peak. However, further testing of the solution in a different chromatography system resulted in a chromatogram with two peaks. Give three possible explanations of these apparently paradoxical results.

SAQ 5.9

Complete the following table of some common separation techniques (you should be able to think of 12-13 techniques).

Technique	Solute properties utilized for separation	Examples of mixtures which can be separated

Summary and objectives

Although we have not covered in any detail the practical aspects of the various methods, we have seen how the physical properties of biological molecules may be used as 'handles' for extracting them from a mixture. These properties include solubility, size, density, shape, charge and adsorption which together with the specificity of interaction of many sets of biological molecules, provide the basis of separation. Some properties are the basis of more than one technique, eg gel electrophoresis and size exclusion chromatography, and some techniques are more suited to analytical purposes rather than to the recovery of the purified compound. Many of the properties described are employed in the various forms of chromatography in which compounds distribute themselves in a characteristic way between two phases.

Whilst theoretically there are many ways of purifying a given compound from a mixture, research and experience has led to the establishment of many routine approaches for the purification of particular groups of biological molecules. Many of these are described in the other books in the BIOTOL series.

Now that you have completed this chapter you should be able to:

- explain how the properties of molecules, namely adsorption properties, solubility, charge, size, biological specificity and absolute stereochemistry, may be exploited in order to extract them from mixtures.
- explain the principles involved in the choice of a separation method for a particular mixture;
- appreciate the importance of the various intermolecular forces in giving rise to the properties of compounds;
- describe the principles which underly all forms of chromatography;
- classify chromatographic techniques;

Enzyme assay methods

Enzyme assay methods

6.1 Introduction

In this and the next chapter we focus attention on detecting and measuring biomacromolecules with particular exphasis on techniques useful in monitoring their purification. We begin by examining the techniques for detecting and measuring enzymes.

specific test

During the purification of an enzyme, it is essential to have a quick, sensitive and specific test or assay for the enzyme that will tell you if the enzyme is present, and if so, in what quantity. Faced with over 100 fractions from, say, an ion exchange column, it is essential to have an assay that will measure the amount of enzyme present in a sample within, say, 2 min, so that you can quickly screen these fractions, pool the enzyme-containing fractions and then move on to the next purification step. An assay that takes 20 min per sample is of little use at this stage. We require a quick, reliable and reproducible assay that will relate the amount of one enzyme in any fraction to that in another. By developing an appropriate assay, one can quickly determine the amount of enzyme (measured as the number of enzyme units see 6.2) present in a small aliquot of the sample, and hence the total amount of enzyme units present in the original sample.

Assay methods can be either continuous or discontinuous. Continuous assay methods monitor continuously property changes such as absorbance, (eg in a spectrophotometer cell) whereas discontinuous methods require samples to be withdrawn from the reaction mixture at various times and analysed by some convenient technique. It should be obvious that continuous methods are far more convenient and are generally used wherever possible.

Continuous assays are normally achieved by mixing the enzyme-containing sample with a solution of enzyme substrate in a cuvette (spectrophotometer cell) and incubating the cuvette in a spectrophotometer and measuring the change in absorbance (as a result of substrate utilisation or product formation) with time. This gives a measure of either the rate of disappearance of substrate or rate of increase in product. Either way, the measured rate is a measure of the amount of enzyme present. This direct method is the most basic approach, although as we will see below, there are a number of variations on this.

Π The above discussion has centred on the use of enzyme assays to monitor enzyme purification. Can you think of any other instances where one may need a quick and accurate assay for a particular enzyme?

The measurement of enzymes in body fluids, eg serum, is frequently used as an aid to the diagnosis of disease. Some enzymes are present in particularly high levels in certain tissues or organs and damage to these can be inferred by the presence of elevated levels of the enzyme in the serum. For example, the enzyme aspartate aminotransferase (AAT) is present in fairly high levels in heart tissue, and increased blood levels of AAT, taken together with other symptoms, is diagnostic of a heart attack. In a heart attack, or myocardial infarction, part of the heart tissue is starved of oxygen, the tissue dies and the cell contents, amongst which is AAT, are released into the blood stream. Similarly, elevated serum levels of the enzyme lactate dehydrogenase is diagnostic for liver damage.

We could also add, the measurment of enzymes used as labels for example in ELISA techniques. Again we need a quick assay procedure. We will examine these techniques in the next chapter.

6.2 Enzyme units

International Units

We express the amount of an enzyme present in a solution, not in terms of mass or moles, but in terms of units based on the rate of the reaction that the enzyme catalyses. The activity of an enzyme is the amount of conversion of substrates to products that a certain amount of enzyme will produce in a specific period of time. The International Unit (U) of activity is defined as the amount of enzyme which will convert one micromole of substrate to products in 1 min, under defined conditions of temperature, pH and substrate concentration.

Although this is the internationally accepted unit, you will find that many workers use a more convenient definition to simplify their calculations. Take, for example, the enzyme aspartate aminotransferase which catalyses the reaction shown in Figure 6.1.

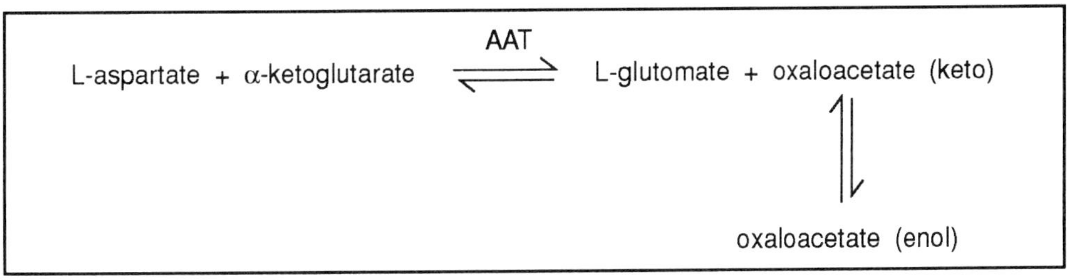

$$\text{L-aspartate} + \alpha\text{-ketoglutarate} \underset{}{\overset{\text{AAT}}{\rightleftharpoons}} \text{L-glutomate} + \text{oxaloacetate (keto)}$$

$$\text{oxaloacetate (keto)} \rightleftharpoons \text{oxaloacetate (enol)}$$

Figure 6.1 The reaction catalysed by aspartate aminotransferase.

To assay for the enzyme, an aliquot of the enzyme solution is added to a substrate solution containing L-aspartate and α-ketoglutarate. One of the products of the reaction, oxaloacetate can exist in either the keto or enol form (keto-enol tautomerism) which are in rapid equilibrium with each other. Since the enol form absorbs at 260 nm, the rate of increase in absorbance at 260 nm is a measure of the reaction rate and hence the amount of enzyme present. For this enzyme many workers define one unit of AAT activity as the amount of enzyme which, in 3.0 cm^3 (ml) of substrate solution, produces a change in absorbance at 260 nm of 0.1 per min at 22 °C.

SAQ 6.1

Using the assay for the AAT described above, 50 µl of a tissue extract was added to 3.0 cm^3 of substrate solution and an absorbance change of 0.26 per min at 260 nm was recorded. Using the above definition for AAT activity, determine the number of enzyme units per cm^3 (ml) in the original extract.

Π Can you think of any enzymes where it would not be possible to define the number of enzyme units?

This would be the case for enzymes where the substrate is a macromolecule of uncertain molecular mass, eg. amylase (starch as substrate), DNase (DNA as substrate) or proteases (protein as substrate). This problem is overcome either by using synthetic or artificial substrates (see Section 6.5.2) or by defining an arbitrary unit based on some observable change in the property of the substrate. For example, α–amylase activity can be measured by recording the decrease in viscosity of a starch solution.

yield

Another term that we can introduce here that relates to the use of enzyme units is the idea of the yield of a particular purification step. Obviously after any purification step one hopes to recover as much as possible of the enzyme of interest, i.e one requires a high yield. The yield following any particular step (given as a percentage) is defined as:

$$\text{Yield} = \frac{\text{total enzyme activity of fraction}}{\text{total enzyme activity of starting solution}} \times 100$$

SAQ 6.2

A homogenate of pig heart tissue was centrifuged to give 100 cm^3 of supernatant. An assay of an aliquot of this solution for the enzyme AAT showed there were 23 enzyme units per cm^3. Following ammonium sulphate fractionation, 30 cm^3 of solution was obtained which was shown to contain 21 enzyme units per cm^3. Calculate the yield of enzyme from this step.

6.3 Enzyme purity - specific activity

As well as using your assay to monitor the purification of the enzyme, the assay can also be used in a calculation that measure the purity of the enzyme. This introduces the idea of the specific activity of an enzyme which is defined as follows:

$$\text{specific activity (Sp.Ac.)} = \frac{\text{total units of enzyme (U)}}{\text{total amount of protein (P)}}$$

Obviously in an initial crude tissue extract the specific activity of the enzyme will be low. There will be a certain number of enzyme units (U) present, but a large amount of contaminating protein (P), hence the ratio U/P will have a low value.

Π What do you think will happen to the specific activity of the enzyme as we start to purify the enzyme?

The specific activity should increase at each purification step. Hopefully the number of units will remain essentially the same (although in practice on average a 10-15% loss of enzyme at each step is not uncommon). However, if it is a meaningful purification step, the total amount of protein (P) should decrease considerably at each step, since the aim is to remove contaminating proteins.

For example, an initial crude extract may have a specific activity of 5 Iμ mg^{-1}, then following an initial fractionation step (say ammonium sulphate fractionation) it may have a specific activity of 25 Iμ mg^{-1},. We can therefore say that this purification step produced a five-fold purification, where fold purification (the degree of purification) is

fold purification

$$\text{fold purification} = \frac{\text{Sp.Ac. at step 2}}{\text{Sp.Ac. at step 1}}$$

$$= 25/5 = 5, \text{ in this example}$$

defined as:

In a recent survey of 100 published purification protocols, it was shown that the average fold purification per step was 8, and that the overall fold purification (from crude extract to purified protein) was 6380.

Π Since the specific activity should increase at each step, do you think there is a maximum value for the specific activity of an enzyme solution?

Yes, every enzyme has a maximum specific activity. When the enzyme is pure, the total amount of protein (P) is due solely to the enzyme. The ratio U/P should therefore have reached a maximum value. If we reduce (P) we will only be throwing away enzyme and therefore (U) will also reduce, leaving the ratio U/P and hence the specific activity the same. In fact, the inability to increase the specific activity of an enzyme solution is often taken as an indication that the enzyme is pure.

It is worth noting that the concept of specific activity does not apply only to enzymes. Any protein or peptide that has a measurable activity e.g a hormone, antibacterial or antiviral peptide) can be measured in terms of its specific activity.

We will now move on to look at how we can design an assay for a particular enzyme, such that the assay is user friendly, i.e quick, sensitive and specific.

6.4 Design of enzyme assays

6.4.1 General considerations

When measuring an enzyme we generally aim at having conditions optimised so that the enzyme exhibits maximum activity. The subject of enzyme kinetics will not be treated in any depth here (the subject is worth at least a chapter if not a whole book on its own!) but there are some factors that affect enzyme activity that you need to at least be aware of. (Enzyme kinetics are dealt with in depth in the BIOTOL texts 'Principles of Enzymology for Technological Application' and in 'Principles of Cell Energetics'). In this text we have provided additional material in the form of an appendix. This appendix explains the importance of measuring initial reaction velocities and examines some numerical relationships between reaction velocity and substrate concentration. Particular emphasis is placed on Michaelis-Menten kinetics. You should find this a useful reminder.

Π What factors can you list that you think might affect enzyme activity and that we need to take account of when designing an enzyme assay?

Your list should have included pH, temperature and substrate concentration. We will consider each of these in turn.

6.4.2 Effect of pH on enzyme activity.

The activity of most enzymes changes with pH. A typical pH/activity profile is shown in Figure 6.2. Here we see the enzyme has maximum activity at about pH 8 but above and below this pH value the activity drops off. Each enzyme has its own characteristic optimum pH, eg for pepsin(secreted in the stomach) the optimum is pH 2, trypsin (found in the intestine) has an optimal activity at pH 8 and alkaline phosphatase (kidney, liver) at pH 9 - 10. It is usual, therefore, to carry out an assay in a buffer solution at a pH value that is optimal for the enzyme.

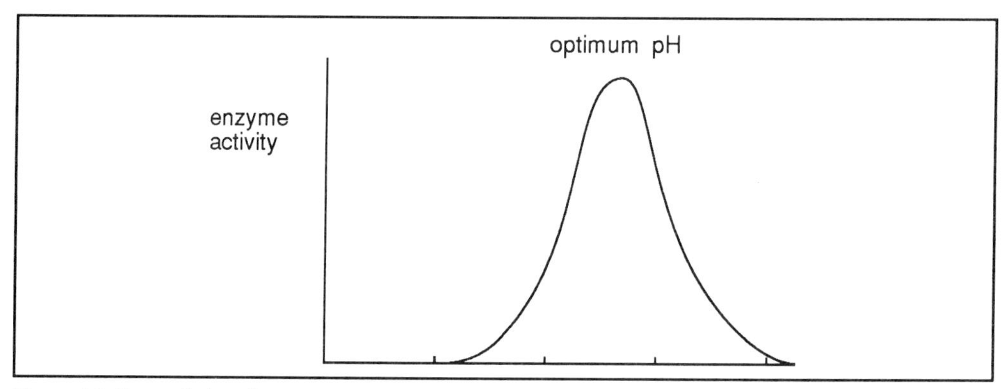

Figure 6.2 The variation of enzyme activity with pH. Note that the enzymes has a distinct pH at which the enzyme is most active (the optimum pH). The actual shape and position of the curve very much depends on the enzyme. Some enzymes have low optimum pHs, other have greatest activity at high pHs. Some enzymes are active over a narrow range of pH's, others are active over a wide range.

6.4.3 Effect of temperature on enzyme activity.

Enzyme catalysed reactions require enzyme and substrate to meet to allow reaction to take place. As the temperature is increased, so the collisions will become more frequent and therefore the rate of reaction will increase. In general, a 10 °C rise in temperature approximately doubles the rate of reaction. Therefore measuring an enzyme at 30 °C will give a rate approximately double that when measured at 20 °C.

Proteins are not very resistant to increased temperature and so the greater rate of an enzyme catalysed reaction is soon offset by the increased denaturation of the enzyme. Thus there is an optimum temperature (T) for a given enzyme system, usually in the range 35-45 °C. (Figure 6.3).

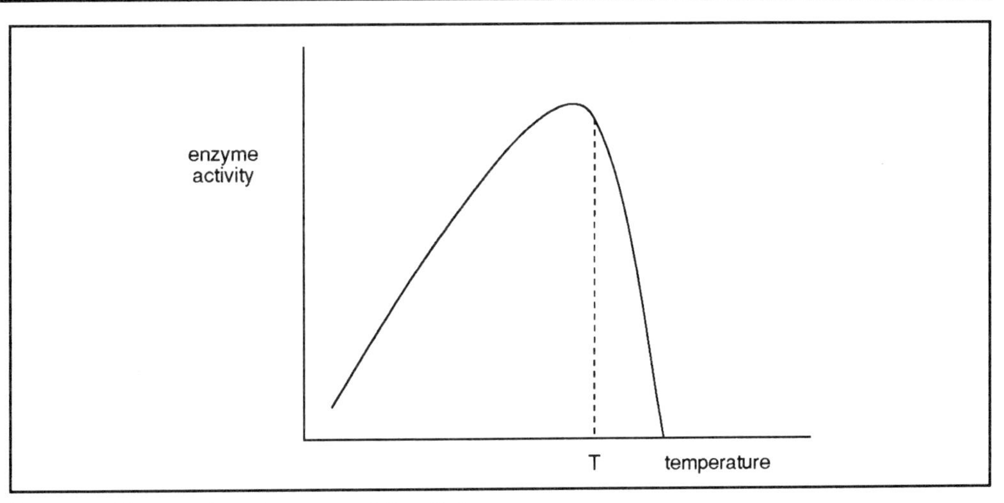

Figure 6.3 The variation of enzyme activity with temperature.

If for example you are monitoring a series of column fractions, temperature is not an important factor since they will all be measured at approximately the same time and same temperature, (ie room temperature). However, temperature is an important factor if you want to make careful measurements of enzyme activities, particularly if these are to be made at different times when room temperature may vary.

6.4.4 Effect of substrate concentration on activity.

Obviously we are measuring the amount of our enzyme in a indirect way by measuring the number of substrate molecules converted to product(s) in a given time. Supposing, however, that there were very few substrate molecules around; would we get a true measure of the amount of enzyme present? Obviously not, because in very simplistic terms we can imagine many enzyme molecules, wandering around aimlessly looking for a substrate molecule to react with! In this case we would not 'see' these enzyme molecules because they would not be converting substrate into product.

To get an accurate measure of the amount of enzyme present, we must ensure that there is a considerable excess of substrate molecules, so that all enzyme molecules are kept busy converting substrate molecules into product (Figure 6.4).

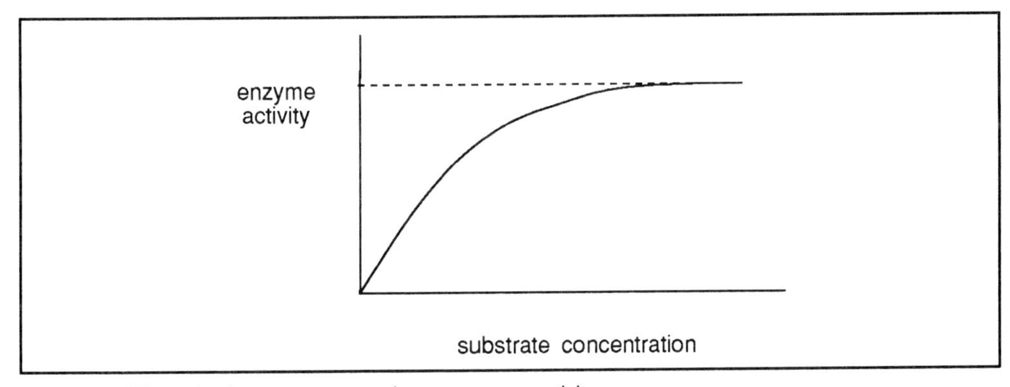

Figure 6.4 Effect of substrate concentration on enzyme activity.

Only then will we get a true measure of the amount of enzyme present. An important rule for any enzyme assay therefore is that substrate must be in excess. If it is not, we say that the substrate concentration is rate limiting.

All of the above parameters are important in ensuring that your assay is reproducible.

In many cases, if an enzyme is mixed with the substrate under the appropriate conditions, either no catalysis occurs or there is only slight activity. This may be due to any of the following:

- absence of a coenzyme, a low molecular mass organic compound which acts as an acceptor or donor for specific chemical groups, eg NAD accepts and donates hydrogen atoms for many dehydrogenases;
- absence of an activator, metal ions, eg Mg^{2+}, are activators for a wide range of enzymes;
- presence of an inhibitor, eg heavy metal ions, cyanide.

6.4.5 Specificity of assay.

specific

We now turn to the problem of the specificity of the assay. The need to design an assay that is totally specific for the enzyme is illustrated by the much quoted example of the assay for the enzyme phosphoenolpyruvate carboxykinase (PEP carboxykinase). This enzyme converts PEP to oxaloacetic acid (OAA) as follows:

$$\text{PEP} + CO_2 + \text{GTP} \rightleftharpoons \text{OAA} + \text{GTP}$$

(GTP is guanosine-5'-triphosphate)

An obvious assay for this enzyme would therefore seem to be to make a solution of PEP and GTP then add the test solution (perhaps a crude tissue extract) and measure ether the rate of loss of PEP or the rate of production of OAA. However, this could be highly misleading, since there are a number of other enzymes that may be present in your extract that either deplete PEP or produce OAA (or both).

eg

$$\text{PEP} + HCO^-_3 \rightleftharpoons \text{OAA} + P_i \text{ (PEP carboxylase)}$$

$$\text{PEP} + \text{ADP} \rightleftharpoons \text{pyruvate} + \text{ATP (pyruvate kinase)}$$

$$\text{PEP} + CO_2 + P_i \rightleftharpoons \text{OAA} + PP_i \text{ (PEP carboxy-transphosphorylase)}$$

(P_i is inorganic orthophosphate, PP_i is inorganic pyrophosphate)

Therefore, using the assay indicated above, one could be measuring the total effect of as many as four different enzymes. Indeed, there might be a reasonable measure of enzyme activity even if PEP carboxykinase was totally absent!

The design of an appropriate enzyme assay that is specific, quick and sensitive can be quite an intellectual challenge to the biochemist.

We must now look at the different ways in which enzyme reactions can be monitored.

6.5 Spectroscopic method of monitoring enzyme reaction

6.5.1 Direct methods.

Many substrates and products absorb radiation in the visible or UV region of the spectrum and provided the substrate and product do not absorb at the same wavelength, the change in absorbance can be used as the basis for the assay. This is the simplest possible approach and is referred to as a 'direct' method. For example, the enzyme xanthine oxidase catalyses the following reaction:

direct method

$$\text{xanthine} + O_2 \rightleftarrows \text{uric acid} + \text{superoxide}$$

The rate of reaction is measured by monitoring the increase in absorbance at 295 nm due to the production of uric acid.

In many reactions, however, neither the substrate nor the product have useful, or sufficiently different, UV spectra. It is fortunate, therefore, that nicotinamide adenine dinucleotide is used as a cofactor (coenzymes) by many dehydrogenase enzymes. This cofactor can exist in either the oxidized (NAD^+) or reduced form (NADH). The spectra of these two forms are shown in Figure 6.5.

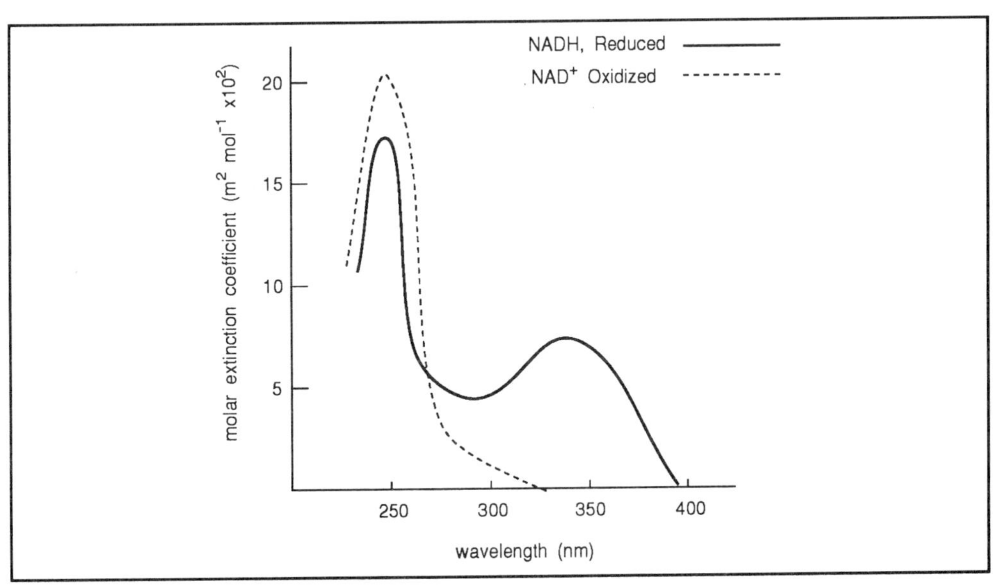

Figure 6.5 The absorption spectra of the oxidized and reduced forms of NAD

It can be seen that NAD^+ does not absorb above a wavelength of 300 nm whereas NADH has an absorption maximum at 340 nm. The interconversion of one form into the other can therefore be measured by the change in absorbance at 340 nm. This in turn can be related to the rate of enzyme reaction. The phosphorylated form of the cofactor, $NADP^+$ and NADPH also show similar spectra and can be used in the same fashion to monitor enzymes that use them as cofactors.

For example, the enzyme lactate dehydrogenase (LDH) catalyses this reaction:

$$\text{lactate} + NAD^+ \underset{}{\overset{LDH}{\rightleftharpoons}} NADH + H^+ + \text{pyruvate}$$

Neither lactate nor pyruvate have useful absorption spectra which would allow the reaction to be monitored. However, for every molecule of lactate that is converted to pyruvate, one molecule of NAD^+ is converted to NADH. The rate of reaction is therefore monitored by the rate of increase in absorbance at 340 nm which is a direct measure of the rate of reaction.

The two examples given above both use the natural substrate of the enzyme. However, for some enzymes conversion of the natural substrate to the product results in no obviously measurable change. There are a number of ways in which this problem can be overcome, and these are described in the following sections.

6.5.2 Direct methods using synthetic substrates

alkaline phosphatase

For some enzymes it is possible to use an artificial substrate that is converted into a measurable product. A fine example of this is the assay for the enzyme alkaline phosphatase. This enzyme is much used as a marker enzyme in molecular biology, particularly when linked to an antibody molecule, in such techniques as ELISA (enzyme linked immunosorbant assay) and protein blotting. As the name suggests, the role of alkaline phosphatase is to cleave phosphate groups from substrate molecules. However, this simple cleavage reaction results in no obvious spectral change that would allow the reaction to be monitored. To overcome this the artificial substrate 4-nitrophenol phosphate is used. This colourless substrate is converted to the coloured (yellow) product 4-nitrophenol by cleavage of the phosphate group by alkaline phosphatase. (Figure 6.6). The rate of increase in absorbance at 405 nm (due to the appearance of the yellow colour) at pH 10 is therefore a measure of the amount of enzyme present.

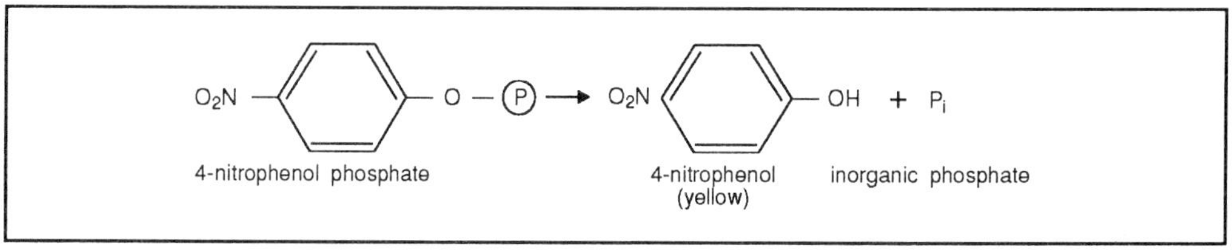

O_2N–C_6H_4–O–(P) → O_2N–C_6H_4–OH + P_i

4-nitrophenol phosphate → 4-nitrophenol (yellow) + inorganic phosphate

Figure 6.6 Hydrolysis of 4-nitrophenol phosphate by alkaline phosphatase.

α-glucosidase

The enzyme α-glucosidase (maltase) can be assayed in a similar fashion. α-Glucosidase cleaves maltose at the α-glycosidic bond to release two molecules of glucose (Figure 6.7). This simple cleavage does not results in an obvious change in the UV spectrum and therefore cannot be monitored directly.

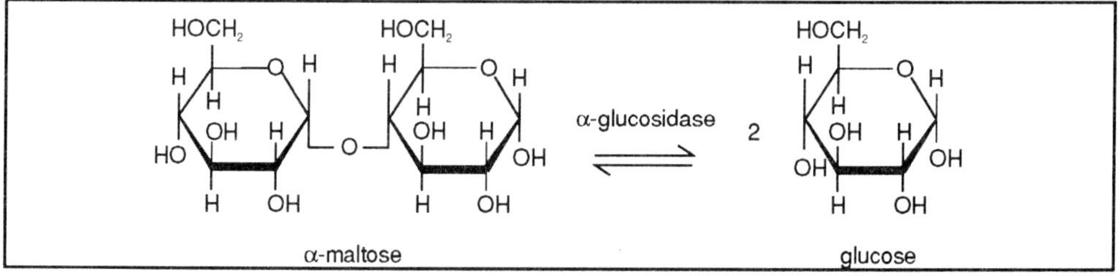

α-maltose ⇌ (α-glucosidase) 2 glucose

Figure 6.7 Cleavage of maltose to two glucose molecules by α-glucosidase.

The artificial substrate 4-nitrophenol-α-D-glucopyranoside can be used far more conveniently (Figure 6.8).

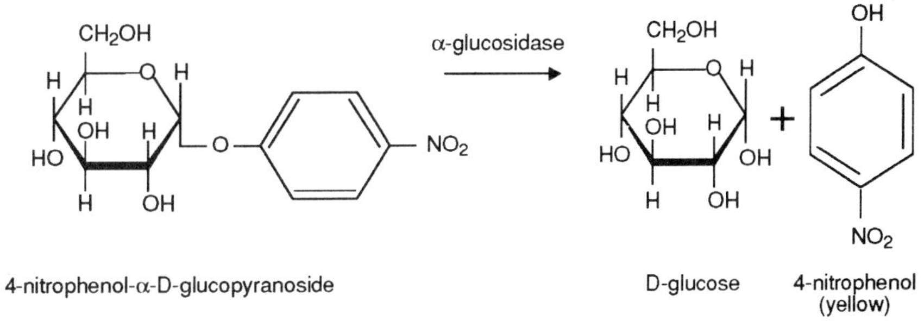

Figure 6.8 Use of the synthetic substrate 4-nitrophenol-α-D-glucopyranoside to assay α -glucosidase (maltase)

The α-glycosidic bond between the sugar group and the benzene ring in the colourless artificial substrate is readily recognised and cleaved by the enzyme, to release yellow 4-nitrophenol. The rate of reaction is therefore again monitored by recording the rate in increase of absorbance at 405 nm.

Π Many proteases are assayed using artificial substrates. Can you remember what proteases do?

They cleave peptide bonds in proteins (they are said to have peptidase activity).

However, if we put a protein solution in a cuvette and add a protease, although many peptide bonds will be cleaved there would be no spectral change that can be monitored to record the enzyme activity. Fortunately, however, many proteases, in addition to their peptidase activity also have esterase activity, ie they cleave ester bonds. We can therefore use this ability to monitor enzyme activity using artificial substrates.

chymotrypsin

For example, chymotrypsin can be monitored using 4-nitrophenylacetate as substrate (Figure 6.9). This substrate is colourless, but the esterase activity of the enzyme cleaves the ester bond to release yellow 4-nitrophenol. The rate of reaction is therefore monitored by the increase in absorbance at 405 nm.

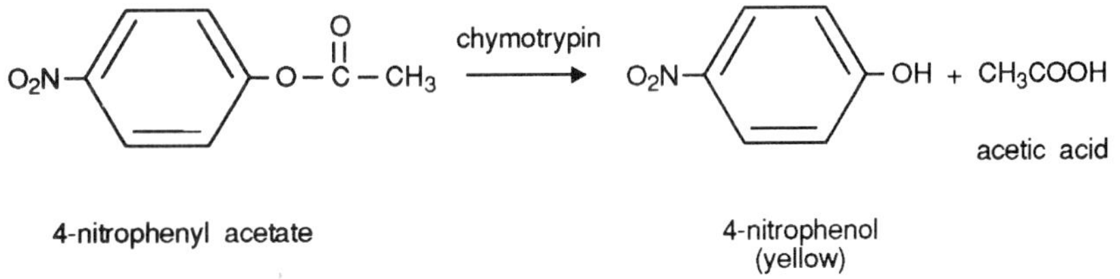

Figure 6.9 The use of 4-nitrophenyl acetate to assay chymotrypsin.

Π Why is the buffer solution for this assay particularly important?

You will have noticed in Figure 6.9 that acetic acid (CH_3COOH) is one of the products of this reaction. If the assay mixture was not well buffered, the pH of the reaction mixture would drop as the reaction proceeded and the enzyme activity would consequently change.

6.5.3 Enzyme-coupled methods

When there is no obvious spectrophotometric change that occurs in a catalysed reaction, it may be possible to link the product to another reaction that is measurable. Take for example the enzyme triose phosphate isomerase (TPI). This catalyses the reaction shown in (Figure 6.10).

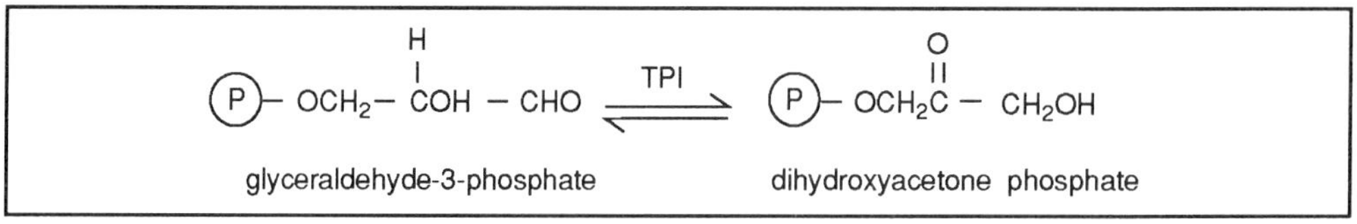

$$\text{(P)}-OCH_2-\overset{\overset{H}{|}}{C}OH-CHO \underset{}{\overset{TPI}{\rightleftharpoons}} \text{(P)}-OCH_2\overset{\overset{O}{||}}{C}-CH_2OH$$

glyceraldehyde-3-phosphate   dihydroxyacetone phosphate

Figure 6.10 Reaction catalysed by triose phosphate isomerase (TPI).

For this reaction there is no obvious spectrophotometric change and therefore the reaction cannot be measured directly.

However, if the enzyme glycerol-1-phosphate dehydrogenase and NADH are included in the substrate solution, the dihydroxyacetone that is formed by TPI is immediately converted into glycerol-1-phosphate. (Figure 6.11).

$$\text{(P)}-OCH_2\overset{\overset{O}{||}}{C}-CH_2OH + NADH + H^+ \rightleftharpoons \text{(P)}-OCH_2CH(OH)CH_2OH + NAD^+$$

dihydroxyacetone phosphate   glycerol-1-phosphate

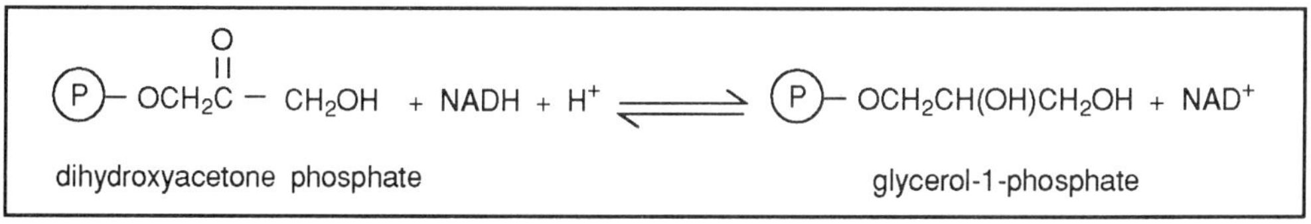

Figure 6.11

SAQ 6.3

How would you monitor this linked reaction?

SAQ 6.4

The enzyme malate dehydrogenase (MDH) catalyses the following reaction:

$$\text{oxaloacetate} + NADH^+ + H^+ \overset{MDH}{\rightleftharpoons} \text{malate} + NAD^+$$

Referring back to Figure 6.1, see if you can design a linked assay for AAT.

6.5.4 The use of chemicals to detect the product

This is really another example of a coupled method, where, instead of reacting the product of an enzyme catalysed reaction with another enzyme to give a measurable product, we react the product with a chemical to give a measurable product, usually a coloured derivative. Ideally we would like to react the product of the enzyme reaction with the chemical directly in the spectrophotometer cell so that the increase in colour can be measured continuously. A good example of this is the use of 5,5′-dithiobis-(2-nitrobenzoic acid), (DTNB). DTNB reacts with compounds containing free thiol groups (-SH groups) resulting in the release of 5-thio-2-nitrobenzoic acid (TNB) which is yellow and can be monitored at 410 nm (Figure 6.12).

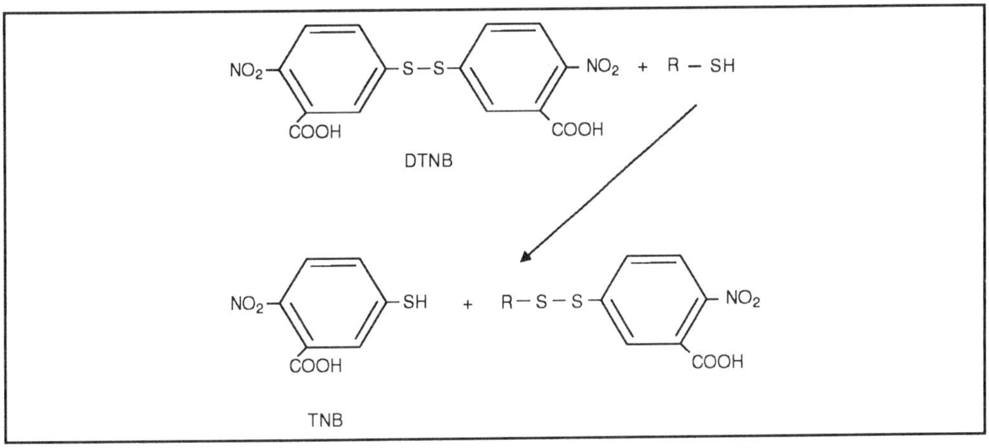

NO_2–C$_6$H$_3$(COOH)–S–S–C$_6$H$_3$(COOH)–NO_2 + R – SH
DTNB
→
NO_2–C$_6$H$_3$(COOH)–SH + R–S–S–C$_6$H$_3$(COOH)–NO_2
TNB

Figure 6.12 The reaction of DTNB with a sulphydryl compound to release the yellow product TNB.

An example of how this can be used is in the assay of citrate synthase which catalyses the condensation of acetyl-CoA with oxaloacetate to form citrate (Figure 6.13).

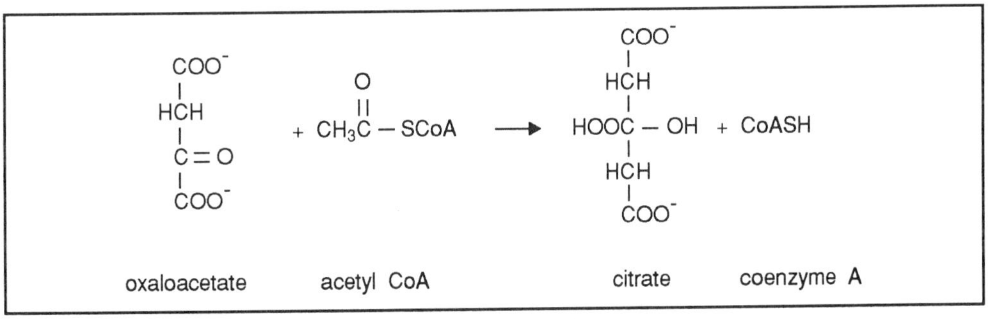

$$^{-}OOC{-}CH_2{-}CO{-}COO^{-} + CH_3CO{-}SCoA \longrightarrow {}^{-}OOC{-}CH_2{-}C(OH)(COOH){-}CH_2{-}COO^{-} + CoASH$$

oxaloacetate | acetyl CoA | citrate | coenzyme A

Figure 6.13 The reaction catalysed by the enzyme citrate synthase.

Coenzyme A is a fairly complicated molecule containing a thiol group which is important for its activity. For this reason coenzyme A is often written as CoASH. As you can see from Figure 6.13, for each citrate molecule formed, one coenzyme A molecule is released. The thiol group in this molecule will immediately react with a DTNB molecule with the release of yellow TNB. The rate of increase in yellow colour (measured at 410 mn) is a direct measure of the enzyme activity.

Π Can you think of any problem that might arise in an assay of this type?

Obviously a direct link of this kind cannot be used if the chemical being used reacts with and inactivates the enzyme being measured. In the case of DTNB this would occur if the enzyme had a thiol group that was important to its functions. Also, of course, the chemical must be capable of reacting instantly with the product under the same conditions (pH, temperature, etc) as the enzyme being measured.

Despite these restrictions, a number of assays, like the one described here, have been developed where the product of an enzyme reaction is converted to a coloured derivative *in situ* without interfering with the enzyme reaction itself.

isocitrate lyase

Figure 6.14 shows an assay for isocitrate lyase. In this case the aldehyde group in one of the products, glyoxylate, is reacted with dinitrophenyl hydrazine to give an orange dinitrophenylhydrazone derivative which can be measured spectrophotometrically. However, in this case the dinitrophenyl hydrazine cannot be included *in situ* as it will react with the enzyme.

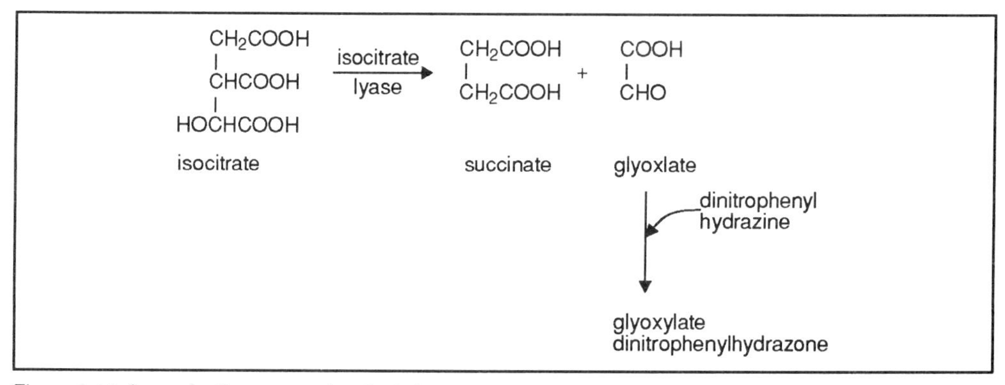

Figure 6.14 Assay for the enzyme isocitrate lyase.

Π How could you use this colourimetric reaction to design an assay for isocitrate lyase?

discontinuous assay

This is an example of a discontinuous assay. Aliquots of the reaction mixture are taken at timed intervals and added to dinitrophenyl hydrazine. The colour developed at each time interval is measured on a spectrophotometer and a graph is produced of intensity of colour (absorbance) formed against time. From this graph the rate of colour production can be determined; this is a direct measure of the rate of the enzyme reaction.

SAQ 6.5

The enzyme biotolase catalyses the reaction:

$$A \rightleftarrows B$$

The spectra of pure compound A and pure compound B are shown in Figure 6.15. Which of the following wavelengths could be used to monitor the reaction? Explain your reasons for your choice.

(a) 220; (b) 250; (c) 350; (d) 420 nm.

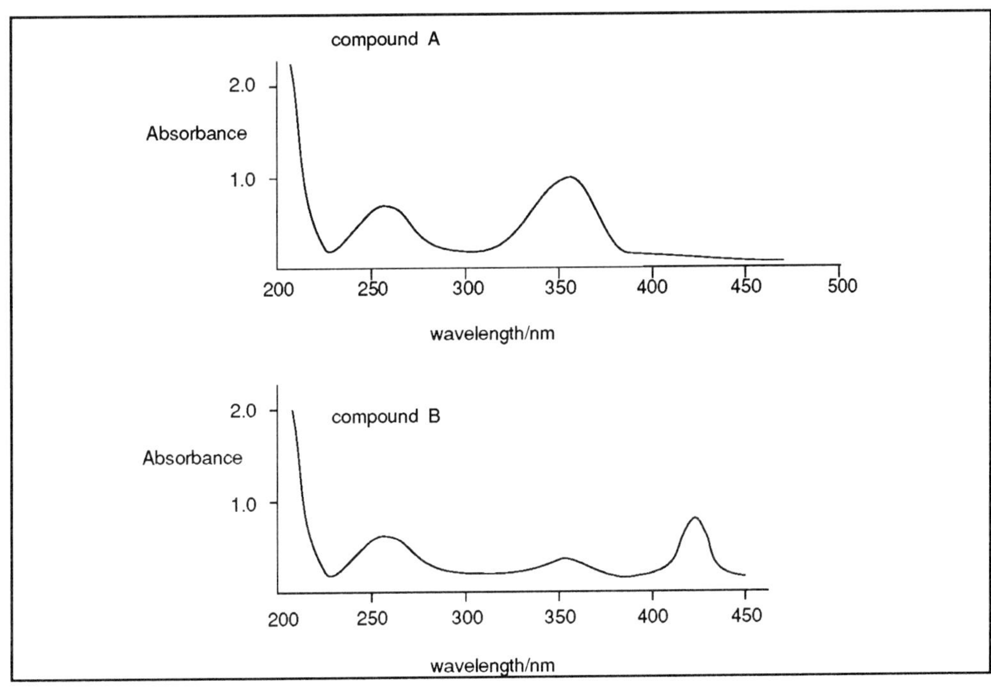

Figure 6.15 A comparison of the spectra of compounds A and B.

SAQ 6.6

The following are two examples of enzyme assays. Explain the principle behind each assay (eg enzyme-linked assays, direct), and where possible explain what wavelength you would use to monitor the reaction.

1) Hexokinase

D-glucose + ATP $\rightleftharpoons$ D-glucose-6-phosphate + ADP

glucose-6-dehydrogenase

D-glucose-6-phosphate + $NADP^+$ $\rightleftharpoons$ D-glucose-δ-lactone-6-phosphate + NADPH + H^+

2) Glucose oxidase

glucose + O_2 + H_2O $\rightleftharpoons$ gluconic acid + H_2O_2

peroxidase

o-dianisidine + H_2O_2 (reduced, colourless) $\rightleftharpoons$ o-dianisidine + H_2O (oxidised, yellow)

SAQ 6.7

1) Give a reason why the direct measurement of the product of an enzyme catalysed reaction might not be possible;

2) describe a potential disadvantage of the measurement of the product of an enzyme reaction by chemical derivatization;

3) explain a disadvantage of using artificial substrates.

6.5.5 Fluorimetric method of enzyme assay

This method of assay is capable of much greater sensitivity than the corresponding absorptiometric methods previously described.

In fluorimetry the radiation emitted by the sample is collected at right angles to the incident ray; in spectrometry the incident radiation passes through the sample to the detector (see Figure 6.16). Note the equations relating to absorbance and fluorescence, we will develop these further in Chapter 8, when we discuss spectroscopy.

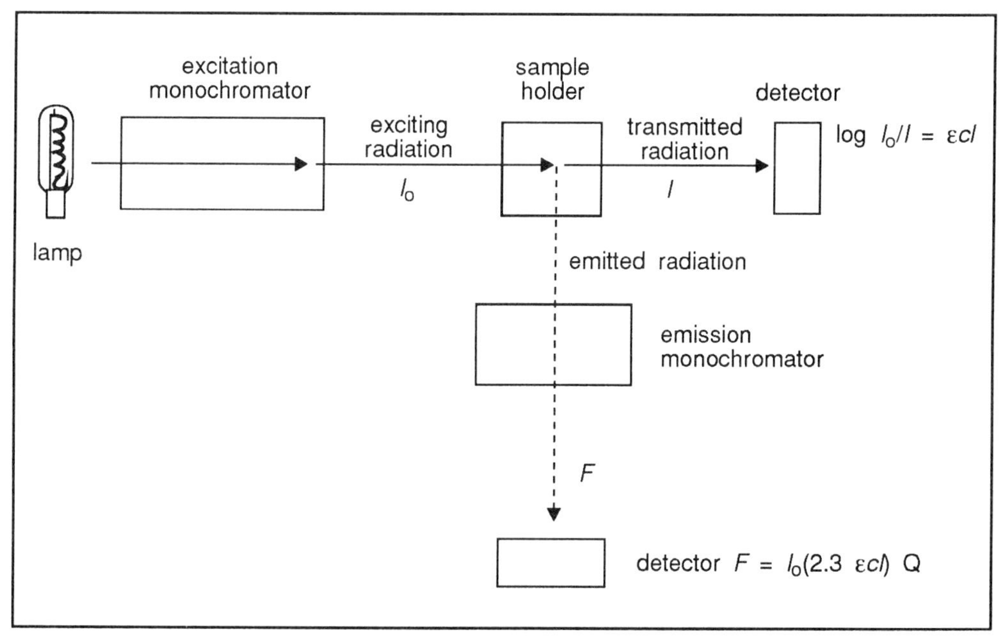

Figure 6.16 A spectrofluorimeter. Fluorescence is emitted in all directions by the sample, but most systems look at only that emitted at 90° with respect to the exciting light. Note that this differs from an absorption spectrophotometer only in that, with that instrument, the transmitted light would be detected.

The intensity of fluorescence, F, is directly proportional to the concentration, c, of the fluorescing material provided that $\varepsilon\ cl$ (ie the absorbance, A) is in the range 0.01 to 0.1 (Q is the quantum efficiency and ε the molar extinction coefficient). This is the region where measurement of the absorbance is not very accurate; this method thus complements the absorptiometric methods.

The principle of fluorimetric assays is analagous to that of the corresponding chromogenic system, based on 4-nitrophenol conjugates. The artificial substrate is based on either methylumbelliferone (Figure 6.17) or a naphthol.

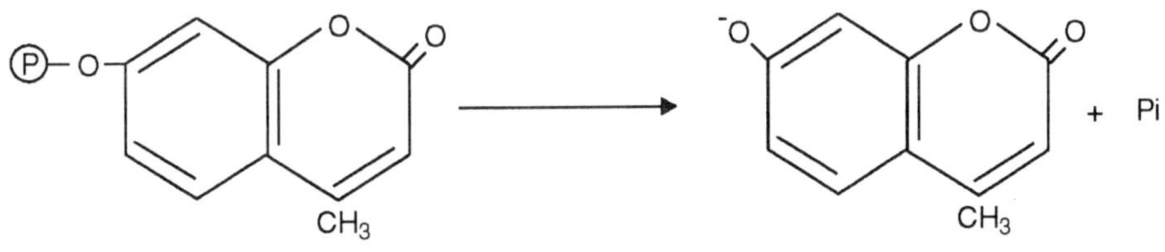

4-methylumbelliferyl phosphate (conjugate) → 4-methylumbelliferyl anion + Pi

Figure 6.17 Hydrolyses of a methylumbelliferyl conjugate.

This conjugate on hydrolysis by the enzyme gives a product which at high pH values fluoresces at a higher wavelength than the original conjugate. The progress of the reaction is monitored by measuring the intensity of fluorescence (proportional to the concentration of the product) at frequent intervals.

Π Make a list of the factors which determine the sensitivity of this method? (Try this before examining our list below).

The following factors are important:

- the intensity of the incident radiation, I_o;
- the sensitivity of the detecting system;
- the proportion of the emitted radiation which can be collected and transmitted to the detector;
- the quantum efficiency, and the molar extinction coefficient.

For a given instrument and enzyme system these factors are constant.

A wide range of artificial substrates is available commercially; eg

4-methylumbelliferyl-β-D-galactopyranoside for β-galactosidase,

4-methylumbelliferyl-acetate for esterases,

4-methylumbelliferyl-phosphate for alkaline phosphatase.

6.6 Non-spectroscopic methods of enzyme assay

Although the vast majority of enzyme reactions can be monitored by following the appearance of a product (or disappearance of substrate) by a spectroscopic method there are a number of cases in which alternative techniques which have specific applications are used.

Π Make a list of any reasons why an alternative technique might be more appropriate?

You might have included the following reasons:

- the chemistry of the reaction or the nature of the product is such that none of the spectroscopic approaches gives an acceptable measurement;
- for some systems a spectroscopic method may be feasible but it is time consuming, requiring expensive or unstable reagents and prone to error on account of its complexity;
- samples which are turbid or highly coloured present difficulties in spectroscopic methods due to their high absorbance.

We will consider briefly some other techniques which can be applied, maybe to only a very limited number of assays.

6.6.1 Chemical methods of assay

These methods tend to be simple and do not require sophisticated instrumentation. In this type of assay the concentration of either the substrate or the product(s) is estimated by a chemical method in a discontinuous manner (ie samples removed at frequent times are analysed).

yeast invertase

As an example consider the assay of yeast invertase using sucrose as substrate. The products, a mixture of the reducing sugars glucose and fructose, can be estimated by titration using the Benedict reagent.

SAQ 6.8

Describe a physical method for the assay of catalase which causes the decomposition of hydrogen peroxide:

$$2\ H_2O_2 \rightarrow 2H_2O + O_2$$

6.6.2 Electrical methods of assay

A number of experimental techniques are based upon changes in the charged states of the system as a result of an enzymatic reaction.

Conductance method

Conductance method. If the reaction produces a significant change in the number of ions then there will be a change in the electrical conductance of the system; this can be used as a measure of enzymic activity.

Π Can you think of any problems which may arise using conductance measurements to monitor the enzyme reaction?

The main problems include:

- the presence of high concentrations of buffer ions means that it can be difficult to detect small changes in conductance as a result of small changes in the total ion concentration;
- changes in the degree of dissociation of weak electrolytes due to the high concentration of buffer ions;
- careful control of temperature is essential since conductance is very sensitive to temperature changes.

The platinum electrodes used to measure the conductance are inert and will therefore have no effect on the enzyme activity.

A good example of the use of this method is the assay of urease using urea as the substrate, in which an uncharged molecule is converted into three ionic species:

$$CO(NH_2)_2 + H_2O \rightarrow 2NH_4^+ + CO_3^{2-}$$

electrometric method

Electrometric method. In this method the change in the concentration of hydrogen ions, ie the pH, of the reaction is monitored. A typical example is the hydrolysis of trialkylglycerols by gastric enzymes, such as lipase, to glycerol and fatty acids. The progress of this reaction can be monitored directly using a glass electrode assembly.

Π Perhaps you would like to consider reasons why this method is not satisfactory.

The major problems include:

- the continual change in pH, unless the mixture is buffered, will result in a progressive decrease in activity (see Figure 6.2);
- any protons released into a buffered reaction will be immediately be taken up by the buffer anion and there will be no detectable change in pH;
- possible contamination of the reaction mixture with metal ions from the reference calomel electrode.

To avoid the adverse effects of pH an automatic titrator ('pH-stat') can be used; in this the pH value is kept as constant as possible by the continual addition of acid or base. The amount added in a given time is a measure of the reaction rate. We will examine electrometric methods of analyses more thoroughly in the final chapter of this text.

6.6.3 Miscellaneous techniques for enzyme assay

There are some other techniques which can be used for specific assays; these have limited application and so will be discussed briefly.

Polarimetry

Polarimetry in which the rate of change of optical rotation of the substrate gives a measure of enzyme activity, eg the assay of yeast invertase using sucrose as the substrate to give a mixture of the optically active glucose and fructose.

Manometry

Manometry in which changes in the volume of gases evolved or taken up are measured. This classical technique, which requires elaborate equipment, was used originally by Krebs and others to elucidate the metabolic pathways of respiration.

Radiotracer method

Radiotracer method. The use of radioactively labelled substrate lowers the detection limits dramatically and is a valuable tool if the enzyme has a fundamentally low activity or is present in small quantities, eg the assay of glutamic acid decarboxylase which can be followed by monitoring the $^{14}CO_2$ emitted fromspecifically labelled glutamic acid:

$$\begin{array}{c} ^{14}COOH \\ | \\ HCNH_2 \\ | \\ (CH_2)_2 \\ | \\ COOH \end{array} \longrightarrow \begin{array}{c} NH_2 \\ | \\ CH_2 \\ | \\ (CH_2)_2 \\ | \\ COOH \end{array} + \ ^{14}CO_2$$

See Chapter 9 for a fuller discussion of techniques using radiotracers.

Microcalorimetry

Microcalorimetry. Enzyme catalysed reactions are accompanied by thermal changes. The greater the reaction, the greater the heat effect. We can therefore use measurement of the heat effect as a device for measuring enzyme activity. This technique should, in theory, be applicable to virtually every reaction; it is particularly useful with turbid or coloured samples (eg cells, cell extracts, sera) which require excessive and often error prone processing for spectroscopic analysis. It suffers from the disadvantage that high cost instrumentation is needed to measure the small amounts of heat evolved; it is, however, rapid and suitable for routine analysis. We will discuss microcalorimetric procedures in more detail in Chapter 11.

SAQ 6.9

List the advantages and disadvantages with enzyme assay using each of the following techniques:

1) direct spectroscopic analysis;

2) microcalorimetry;

3) conductance measurements.

Summary and objectives

The terms used to express enzyme activity with special reference to changes during purification processes are defined. The factors which must be controlled during enzyme assay, such as the pH of the mixture, temperature, and substrate concentration are discussed. Enzyme assays can be made either in a continuous or a discontinuous manner. The principles underlying the design of different spectroscopic assays are discussed, these include direct methods, the use of synthetic substrates, enzyme coupled reactions and the use of chemicals to detect the product. Finally there is a brief review of alternative methods of assay.

Now that you have completed this chapter you should be able to:

- define the terms enzyme unit, yield, specific activity and fold purification and calculate their values from experimental data;
- differentiate between a continuous and discontinuous assay method;
- describe the factors which must be controlled in any measurement of enzyme activity;
- explain the principles underlying enzyme assay by spectroscopic methods including the use of synthetic substrates, enzyme coupled and chemically coupled reactions;
- explain the advantages and applications of spectroscopic and non-spectroscopic methods of assay.

Detection and measurement of specific biomacromolecules: Immunological and hybridisation techniques

Detection and measurement of specific biomacromolecules: Immunological and hybridisation techniques

7.1 Introduction

In the previous chapter, we examined the measurement of enzyme activity predominantly as a method for monitoring the purification of enzymes from cell homogenates. There are, however, a large number of cellular macromolecules which do not have enzymic activity. Nor do they have any specific chemical reactivities. For example, one messenger RNA (mRNA) is chemically and physically very much like that of any other mRNA. There are however many reasons why we need to be able to distinguish between such compounds and to be able to map their distributions in cells and to quantify their occurrence. How then do we measure such complex, chemically more-or-less indistinguishable, macromolecules?

antibody specificity

In this chapter we examine two types of techniques which enable us to carry out such investigations. The bulk of the chapter describes techniques which make use of the abilities of antibodies to identify and bind to specific molecular structures. The key to the success of this strategy lie both in the specificity of the antibodies used and in our ability to detect interaction between these antibodies and their target structures.

molecular hybridisation

The remainder of the chapter describes techniques which depend upon the ability of the nucleotide bases in nucleic acids to specifically basepair with each other. This property of the nucleotide bases enables us to use particular sequences of nucleotides to identify and measure complementary sequences. These techniques are usually referred to as hybridisation as hybrid nucleic acid molecules are produced as the key process in molecular identification.

The specificity and diversity of antibodies enables us to measure a wide variety of complex molecular structures but is especially employed to detect and assay proteins and glycoproteins. Clearly the hybridisation techniques are only applicable to the analysis of nucleic acids.

This chapter has been written on the assumption that the reader is familiar with the general properties of antibodies and nucleic acids. Appropriate background knowledge is provided by the BIOTOL text 'Molecular Fabric of Cells'.

7.2 Antibodies used in immune-assay procedures

7.2.1 Polyclonal antibodies

The antibodies used in immune-assays are either polyclonal antibodies (antibodies generated by administering antigen to an animal) or monoclonal antibodies (antibodies produced by pure, cloned cultures of antibody-producing cells - B cells). It is beyond the scope of this text to describe the 'manufacture' of antibodies in detail (The BIOTOL text 'Technological Application of Immunochemicals' gives a thorough account of this aspect). However, in outline the production of polyclonal antibodies is achieved by injecting an animal (for example, rabbit, goat etc) with the compound (antigen) of interest. After a few weeks, the animal is given an addition dose of the antigen This process may be repeated.

immunisation by means of a strong antigen

The timetable to generate antibodies against a good antigen (large molecule with repetitive units) looks as follows: the antigen is subcutaneously administered to the animal in an isotonic solution. The first antibodies will appear in the serum after approximately 10 days: the IgMs. The maximum concentration is reached after 15 - 20 days. Then the concentration will decrease within several weeks. A secondary response can be induced by means of a second injection with antigen administered twenty days after the first.

immunisation by means of a weak antigen

Because memory cells have been produced during the first (primary) response, the differentiation of these memory cells into antibody-producing plasma cells will proceed much faster than after the first injection. After approximately 10 days the concentration of antibodies (in this case the IgGs) is at its highest level (3 - 10 times higher than the amount of IgMs in the first response). By injecting the animal every 14 days with antigen the concentration of antibody will rise to 1 - 5 mg of antigen-specific IgG per cm^3(ml) of serum. In cases where we have to use smaller quantities of antigenic compounds another technique can be applied. Efforts can be made to prolong the time during which the animal is exposed to the antigen. This can be realised by means of repeated injections or by administering the antigen in a precipitate containing for example aluminium potassium phosphate. The antigen will be released slowly from these precipitates and will have the same effect as in the case of repetitive administration of free antigen.

Most antibodies that are used in immune chemistry are generated by administering an antigen solution to a rabbit. 5 to 30 cm^3 of blood is taken from the ear vein and the blood is left to clot. After about 1 hour at 30°C the clot is separated from the serum. After the serum has been drained from the clot, the clot is incubated at 4°C so that is shrinks to half its size and an additional equivalent amount of serum becomes available. The serum is centrifuged at 10 000 x g for 15 minutes in order to remove cells and the clot debris. After centrifugation the serum is heated for 45 minutes at 46°C in order to inactivate proteases and the complement system. Subsequently it is frozen in small amounts and stored at -20°C or lower temperatures. It is wise to take some blood from the rabbit prior to immunization and use this as control serum. Rabbits, mice and rats are used to generate smaller quantities of antibodies whereas sheep, goats and horses are used to harvest larger quantities of antibodies.

Π Why are such preparations called polyclonal antibodies?

The answer is that serum produced in this way contains many different antibodies each produced by different clones of antibody-producing cells (B-cells). There may be many different antibodies present which will react with the antigen of interest. Each will bind to particular structures (epitotopes) on the antigen.

immunisation with small molecules

If we wish to generate antibodies against small organic molecules, such as drugs or non-peptide hormones, the molecule (the hapten) must be coupled to a strong antigenic material, like proteins, polysaccharides or (sheep) erythrocytes (the carrier). The material can be administered after it has been coupled. Note that a hapten is a molecule against which antibodies can be produced but which, by itself, cannot stimulate the production of antibodies.The antiserum which is produced can simply be freed of the antibodies that were produced against the carrier of the hapten by titrating the antiserum with the carrier and centrifuging the precipitate. Many different antibodies are produced against large molecules, such as proteins. If it is known that about six different immunoglobulins (antibodies) are produced against a hapten with a molecular weight of 200, it can be expected that in the case of a protein of 20 kDa a very large number of antibodies will be produced which all recognize a different part of the surface of the antigen. These different antibodies have all been produced by just as many different clones of B lymphocytes. This explains why the serum which is induced in the manner described above is called a polyclonal antiserum.

SAQ 7.1

Will an antiserum that has been generated against bovine serum albumin in rabbit A be the same quantitatively as an antiserum generated against the same protein in rabbit B? (Give reasons for your answer).

7.2.2 Monoclonal antibodies

myeloma
hybridoma

ascites
tumours

Monoclonal antibodies are produced in cell cultures. In this case, a single type of B-cell is cultured and thus only one type of antibody is produced. A detailed description of how the appropriate B-cell is identified, isolated and cultured is beyond the scope of this text. (The reader is referred to the BIOTOL text 'Technological Applications of Immunochemicals' for details), In outline, the procedure is to first immunize an animal with the antigen of interest and the antibody-producing B cells are subsequently isolated from the spleen. These cells are fused with myeloma (tumour) cells which grow *in vitro*. The right antibody-producing cells (hybridomas) are selected from the fused cells. As soon as the right cell line has been found, it can be cultured and the monoclonal antibody can be isolated from the extracelluler fluid. Alternatively the desired B-cells may be cultured in the peritoneal cavities of mice where they form ascites tumours. The serum of such mice contains very high levels of the required antibody.

Π Make a list of the advantages and disadvantages of using monoclonal antibodies.

The application of monoclonal antibodies has important advantages. It is absolutely certain that only one epitope is recognized. If the antibody reacts with a different protein, it is certain that the same epitope is present on that protein. Monoclonal antibodies are used for the detection of hormones, drugs, serum proteins, enzymes and proteins at the cell surface. The detection of determinants at the cell surface especially offers many possibilities for application. It is possible to distinguish between different cell types, such as tumour cells and normal cells.

The disadvantage of monoclonal antibodies is that production is very expensive and labour-intensive. These economic aspects will prevent the application of monoclonal antibodies in cases where it is not absolutely necessary. There are many possibilities in which polyclonal antibodies can be applied just as well. In certain cases it is even better to use them. An example is the case where a viral infection must be demonstrated by detecting the viral coat protein by means of an immune reaction Here it is better to use a polycolonal antibody because of the risk that the epitope for the monoclonal antibody will be changed when something changes (mutates) in the viral coat protein gene.

7.3 Immune precipitation and immune reactions in gels

7.3.1 The immune precipitation reaction

When antigen is added to particular amounts of antibodies, several types of complexes can be formed between antibody and antigen (see Figure 7.1).

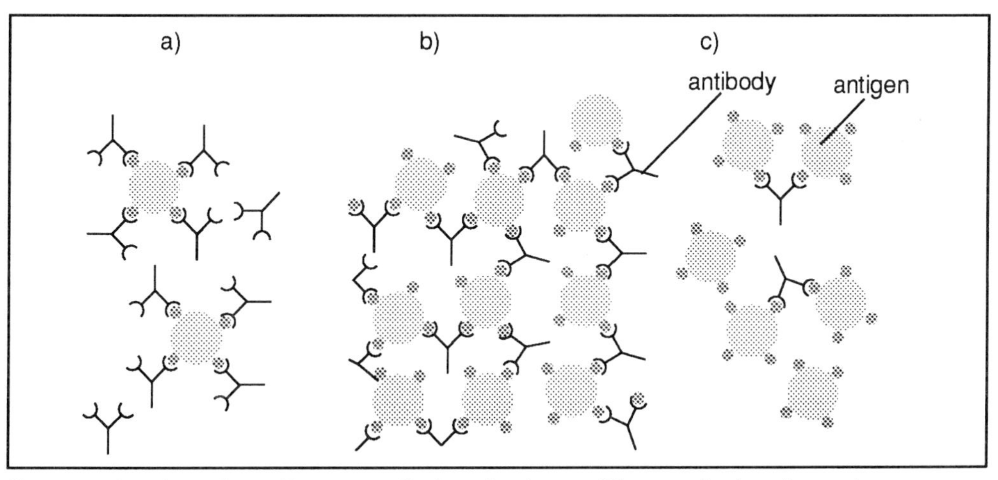

Figure 7.1 Complexes formed between antibody and antigen at different antibody:antigen ratios. a) Complexes formed in the presence of excess antibody; b) compelxes formed in presence of equivalents amounts of antigen and antibodies; c) complexes formed in the presence of excess antigen.

In the case of an excess of antibodies all epitopes on the antigen will be occupied by an antibody. Depending on the source of the serum the antibody (IgG) molecule-antigen complexes will be soluble or insoluble. With rabbit IgGs these complexes are insoluble, with horse IgGs they are soluble. So the serum that is used is the factor which determines whether precipitation will or will not occur when we use a large excess of antibodies. On the addition of more antigen, the relative excess of antibodies will decrease and the binding sites of the antibodies will become increasingly occupied. Hence, cross-linking between the different antigen molecules by means of the antibody molecules will increase. In this manner larger complexes are formed and precipitation will take place. In the area with the maximum precipitation the number of epitopes on the antigens and the number of binding sites of the antibodies are more or less equal. This area is designated the equivalence area (point).

equivalence area

In the case where we have an excess of antigen, large three-dimensional structures will not be formed. In this case complexes of two antigen molecules are created which are linked by means of an antibody. These complexes will generally be soluble. (Case c in Figure 7.1).

Π The antibody molecules shown in Figure 7.1 are shown as being bivalent (two sites at which they bind antigen) whilst the antigen is shown as being tetravalent. Each antigen has four sites which interact with the antibody molecules (in other words, each antigen carries four epitopes recognised by the antibody). Use this information to calculate the ratio of antigen: antibody in the area of equivalence.

You should have calculated that the ratio would be 1:2. When antigen and antibody are present in this ratio, the number of epitopes on the antigen and the number of binding sites on the antibody molecules are equal.

Figure 7.2 shows the result of an experiment in which a variable amount of antigen is added to a constant amount of antiserum. Such experiments enable us to calculate the concentration of the IgG molecules in the serum which react with the antigen and the number of determinant sites on the antigen. This will be shown by means of the precipitation of human serum albumin (HSA) with an antiserum generated in a rabbit.

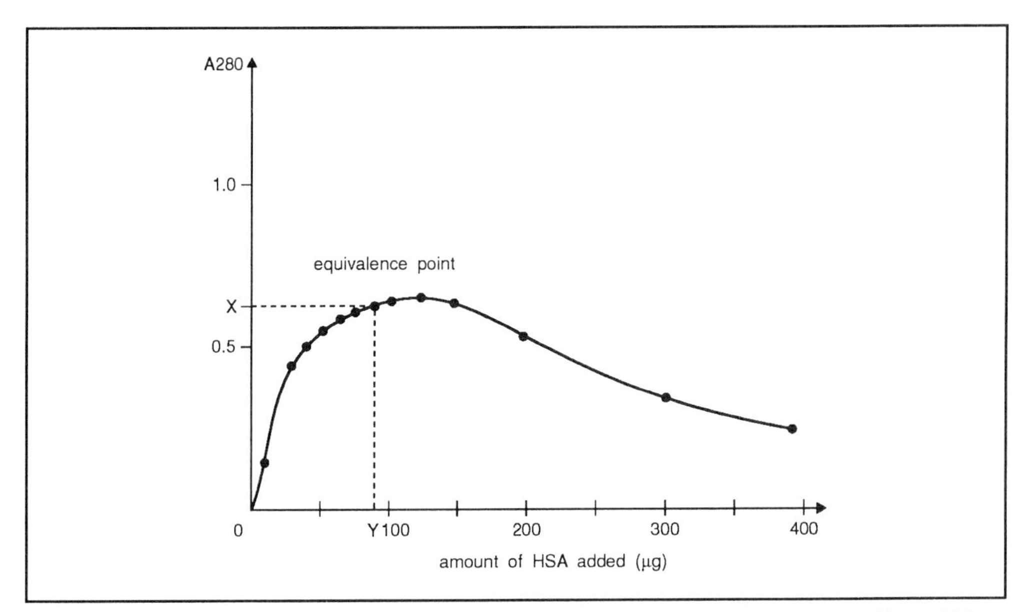

Figure 7.2 The precipitation curve of human serum albumin by using an antiserum generated in a rabbit. An increasing amount of HSA is added to a constant amount of antiserum. The turbidity (absorbance) of the precipitate is determined at 280 nm. It was also determined by means of an ELISA test to check whether any antigen or antibody molecules were left in the supernatant after centrifugation. (See text for further details).

The test of which the results are given in Figure 7.2 was carried out as follows. Increasing amounts of HSA were added to 0.1 cm^3 portions of antiserum (rabbit). The precipitates formed in this manner were separated from the solutions by means of centrifugation. Each precipitate was washed twice and, after it has been dissolved in 1.0 cm^3 NaOH (0.1 mol dm^{-3}), its light adsorbance was measured at 280 nm. After removal of the precipitates by means of centrifugation, the supernatants were examined for the presence of antibodies against HSA or the presence of HSA. (This was done with the relatively crude Ouchterlony test or by means of the more sensitive ELISA test. We will discuss these later). There were some measurements in the titration curve in which neither antibodies against HSA nor HSA could be demonstrated in the supernatant. This area is the equivalence area. In the experiment shown in Figure 7.2, this was at about 87.5 μg HSA. Because it is known that 87.5 μg HSA was added, we may use the extinction coefficients of IgG and HSA to calculate how much IgG reacting with HSA was present in the antiserum. A solution of 1 mg per cm^3 IgG gives an extinction of 1.439 at 280 nm with a path length of 1 cm. For HSA, the value is 0.530. 0.0875 mg HSA dissolved in 1.0 cm^3 NaCH yields an extinction of 0.0875 x 1 x 0.530 = 0.046 at a path length of 1 cm. Total extinction of the complex at the point of equivalence was 0.65 (see Figure 7.2), thus IgG contribution was 0.65 - 0.046 = 0.604. This corresponds with

$0.604/1.439 = 0.419$ mg per cm^3. 0.1 cm^3 of serum was used in the experiment, so 4.19 mg IgG which reacts with HSA is present per cm^3 of serum. Now we know how much HSA reacting antibody is present in the serum, we can in principle use this to determine HSA in samples containing HSA.

determination of number of epitopes

In the case of rabbit serum, the number of epitopes per antigen molecule can be calculated on the basis of observations in the area in which we use excess antibodies because all antigen molecules precipitate despite this excess of antibody. Using the technique described above, the amount of IgG present in the precipitate can be calculated from the absorbances of solutions prepared from the precipitates. Because the molecular weights of HSA (68 kDa) and IgG (150 kDa) are known, the molal ratio of IgG and HSA can be calculated for each precipitate. If this ratio is plotted against the added amount of HSA in a curve, it is possible to calculate the number of epitopes on HSA by means of extrapolation to an infinite excess of IgG (all epitopes on the antigen bind to an IgG molecule). (See Figure 7.3).

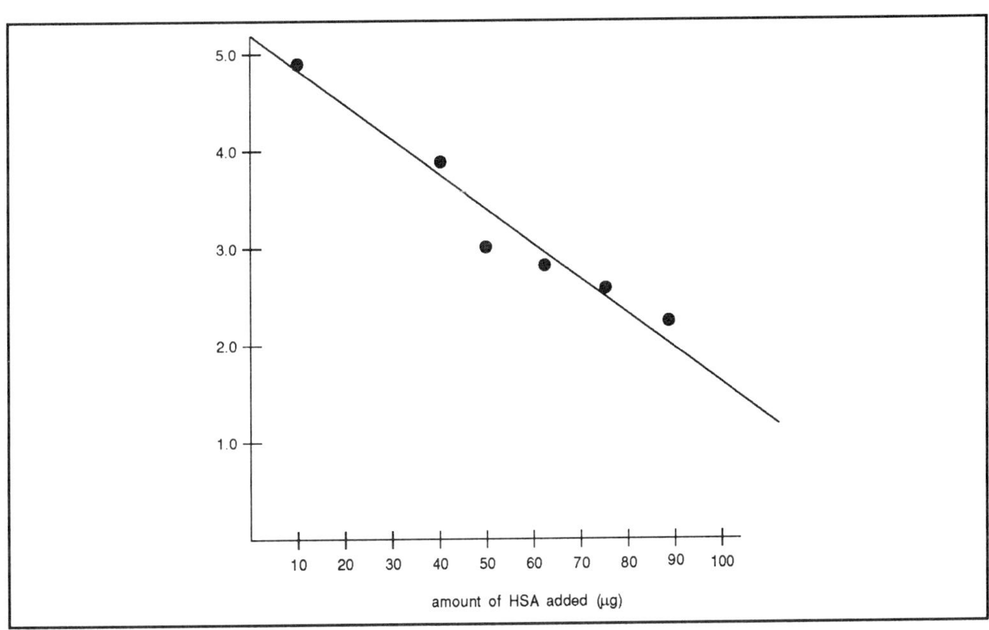

Figure 7.3 Ratio antibody/antigen precipitated at different amounts of antigen. The line, drawn by means of linear regression, yields a ratio of 5.2 when extrapolated back to HSA (= 0μg). This indicates that there are five antigen-binding sites on the HSA molecule for the applied serum).

We provide a sample calculation for the type of experiment depicted in Figure 7.3

Calculation of the number of epitopes per human serum albumin molecule.

The precipitate which is produced on addition of 10 μg HSA has an extinction of 0.15. 10 μg HSA yields an extinction of 0.0053 per cm^3. Hence, the contribution to the extinction of the antibodies is $0.15 - 0.0053 = 0.1447$ and this is caused by 0.101 μg cm^{-3} IgG. Therefore the molal ratio IgG/HSA is $(0.101/150\,000)/(0.010/68\,000) = 4.6$.

So if 10 μg HSA is added to 0.1 cm^3 serum, 4.6 antibody molecules are bound per HSA molecule.

Quantitative precipitation is quite a crude method to demonstrate the presence of antigen in a preparation. It is moreover important to make sure that measurements are carried out around the point of equivalence. If the antigen is a carbohydrate, the absorbance measurement at 280 nm can no longer be used to determine the amount of precipitated antigen since carbohydrates do not absorb at 280 nm. A carbohydrate analysis must be performed in this case.

You should realise that the precipitation reaction of antigen solutions is no longer applied as a routine quantitative measurement (precipitation reactions in gels and the ELISA technique are more suitable for this aim). It does, however, still have some uses. It is a convenient method for determining small quantities of radioactive material., for example in haemoglobin synthesis with mRNA extract. When mRNA, isolated from the red blood cells of chicken, is added to *E.coli* ribosomes, radioactive haemoglobin will be synthesized if radioactive amino acids and other necessary cofactors for protein synthesis are present. The rate of synthesis can be determined by measuring the incorporation of radioactivity into the synthesized protein. The haemoglobin can be induced to precipitate effectively by adding additional non-radioactive haemoglobin to the extract. The specificity of the immunoprecipitation and protein synthesis can be checked by means of an SDS gel which is followed by autoradiography. The film should only turn black at the sites of the haemoglobin subunits. (We discuss autoradiography in Chapter 9).

SAQ 7.2

The quantities of bovine serum albumin (BSA) listed below are added to 1 cm^3 of an antiserum solution. The pellet is washed twice after centrifugation and dissolved in 1 cm^3 NaOH (0.1 mol dm^{-3}). The extinction at 280 nm has been determined. It is also known that 1 mg per cm^3 BSA at 280 nm yields an absorbance of 0.667 and 1 mg of IgG an absorbance of 1.439. The molecular weight of BSA is 68 000 Da and that of IgG 150 000 Da. Calculate the number of epitopes per BSA molecule.

BSA added (μg)	absorbance at 280 nm
5	0.097
10	0.184
15	0.272
25	0.429
30	0.557
50	0.700
75	0.859
100	0.894
150	0.865
200	0.845
400	0.655

7.3.2 Precipitation reactions in gels

Immune reactions in gels should always proceed in a humid environment to prevent the gel from drying out.

immune precipitations in gels

The principle of precipitation reactions in gels. If a solution with an antigen is put into an agar gel, the antigen will diffuse radially in this gel. A concentration gradient is created,decreasing to zero at some distance from the point of application.If the gel contains antibodies and the antigen concentration was high enough at the start, there will be an area around the point of antigen application in which good precipitation will occur (the area of equivalence). We can extend this approach. Both partners (antibody and antigen) of the precipitation reaction can be induced to diffuse. If they diffuse towards each other they will eventually meet, react and precipitate. This approach is designated double diffusion. The driving force behind the motility of the reacting molecules can also be assisted by electrophoresis. In this manner the separating force of electrophoresis can be combined with an immune reaction, which is called electroimmune diffusion.

single immune diffusion

Single and double immune diffusion. Single immune diffusion can be carried out vertically in a narrow tube (one-dimensional diffusion) or horizontally (two-dimensional diffusion). In the first case an antigen-containing solution is put onto a gel which contains antibody. The antigen diffuses into the gel and precipitation will take place in the area where antigen and antibody are present in equal quantities. Because antigen will continually diffuse into the gel, the precipitate which is formed at the start of the reaction will dissolve again due to the excess of antigen. In most cases the precipitation line will be visible to the naked eye. The line of precipitation will move as long as the supply of antigen continues. However, after a time the concentration gradient of the antigen in the gel will be so small that the transport of antigen will almost come to a halt. So will the movement of the precipitation line. This technique is called the **Oudin tube technique**.

Oudin tube technique

Mancini technique

This experiment can also be performed in a horizontal gel in which antibody is present (the **Mancini technique**). In this case, radial diffusion will take place. A precipitation circle will become visible around the site where the substance was introduced. If a series of standard solutions is administered to the same horizonal gel, A semiquantitiative concentration determination can be performed for an unknown antigen solution. In this case the diameters of the precipitation rings of the known antigen solutions should be plotted against the concentration of antigen. Figure 7.4 gives an example of the single radial diffusion technique. This method is applied in clinical chemistry to assess the concentration of plasma proteins in the serum of patients suffering from diseases which lead to an overproduction of some serum proteins.

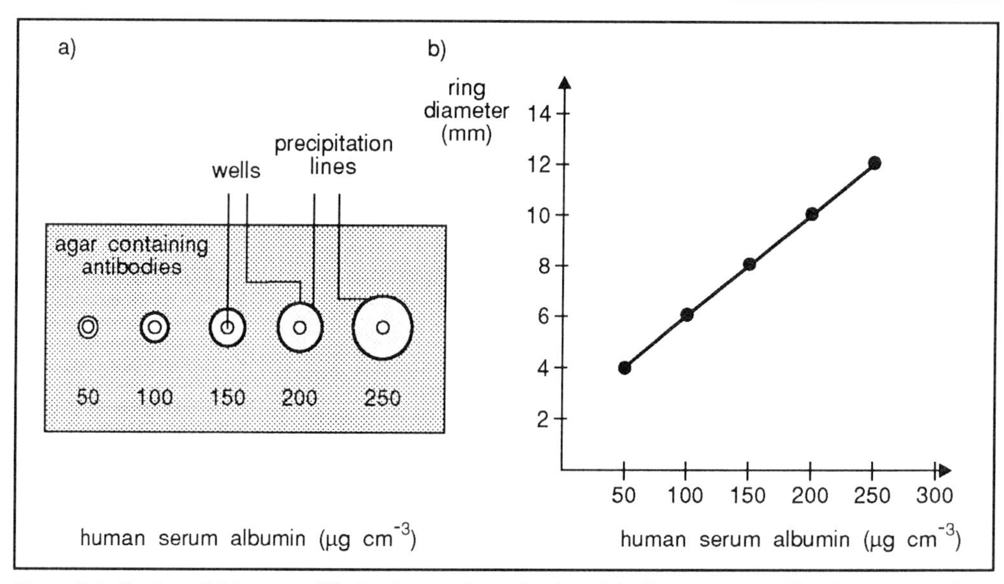

Figure 7.4 Single radial immune diffusion for the determination of the human serum albumin concentration (HSA). a) a slide is coated with a layer of agar containing anti-HSA rabbit serum. The given concentrations of HSA are placed in the wells. The precipitation reaction is performed overnight; b) standard graph (not drawn to scale) for determining the HSA concentration in samples using the Mancini technique.

double immune diffusion Ouchterlony technique

In double immune diffusion both the antibody and the antigen diffuse in the gel. This is probably the most frequently applied technique. It is relatively insensitive compared to the modern techniques like RIA or ELISA but it is easy to perform and does not require expensive equipment. The **Ouchterlony technique** - a synonym for double immune diffusion - not only offers the possibility of demonstrating the presence of antigen in a specimen but also tells us something about the immunological relationship between different antigens. This will be explained by reference to Figure 7.5. A thin layer of warm liquid agar is put onto a slide. After it has cooled, wells are punched into the agar and the antibody and antigen solutions are placed in them. Now examine Figure 7.5 and read its legend carefully.

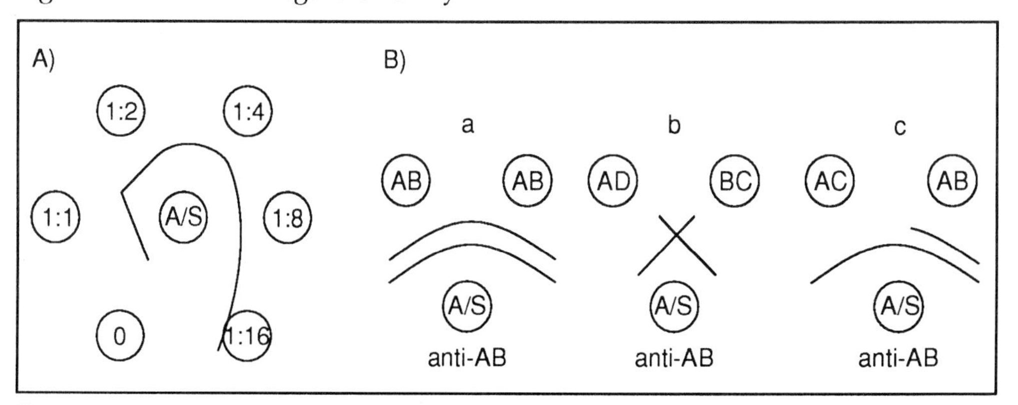

Figure 7.5 Ouchterlony double diffusion experiments A) Antiserum (A/S) is placed in the central well, and different dilutions of the antigen in the outer wells. Note that the position of the precipitation lines depends on the concentration of the antigen. B) Experiments to examine immunological relationships. Note that in a) the two antigens in the two wells are the same. These produce two continuous precipitation lines. In b) note that the two antigen samples do not contain common antigens and that the precipitation lines cross. In c) there is one antigen in common and this gives a continuous precipitation line.

Figure 7.5A shows the result of double diffusion. It is clear that the distance of the precipitation line to the central well containing the antibody depends on the concentration of the antigen in the surrounding wells. The lower the concentration of the antigen in the surrounding wells, the further the antibodies have to diffuse in order to attain an equivalent concentration with the low antigen concentration. Figure 7.5B depicts what the different double diffusion patterns can look like if immunological identity research is done. The antiserum contains antibodies against both the antigenic determinants A and B. If antigens bearing the determinants of A and B are present in both wells at the same time, flowing, curved precipitation lines will appear. This means that common antigens are present in both wells. If the wells have no antigens in common, the double diffusion of the antibodies against antigen A will develop independently of antigen B. Crossing lines will appear (see Figure 7.5B (b)). The pattern with antigens with partial identity yields a pattern as shown in Figure 7.5B(c). This pattern appears if two antigen mixtures have at least one antigen in common and if the antiserum contains antibodies as well against an antigen which occurs in one sample but not in the other.

We can also use this technique to explore whether antigens carry several common determinants. For example if the antigenic determinants A and B in Figure 7.5B(a) are on the same molecule we would of couse get a single precipitation line. From these types of precipitation patterns we can therefore determine whether two samples have common antigens. We can also explore whether molecules carry more than one antigenic determinant and whether different molecules carry common antigenic determinants.

Π Can the Ouchterlony method provide information about the molecular weight of an antigen?

Remember that the position of the precipitation line will be dependant upon the concentration gradient of the antigen (see Figure 7.5A). The concentration gradient is, of course, dependent upon the concentration of antigen in the wells but is also influenced by the rate at which the antigen diffuses. The rate of diffusion is influenced by the size of the molecule. Thus in principle we can use the position of the precipitation line to give us some idea of the size of antigen molecules. The method is not very precise because the exact position of the precipitation line is dependent upon many factors other than the size (molecular weight) of the antigen (eg shape of the antigen, concentration of the antigen, concentration of the antibody).

qualitative immune electrophoresis

Quantitative and qualitative immune electrophoresis. Qualitative immune electrophoresis combines the specificity of immune precipitation with the separating capacity of electrophoresis. Molecules move through an electric field depending on their size and charge (electrophoretic mobility, see Chapter 5). The analysis is usually performed on a slide covered with agar. Two wells and two troughs are made in the agar (see Figure 7.6).

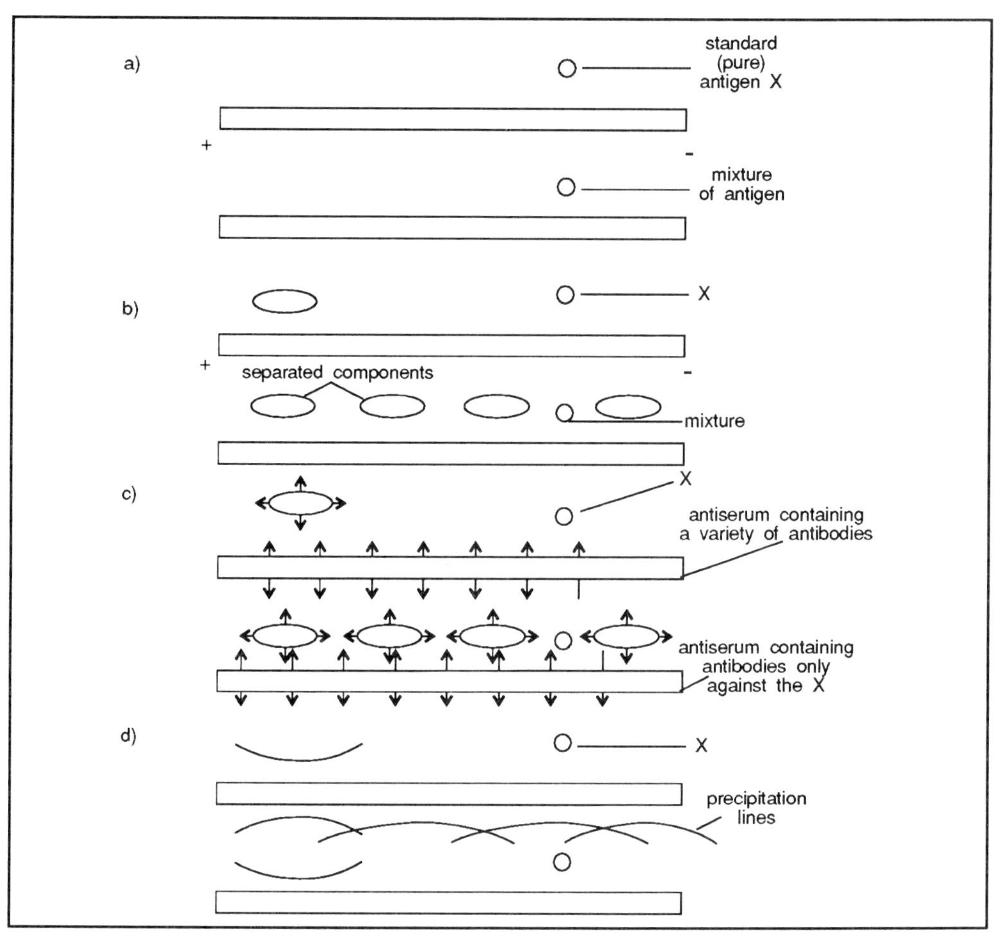

Figure 7.6 The steps of an immune electrophoretic experiment. a) the wells have been punched into the agar and are filled with purified antigen X and with a mixture of antigens and the electrodes have been connected as indicated; b) the proteins have migrated after electrophoresis. This is shown for the sake of clarity, although proteins are usually not visible to the naked eye. The agar is removed from the troughs; c) anteriserum is placed in the troughs and the antigens and antibodies will start to diffuse as indicated; d) the precipitation lines will become visible after some time.

This method allows for the examination of several antigens which have different electrophoretic mobilities. Refer to Figure 7.6 to follow the description. The mixture of antigen is placed in a well in a gel and a purified antigen is applied in another well as a reference. Usually the sample size is about 1-5μl containing 1-100μg antigen. Electrophoretic separation is then carried out as described in Chapter 5 and the antigens are separated. The current is switched off and antiserum is run into the troughs. The antigens and antibodies start to diffuse towards each other. After sufficient time, precipitation lines will become visible. This enables us to compare antigens in a mixture and to identify them by comparing precipitation lines with those achieved with purified antigens. It is possible to test for a variety of antigens either by using specific antisera or by using a polyclonal antiserum containing a number of antibodies.

cross-over electrophoresis

Another application is cross-over electrophoresis. Most proteins (antigens) have a negative charge at pH 8.0 and hence they will migrate to the anode. IgGs have hardly any charge at pH 8.0. Sugar molecules of agarose always contain a number of negatively charged groups, like sulphate, pyruvate and carboxyl groups. The ionised groups of the agarose are surrounded by cations.

electro-endosmosis

If an electric field is created, these cations will move to the cathode together with the water molecules which surround them. This induces a transport of the solvent causing uncharged molecules, among which are antibody molecules, to be dragged along. This phenomenon is designated electroendosmosis. If the antiserum is placed on the anode side, the antibodies and the antigen will move toward each other after the electrodes have been connected. This is shown in Figure 7.7. Specimens of antigen and antibody have been placed in the gel as indicated in Figure 7.7a.

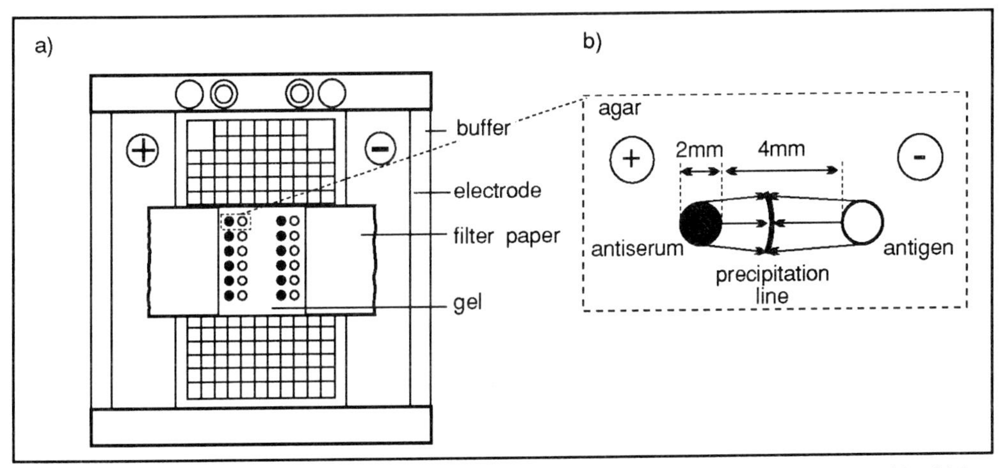

Figure 7.7 The principle of cross-over electrophoresis. a) schematic representation of a set-up with which immune electrophoresis can be accomplished. Antibody is put into the black circles. The open circles are filled with antigen; b) enlarged representation of the site of application.

By placing specimens at different sites on the gel, more specimens can be examined at the same time by means of electrophoresis. For this technique an agar is used which contains a comparatively large number of negatively charged groups in order to stimulate the transport of the antibodies by means of electroendosmosis. The antigens move towards the anode because of their electrical charge and the IgG molecules will move to the cathode because of the electroendosmosis. The advantage is that cross-over electrophoresis is more sensitive than Ouchterlony double diffusion. It requires 0.05 mg cm^{-3} reagent compared to 2 mg cm^{-3} for Ouchterlony double diffusion. Cross-over immune electrophoresis can be carried out quickly: within two hours compared to Ouchterlony double diffusion which sometimes needs more than 24 hours in order to produce a stable pattern.

Π Give one reason why the cross-over electrophoresis technique is more sensitive than the Ouchterlony double diffusion method.

Perhaps the most obvious reason is that because incubation time are shorter in cross-over electrophoresis, there is virtually no thermal diffusion of the antiserum and antigen. The localised concentrations of the reactants therefore remain relatively high and thus a visible precipitate is given even with low initial reactant concentrations.

quantitative immune electrophoresis

Laurel rocket electrophoresis

The quantitative **'Laurel rocket electrophoresis'** is basically the same as radial diffusion (Mancini), except that the driving force for the movements of the antigen is not diffusion but an electric field. Different preparations are placed in wells in an agarose gel which contains an antiserum. The electrodes are connected in such a way that the antigens move into the gel (Figure 7.8). The antigen will meet antibodies during its movement in the agar gel. These will bind to the antigen and the complex will move along and bind more and more antibodies. This will continue until equivalence is attained. At this antigen/antibody ratio the complexes are no longer soluble, they will precipitate and not move any further. The result is a rocket-like precipitation line, in which the enclosed surface matches the amount of antigen in the specimen. The outcome of such an experiment is shown in Figure 7.8. Semiquantitative determination is possible by comparing the surface of an unknown solution to some standard specimens.

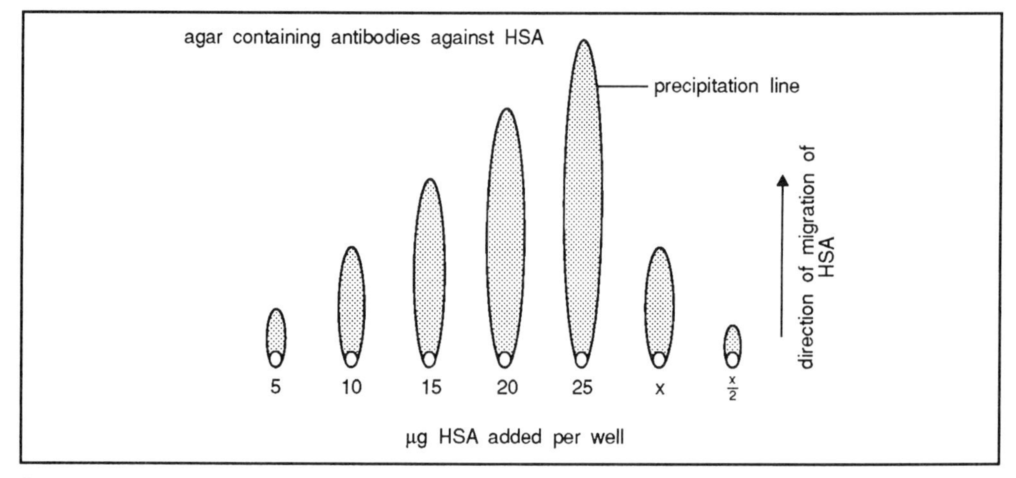

Figure 7.8 The outcome of 'rocket electrophoresis' of human serum albumin (HSA) preparations. Note x = unknown (see text for details).

SAQ 7.3

Which type of agar should be applied for the following techniques: (agar containing relatively many negative ions or agar containing a relatively low concentration of negative ions)?

1) rocket electrophoresis

2) cross-over electrophoresis

3) Ouchterlony double diffusion

7.3.3 Making precipitation lines visible in gels

protein stains

In most cases the precipitation lines are visible to the naked eye. To be able to see less distinct lines, the preparation should be held against a black background, using diffuse light which comes from the side and from below. The white precipitation lines can be distinguished most clearly this way. If this is not sufficient, the precipitate can be stained. The soluble protein must first be removed from the gel. This is possible by incubating the gel for 12 to 24 hours in isotonic saline. The protein which is still present in the gel after washing cannot diffuse through the pores in the gel and hence this must be precipitate. These protein complexes can be stained by means of a dye which binds to protein, (for example Coomasie Brilliant Blue).

7.4 Enzyme-linked immunosorbent assay (ELISA)

7.4.1 Principle of the technique

The term ELISA originates from term Enzyme-Linked ImmunoSorbent Assay. This technique employs the sensitivity of a spectrophotometric assay of an enzyme and the specificity of antibodies. A stable enzyme which can easily be measured and which has a high turnover rate is covalently linked to antibodies or to the antigen. The three methods that are employed most frequently for the performance of an ELISA which will be discussed in this section are:

- the competitive method;
- the double antibody;
- the indirect method.

7.4.2 The competitive ELISA technique

competitive ELISA technique

The competitive ELISA technique is schematically depicted in Figure 7.9. Examine this Figure carefully. The antigen with a covalently linked enzyme is allowed to bind to antibodies. Reference curves can be made by incubating standard solutions which contain known quantities of antigen which is not linked to enzyme together with antigen with covalently coupled enzyme.

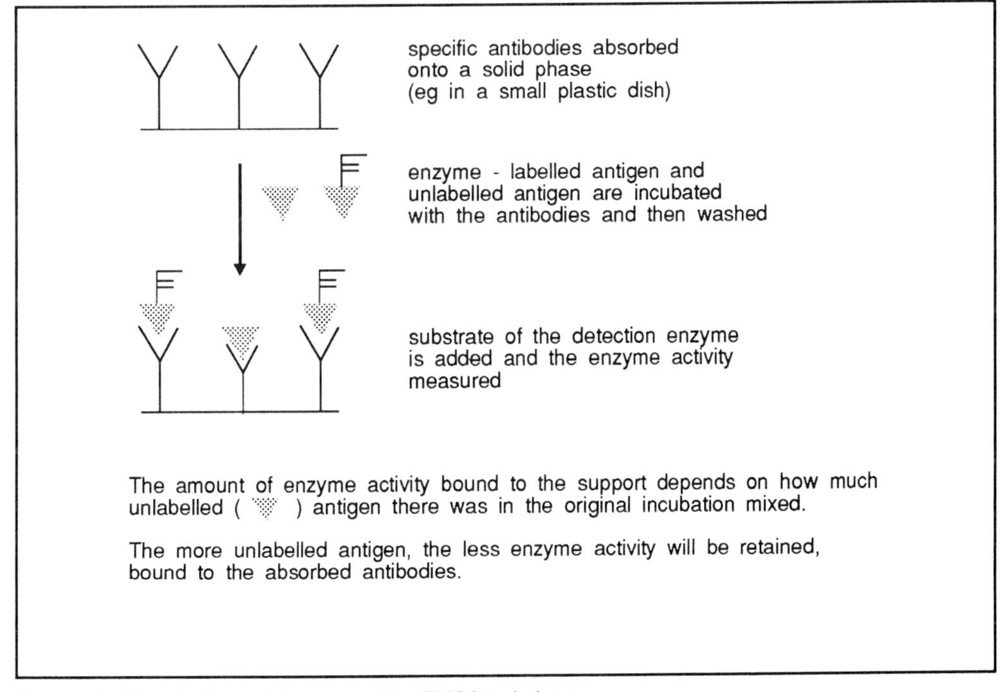

Figure 7.9 The principles of the competitive ELISA technique.

Incubation of mixtures with an unknown quantity of antigen can then be compared to the reference curve (standard curve). An important disadvantage of this method is that it is quite complicated to couple the enzyme to the antigen. Each antigen must be linked to the detection enzyme in a more or less specific manner. This difficulty is overcome by the double antibody and the indirect ELISA techniques described below.

7.3 The double antibody ELISA technique

double antibody ELISA technique

This technique is summarized in Figure 7.10. Antibodies generated against the antigen are attached to the solid surface. Antigen is added and after washing antibody (enzyme labelled) against the same antigen is added. We then measure the amount of enzyme adhering to the complex. A reference curve (standard line) is obtained by adding known quantities of antigen in the second step of the procedure.

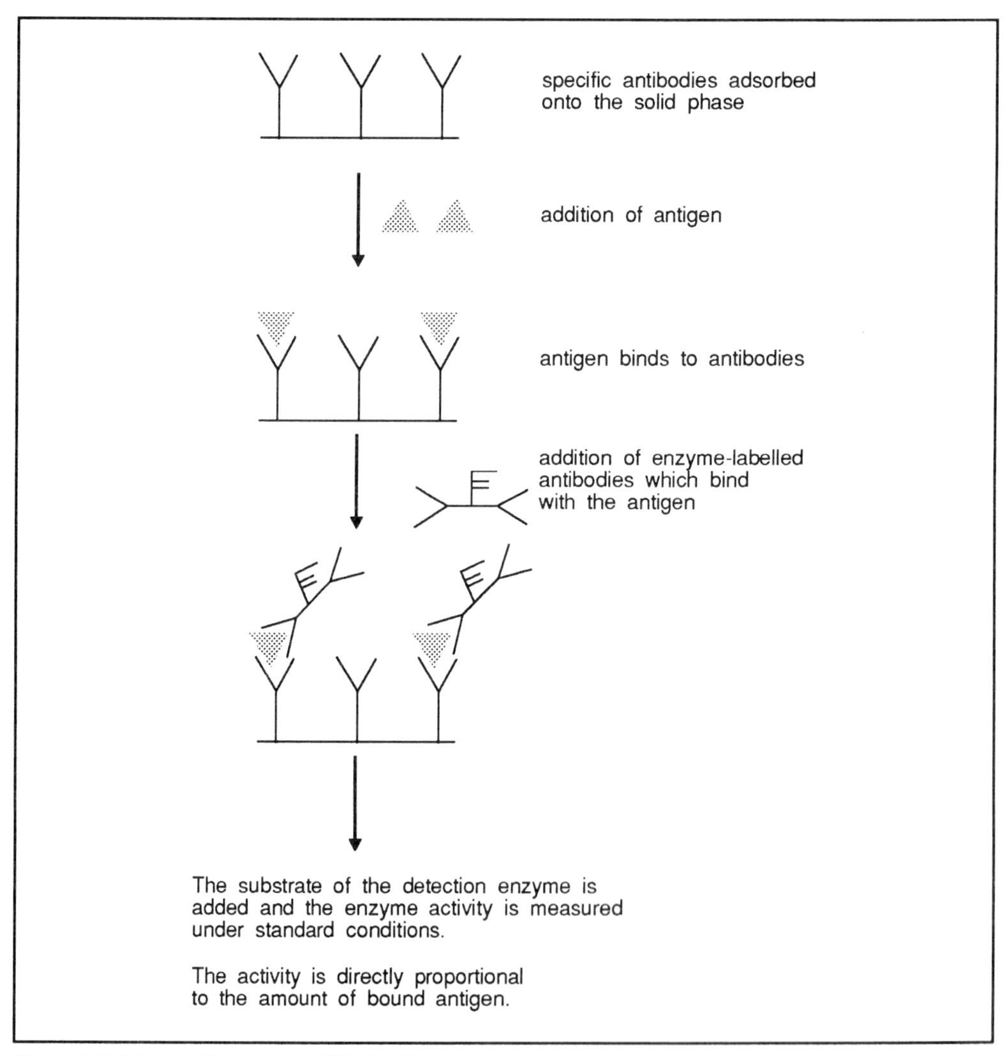

Figure 7.10 The double antibody ELISA technique.

7.4 The indirect ELISA technique

indirect ELISA technique

Use Figure 7.11 to follow the description. In the indirect ELISA technique the antigen is adsorbed onto a solid surface at high pH (approx 10 - 11). The serum is added at neutral pH and all non-binding proteins are removed by washing. The antigen-specific antibodies (IgGs) remain attached to the solid surface by means of the antigen. Then antibodies which will bind with the first IgGs (= anti-antibodies) - to which the detection enzyme is coupled - are added. The enzyme activity is measured after washing. The reference curve is obtained by adding known quantities of antigen in the first step. Rabbit IgGs are often employed as first antibodies and antibodies generated in goats against rabbit IgGs are used as second antibodies. It is to these that the detection enzyme is coupled.

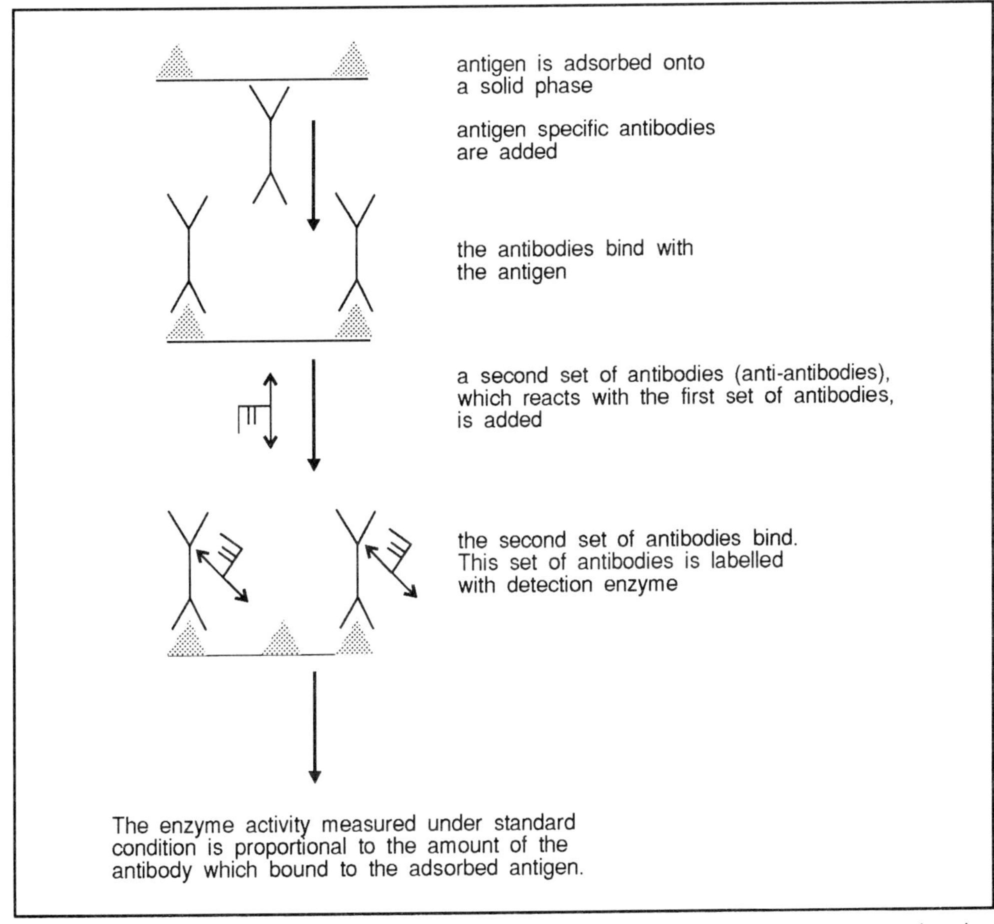

Figure 7.11 The indirect ELISA technique. In this example the antigen specific antibodies were produced in rabbits. The anti-antibodies were produced in goats by immunising goats with rabbit antibodies. (see text for a description).

Π What is the main advantage of the indirect ELISA technique?

advantage of indirect ELISA

You might have had some difficulty with this, if you have limited immunological knowledge. The important point is that the enzyme-linked goat antibodies can be used in a lot of different assays. In the indirect method, these antibodies will bind with rabbit IgGs irrespective of what the rabbit antibodies bind with. Thus we only need to go through the laborious process of linking the enzyme to the antibodies once and then use this in the assay of many different antigens.

7.5 Practical aspects of the ELISA technique

The solid phase

Proteins have successfully been adsorbed onto different surfaces: cross-linked dextranes, polyacrylamide beads, filter paper, nitrocellulose filters, polypropylene tubes and polystyrene microtiter plates. Especially the latter are frequently employed. The antigen or the antibody can bind to the solid surface by means of passive adsorption, but it can also be coupled covalently to a coupling reagent like cyanogen bromide. A mild non-ionogenic soap is added to the solution after coupling or adsorption in order to avoid nonspecific adsorption. The nonspecifically bound protein is washed away after each step.

The detection enzymes

Π What sorts of enzymes are used in ELISA? List some examples.

There is quite a wide choice but the key facts are that we really need stable and easy to assay enzymes. We would obviously prefer enzymes for which there are sensitive assay procedures. The sort of example you might have thought of is alkaline phosphatase (see Chapter 6) which can be easily assayed by merely incubating the enzyme with nitrophenol phosphate. If you glance through Chapter 6, you will probably come up with many other, easy to assay, examples.

single enzymes

The detection enzyme is covalently linked to the antigen or antibody. It is convenient to use detection enzymes which have a broad substrate specificity (and thus can measured by means of a number of different substrates and are detectable in several ways). Two detection enzymes that are frequently employed are alkaline phosphatase and peroxidase isolated from horseradish. As we have learnt alkaline phosphatase converts the colourless substrate p-nitrophenyl phosphate into the yellow p-nitrophenol (maximal absorption at 405 nm after the reaction has been stopped with NaOH) and peroxidase converts a range of compounds. like o-phenylene diamine, which is oxidized to dinitrobenzene by H_2O_2. Dinitrobenzene shows strong absorption at 495nm in an acid environment.

linked enzymes

The sensitivity of an ELISA measurement can be greatly improved if two enzymes are employed in which the first produces a substrate for the second. The advantage is that both enzymes contribute to the amplification of the signal. An example: alkaline phosphatase is the first detection enzyme and in this case it is present in very low concentrations. Alkaline phosphatase can hydrolyze $NADP^+$ into NAD^+ and phosphate. The NAD^+ serves as a cofactor for the second enzyme which can be an NAD^+ dependent redox enzyme. Alcohol dehydrogenase is well suited to this aim. The enzyme specifically reacts with NAD^+ and the NADH which is formed reacts with a tetrazolium dye. The formazan which is formed is coloured and has a high absorption coefficient and can be determined spectrophotometrically.

7.4.6 Applications of the ELISA technique

The ELISA technique can be applied to virtually all immunological assays. It is currently the most important quantitative immunological technique and it has completely replaced the radio-immune technique (Radio Immune Assay, RIA). The latter operates basically in the same way as the competitive ELISA technique but instead of measuring the enzyme, we measure the amount of radioactivity that is bound. The RIA technique employs purified antigen, a radioactively labelled antigen and a specific antiserum. An important disadvantage of the RIA technique is the fact that it requires radioactivity and expensive equipment in order to measure radioactive substances quantitatively. Its sensitivity can be comparable to that of the ELISA technique. Virtually all RIA assays are replaced by ELISA for reasons of cost and safety.

ELISA replacing RIA

Some examples of the application of ELISA in clinical biochemistry are the determination of blood groups, hormones like insulin and oestrogen and the demonstration of pathogens which cause infectious diseases like the Rubella, *Salmonella*, moulds like *Aspergillus* and parasites like trypanosomes.

Good ELISA tests have been described against a large number of vegetable and animal viruses. Hence, ELISA can be applied in each case in which proteins and carbohydrates are involved and in some cases in which DNA must be detected.

7.5 Immune histochemistry

Immune histochemistry employs antibodies which have been generated against cell constituents. These will often be monoclonal antibodies since these reduce the reactions with constituents other than those of interest to a minimum. The monoclonal antibodies can be labelled with gold. The electron beam in the electron microscope is strongly scattered at the sites where gold is present and this indicates the sites of the antigens. Fluorescent dyes are very suitable for the study of the distribution of biomolecules under a light microscope. In this case, a section is incubated with the primary antibody. After washing, the section is incubated with an antibody against the first antibody. An anti rabbit-IgG generated in a goat may serve as an example. A strongly fluorescent substance is covalently linked to the second antibody. If the preparation is illuminated under the microscope by means of ultraviolet light, the site where the antibodies are bound will be visible because of the fluorescence. Fluorescein and rhodamine are frequently used fluorescent chromophores. Fluorescein emits green light and rhodamine emits orange light. The natural fluorescence of biological material is faintly blue.

7.6 Blotting techniques

7.6.1 Principle of blotting

blotting techniques

The principle of this technique is simple. DNA, RNA or protein which is separated in a gel by means of electrophoresis is extracted from the gel by means of diffusion, transport of solvent or electrophoresis and transferred to a solid surface. This is usually a membrane which is permeable to molecules of the solvent and proteins but not to larger structures like bacteria. Despite the fact that the membrane is permeable, protein and nucleic molecules will bind to the surface of the membrane until all binding sites are occupied. The molecules that are bound to the membrane can easily be reached by other molecules so that several reactions are possible, for example, immune reactions with proteins or detection by means of fluorescence or autoradiography. Extraction of the bound molecules is possible after detection. DNA and RNA blotting techniques will be discussed later. We will therefore restrict ourselves here to the blotting of proteins and the immunological reactions on the membrane surface. This technique is often called Western blotting.

7.6.2 Performance of the technique and properties of membrane filters

Proteins can be transferred to a membrane filter from different types of gel (SDS gels, non-denaturing gels and isoelectric focusing gels). The denaturing reagents which are present during electrophoresis are first washed out of the gel. SDS can be removed to a large extent by washing with a buffer which contains urea. With some membranes it is not necessary to remove the SDS completely and washing with a buffer will be sufficient. The SDS must be removed in order to partially renature the proteins. After partial renaturation the protein can be transferred from the gel to the membrane in three ways (see Figure 7.12).

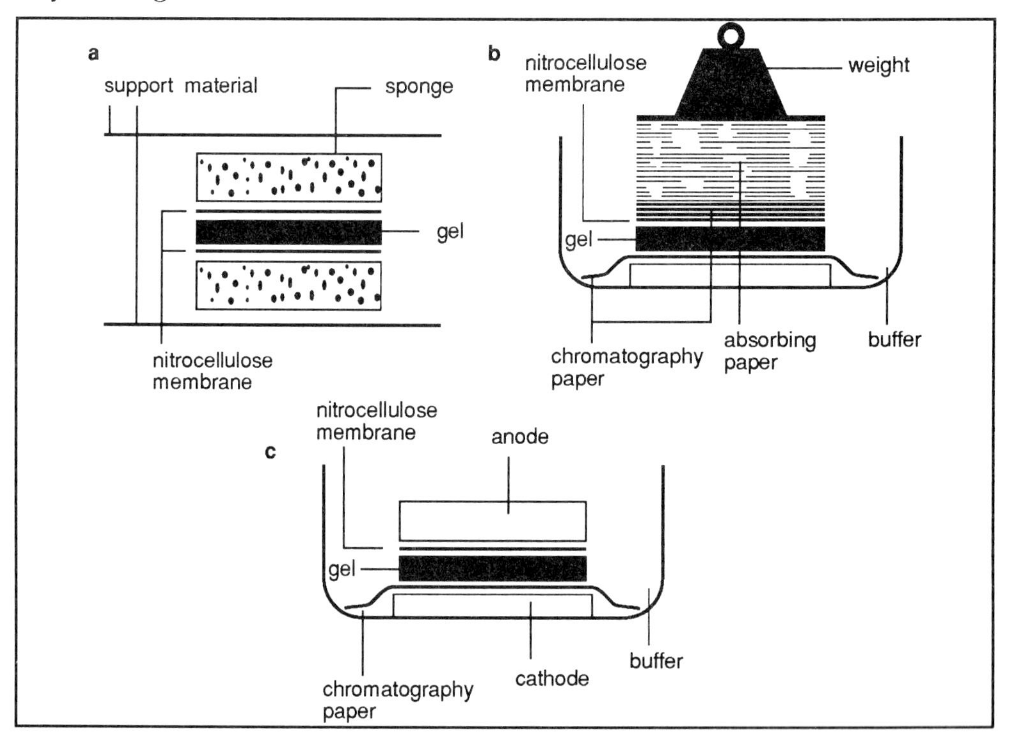

Figure 7.12 The transfer of protein to the blotting membrane. a) the molecules are removed from the gels by means of diffusion and transferred to the blotting membrane; b) the molecules are removed from the gel by means of transport of solvent and transferred to the blotting membrane. The material which absorbs liquid, usually paper, serves to stimulate the transport of solvent; c) electrophoresis is the driving force for the transfer of negatively charged molecules from the gel to the blotting membrane.

a) The first method is diffusion. The gel is sandwiched between two membranes onto which the protein will bind. The drawback is the fact that the transfer takes a long time and that the protein bands broaden slightly due to diffusion. An advantage is that two blots can be produced for each gel.

b) The second method is the extraction of the protein from the gel by letting a large amount of buffer run through the gel and the blotting membrane. One blot can be made for each gel.

c) The method which is currently employed most frequently is the extraction of protein from the gel by means of electrophoretic transfer to the membrane. This method succeeds in most cases but it may happen that there are proteins in the gel of which the pI (pH at the isoelectric point) is equal to the prevailing pH. These proteins will not move. This can be avoided by using another pH during the transfer. The risk is then, however, that other proteins will be uncharged. Another method is to add some SDS so that the proteins carry a negative charge. Care must be taken that the membrane still absorbs proteins under these circumstances. It can also happen that proteins are positively charged under the prevailing pH (as is the case with histones). These proteins will not move to the anode during the electrophoretic blotting, but, contrary to most other proteins, to the cathode. In this case there should be a membrane between the gel and the cathode. The buffers which are used vary depending on the proteins to be blotted and the membrane that is used for the blotting process. Methanol is often added to increase the membrane's polarity and to decrease that of the medium. This will stimulate renaturation and the extraction from the gel but will decrease the hydrophobic binding to the membrane.

properties of membranes

Three types of membrane are used. Their most important properties are listed in Table 7.1. Particularly note their protein-binding capacity and their retention capacity.

type of membrane	protein-binding capacity (μg per cm^2)	protein left behind after washing with a BSA or PBS solution	
		BSA	PBS
nitrocellulose (NC)	249	74.5%	99.1%
nylon	149	81.5%	98.9%
positively charged nylon	176	92.0%	96.8%
PVDF	172	95.0%	99.1%

BSA = bovine serum albumin
PBS = phosphate-buffered salt solution
PVDF = polyvinylidene difluoride membrane

Table 7.1 Protein-binding properties of the different membrane types

a) **nitrocellulose membranes**

The binding of the proteins to these membranes is mainly based on hydrophobic interactions. At pH3 the surface of this type of membrane is negatively charged like most proteins.

b) **nylon membranes**

Mechanically these are the strongest of the membranes which are used. The nylon can be supplied with negatively or positively charged groups. The binding to these membranes is therefore mainly ionogenic.

c) **membranes to which the adsorbed molecules can be coupled covalently**

This has the advantage that, after coupling the membranes can be treated with very strong reagents without removing the proteins bound to them.

You should note that new types of membranes are brought onto the market with increased performance compared with the currently available materials.

Examples of advantages of the newer membranes are great mechanic stability, an open structure which creates a large and well accessible surface, good adsorption capacity and possibilities to carry out specific subsequent reactions without a large background signal. Table 7.1 shows the properties concerning adsorption and desorption of the different types of membranes.

In many immune reactions - after the adsorption of the protein to the blotting membrane - the protein binding sites which have remained open must be occupied. This is necessary to prevent non-specific binding of the antibodies to the membrane. In order to occupy the protein-binding sites which are left over, a bovine serum albumin (BSA) solution, milk or gelatine solution are employed. It is therefore important to know whether treatment of the membrane with a protein solution does not induce the proteins which have already been adsorbed to desorb. (Examine Table 7.1)

7.6.3 Immune reactions after blotting

When the proteins have been transferred to the blotting membrane, all other protein-binding sites have to be occupied. Then a reaction using specific antibodies can take place against the antigens present on the membrane. At this stage there may be a means of directly detecting the antibodies that are bound. Alternatively detection may take place by means of a second reaction with the antibody bound to the antigen (compare with the indirect ELISA). Detection can take place by means of radioactively labelled anti(-anti) bodies or by means of an ELISA reaction in which the detection enzyme produces a strongly stained product at the site of the antigen on the membrane.

In order to investigate rapidly whether an antiserum can be used for a particular immune reaction on a blotting membrane, a drop of an antigen solution can be placed directly on the blotting membrane and after blocking the rest of the protein-binding sites, the desired immune reactions can be carried out. This type of application is known as 'dot blotting'.

7.6.4 Protein staining on a blotting membrane

Staining proteins on a membrane may cause problems because the dyes often bind strongly to the membrane. Nylon membranes are particularly unsuitable for the traditional protein-staining methods. A method has been developed recently to stain proteins by means of an enzyme in a process similar to ELISA. This method can be applied for all types of membranes and it proceeds as follows. Biotin is coupled to the protein which is adsorbed onto the membrane. Then avidin with covalently coupled horseradish peroxidase is added. The binding between biotin and avidin is very strong

biotin labelling

(K_d = 10^{-15} mol dm^{-3}). The site on the membrane where the protein-biotin-avidin-peroxidase complex is present, is made visible by measuring the peroxidase activity. At the sites where peroxidase is present, the product which is formed by peroxidase will stain the membrane purple. Table 7.2 shows which staining methods for proteins can be used for the three types of membrane discussed in the text.

Dye	detection limit	number of steps	Usable with		
			NC	nylon	PVDF
Amido black	1 µg	2	+	-	+
Coomassie blue	1.5 mg	3	-	-	+
Indian ink	100 ng	5	+	-	+
Colloidal gold	3.5 ng	5	+	-	+
Biotin-avidin	30 ng	15	+	+	+

NC = nitrocellulose
PVDF = polyvinylidene difluoride

Table 7.2 Staining methods for proteins adsorbed to blotting membranes.

The number of steps indicates the laboriousness of the technique. The protein-staining methods are sensitive but since very small quantities are concerned in blotting, it is sometimes necessary to apply highly sensitive protein-detection methods.

Π Examine Table 7.2 and decide which type of membrane is likely to have the least use if the proteins transferred by a blotting technique have to be detected by chemical staining.

You probably came to the conclusion that nylon membranes are least suitable. This is because these membranes retain stain as strongly as proteins do.

SAQ 7.4

Without referring to earlier sections, check your knowledge of the membranes used in blotting experiments by answering the following. (Use one of the types of membranes from nitrocellulose, nylon, positively charged nylon; polyvinylidene difluoride).

1) Which type of membrane has the highest protein binding capacity?

2) Which type of membrane has the lowest protein binding capacity?

3) From which type of membrane is it easiest to remove proteins by washing with a solution of bovine serum albumin?

4) Which type of membrane binds proteins more strongly?

5) Which type of membrane is suitable for use with Coomassie blue?

7.7 Detection and measurement of specific nucleic acids

7.7.1 Introduction

Chemically one molecule of DNA is rather similar to any other. Thus all DNA molecules consist of deoxyribose moieties linked to each other by phosphate groups and each sugar moiety is also attached to one of four common nucleotide bases (adenine, guanine, cytosine and thymine). Likewise RNA molecules are all chemically rather similar to each other. This means that chemical (eg by reacting DNA with diphenylamine to give a coloured product; by reaction RNA with orcinol to give a coloured product) and physical (for example UV adsorption or staining with a fluorescent reagent such as ethidium bromide) methods do not enable a distinction to be made between different DNA and RNA molecules. In order to distinguish between different molecular species of RNA and DNA, we have to make use of the fact that the different molecules contain the same nucleotide bases but that these bases are in a different order (sequence). This is achieved by using techniques which are based upon a process known as molecular hybridisation.

In the remaining part of this chapter, we will outline the principles of molecular hybridisation and briefly explain how hybridisation is measured. It is however beyond the scope of this text to give extensive details of the preparation, extraction and purification of nucleic acids which are pre-requists before hybridisation can be carried out. Such details are dealt with in the BIOTOL text 'Analysis of Amino Acids, Proteins and Nucleic Acids'.

7.8 Principles of hybridisation

base-pairing

The bases in nucleic acids will, under favourable condition, form hydrogen bonds with other bases in a very specific way. Thus adenine pairs with thymine (or uracil) and guanine with cytosine. This pairing is of course critical both to the inheritance of genetic information by DNA and it transmission, through RNA synthesis, to the production of proteins.

In principle, if we incubate single-stranded DNA (DNA in which the two strands of the double helix have been separated) in conditions in which hydrogen bonds can form, then the separate strands can combine through base-pairing providing that the two strands are complementary. This process can occur between two strands of DNA or between a strand of DNA and a complementary strand of RNA. Use can be made of this process to help us to identify nucleic acids with particular nucleotide sequences. For example let us assume that we have separated fragments of DNA by electrophoresis and 'blotted' them onto a membrane as described earlier for proteins. Suppose we now 'melt' (separate the strands) the DNA on the membrane and then challenge these resulting single-strands with another sample of single-stranded DNA. If this sample contains complementary nucleotide sequences to those of the DNA on the membrane, hydrogen bonds will be formed and a hybrid double-stranded DNA molecule will be produced. We are effectively using base-pairing as a method of identifying nucleic acids containing particular nucleotide sequence. Usually the challenging nucleic acid is labelled (for example, made radioactive) so that it can be detected and measured by autoradiography or by some other technique. More recently non-radioactive labelling has become more important. A common label using is biotin chemically attached to the challenge nucleic acid (often called a probe). As we have learnt biotin is very tightly bound by a protein called avidin. If we attach an enzyme to the avidin then we can detect where avidin binds by measuring the enzyme. We can represent the sequence of events as follows:

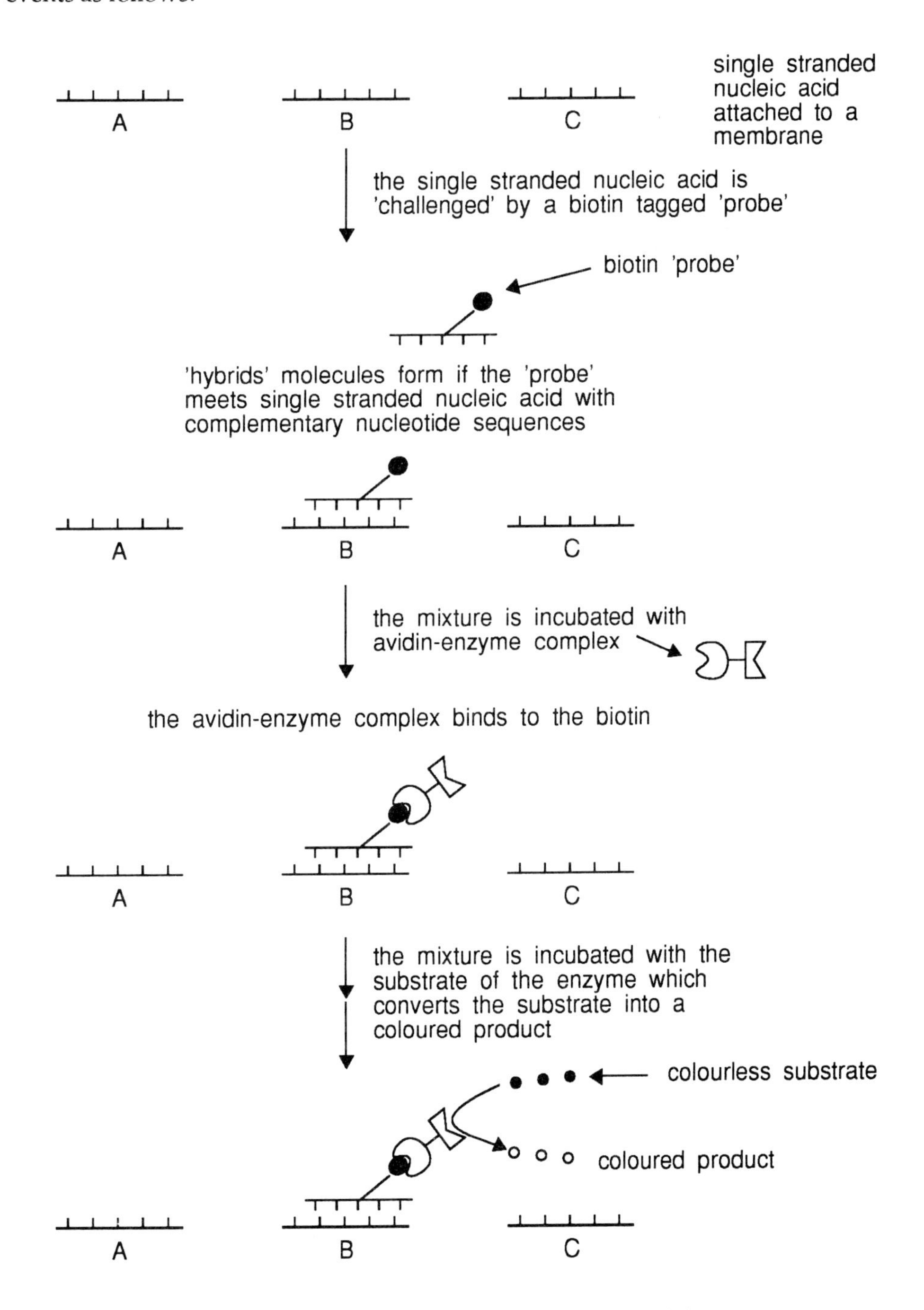

You should note that there are many variations of this general theme.

Π Is the 'probe' used in the flow diagram above specific for a particular nucleotide sequence?

The answer is yes, it only binds with the sequence in molecule B. We can use the amount of enzyme reaction in the final stage to measure the amount of hybridisation that has taken place, which, in turn, may tell us how much of sequence B was present. Hybridisation can therefore be used as both a qualitative (to identify if a particular sequence is present) and as a quantitative (how much of a particular sequence is present) technique. The technique of blotting DNA from electrophoretograms and using DNA hybridisation to detect the DNA on the membranes is called Southern blotting. A similar technique,rather coyly called Northern blotting, can be used to detect RNA separated by electrophoresis.

Southern and Northern blotting

From the description of the technique given above it may appear simple. However, there are many pitfalls. One of these is that if we choose conditions in which hydrogen bonds are very stable (for example at a low temperature), then hybrids may form in which there are many mismatched base pairs. This of course will lower the specificity of the technique.

polymerase chain reaction technique

The production of suitable 'probes' is of course essential for the successful application of this technique. The production of such probes often involves genetic engineering techniques in order to synthesise sufficient quantities of the desired nucleotide sequences. Alternatively a process known as the polymerase chain reaction technique is used to made desired nucleotide sequences. In some cases short chemically synthesised oligonucleotides can be used. It is beyond the scope of this text to discuss these processes in details. They are covered by the BIOTOL text 'Analysis of Amino Acids, Proteins and Nucleic Acids'. The technique of genetic engineering are described in the BIOTOL text 'Techniques for Engineering Genes'.

The important message here is that hybridisation can be used to recognise and measure specific nucleotide sequences.

SAQ 7.5

1) A bacterial genome is cut into fragments and the individual fragments separated by gel electrophoresis. After blotting onto a cellulose nitrate membrane, the DNA was denatured to form single stranded DNA. This DNA was challenged with ^{32}P-labelled DNA prepared from a bacteriophage which contained single stranded DNA. After incubation, the membrane was washed and placed against an X-ray film. After further incubation, the X-ray film was developed and revealed 2 blackened areas. What can be concluded from this experiment?

2) A similar electrophoretogram was incubated with ^{32}P-mRNA which codes for the enzyme β-galactosidase. After washing, an autoradiogram was produced which had one blackened area. What can be concluded from this experiment?

Summary and Objectives

In this chapter, we have described techniques that use the specificity of antibodies and nucleotide base pairing, to enable us to measure a wide variety of biomoleucles. Critical to the success of these techniques is our ability to detect and measure the interaction between our reagent (antibody or nucleic acid probe) and the target compound. In the case of assays involving antibodies we saw that precipitation can be used and requires little in the way of specialist equipment. However, by using (labelled) antibodies, we have seen that it is possible to perform quantitative measurements of very small amounts of material even if closely related molecules are present. Recently the labelling of antibodies using enzymes, rather than radioactive labelling, has dominated this area. In the final part of the chapter, we briefly examined the detection and assay of specific nucleic acids using molecular hybridisation techniques.

Now that you have completed this chapter you should be able to:

- list advantages and disadvantages of using polyclonal and monoclonal antibodies in assays;
- explain the equivalence area (point) of an immune precipitation reaction and interpret data derived from immune precipitation assays;
- explain how immune precipitation in gels can be used to identify and quantify antigens with special reference to the Marcini technique, Ouchterlony plates, immune electrophoresis and the Laurel rocket technique;
- describe a range of techniques which use enzyme labelled antibodies to measure antigens;
- select suitable membranes to use for carrying out a Western blot;
- explain why nucleic acids probes selectively form hybrid molecules with nucleic acids;
- interpret data derived from hybridisation experiments.

Physical methods of structure determination

Physical methods of structure determination

8.1 Introduction

Many biological molecules fall into clearly defined classes, (eg proteins, polysaccharides, nucleic acids), each of which is made up of a relatively small number of 'building blocks'. For each of these groups specific approaches to structural determination have been developed. These are discussed elsewhere in the BIOTOL texts. But for many other biological molecules, particularly relatively small structures, one of the surest methods of structural determination is by one or more of the physical methods described in this chapter.

The development of technology in the last few decades has revolutionised structural chemistry. To elucidate the structure of a compound used to entail a great deal of chemistry and often months or years, of work; today this may be achieved in a fraction of this time.

The main physical methods employed to study structural aspects of biological molecules are ultraviolet, visible and infrared spectroscopy, nuclear magnetic resonance spectroscopy, mass spectrometry and X-ray crystallography. These are not the only methods but for biological materials, the others are used fairly rarely. The principles and applications of the various techniques will be described; for more detailed experimental procedures you are recommended to consult the appropriate book in the list given in 'Suggestions for further reading' at the end of this text.

8.2 Electromagnetic radiation

8.2.1 General features of spectroscopy

wavelength
frequency

The various forms of radiation which include X-rays, ultraviolet and visible radiation, microwaves and radio-waves, are all part of the same phenomenon - electromagnetic radiation. Electromagnetic radiation can be considered as a wave occurring simultaneously in electrical and magnetic fields and is described in terms of its wavelength and frequency. The wavelength (λ) is the distance between the same point on consecutive waves, eg the distance between two adjacent peaks (Figure 8.1) and the frequency (ν) is the number of wave cycles per second (units s^{-1} or Hz). Since all electromagnetic radiation travels at the same speed (2.998×10^8 ms^{-1}), radiation with a long wavelength has a low frequency and vice-versa. This is described by the relationship.

$$\lambda = c / \nu \qquad \text{(E - 8.1)}$$

- where c is the speed of light;

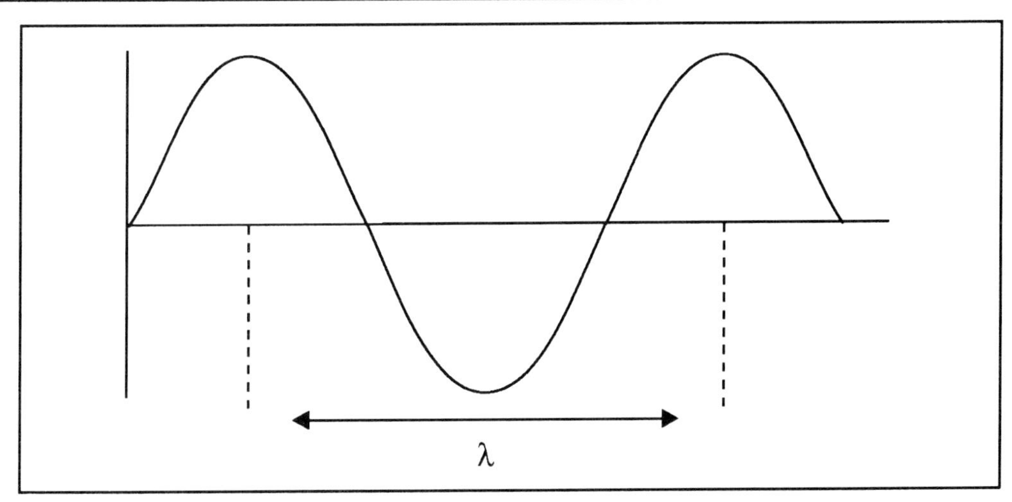

Figure 8.1 The wave nature of radiation.

wavenumber

In infrared spectroscopy the term wavenumber $\bar{\nu}$ is used; this is a reciprocal of the wavelength and the units are usually cm^{-1}. Thus

$$\bar{\nu} = 1/\lambda = \nu/c \qquad \text{(E - 8.2)}$$

Quantum mechanics tells us that electromagnetic radiation can be considered not only as a wave but also as a stream of discrete particles of energy - quanta or photons. The energy E of a quantum is directly related to its frequency:

energy

$$E = h\nu \qquad \text{(E - 8.3)}$$

where h is a constant known as Planck's constant ($h = 6.625 \times 10^{-34}$ J s). For one mole of quanta, we need to multiply this value by the Avogadro constant ($N_A = 6.022 \times 10^{23}$ mol^{-1}).

Thus the energy per mole $E_m = h\,\nu\,N_A$ (E - 8.4)

Figure 8.2 summarizes the energy, wavelength and frequency in all regions of the electromagnetic spectrum. Note that this figure also gives details of the molecular transitions (eg nuclear, electronic, vibration, rotation) involved with the absorption or emission of electromagnetic radiation at different frequencies. It also lists various types of spectroscopy. We will discuss these in more detail later in this chapter.

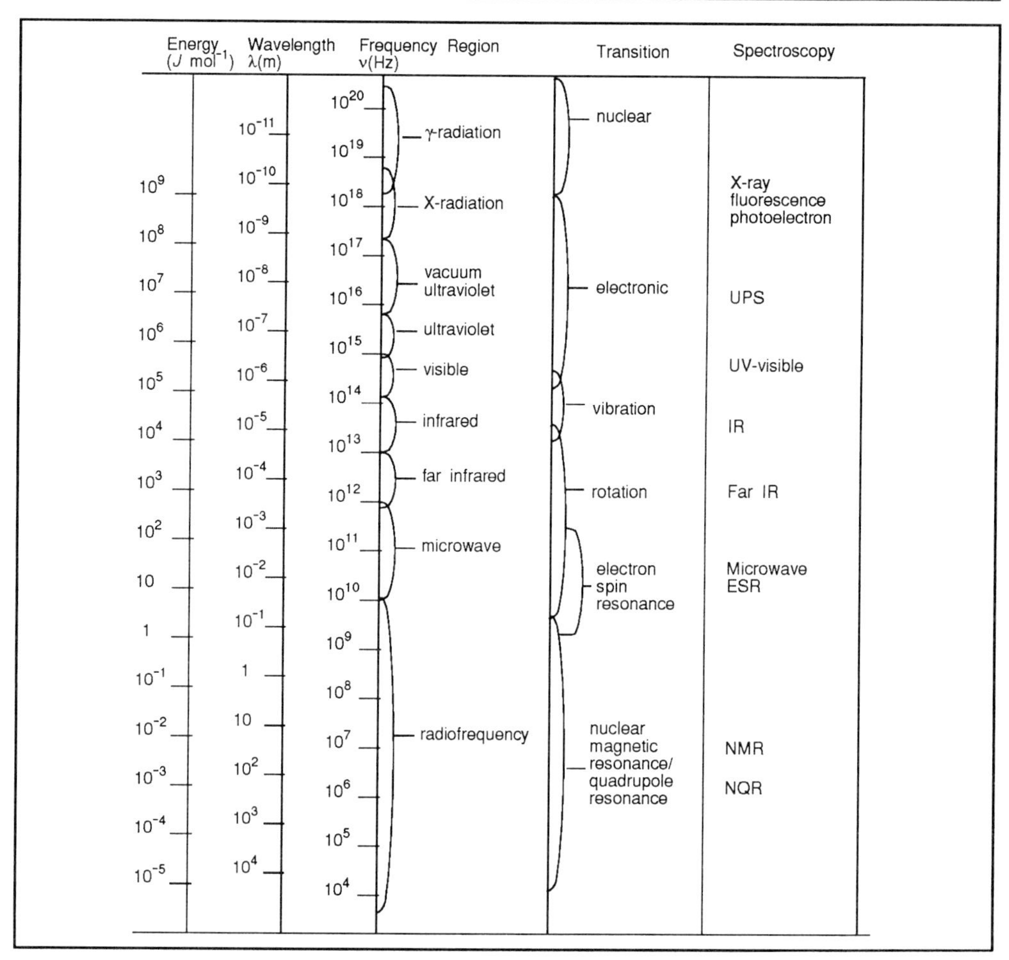

Figure 8.2 The energy, wavelength and frequency of electromagnetic radiation (see text for further details).

Π Bearing in mind that the energy of a quantum is directly related to its frequency, how is it related to the wavelength of the radiation?

If we combine Equations 8.1 and 8.3 we obtain the relationship

$$E = hc/\lambda \qquad \text{(E - 8.5)}$$

From this we can see that the energy is therefore inversely proportional to the wavelength.

SAQ 8.1

Calculate the wavenumber in cm^{-1}, the frequency, the energy in joules per quantum and energy in Joules per mole for wavelengths of:

a) 5×10^{-4} nm (X-ray region);

b) 2500 Å (UV region) and

c) 30 μm (infrared region.)

So, electromagnetic radiation of long wavelength (and hence, low frequency) such as radiowaves, has low energy and that of shorter wavelength such as X-rays, has higher energy.

spectroscopy

Nearly every region of the electromagnetic spectrum has been used to help determine molecular structure, involving a battery of techniques known generally as spectroscopy. Spectroscopy is the study of the interaction of electromagnetic radiation with matter. As a means of elucidating structures, this interaction can be very revealing.

8.2.2 Molecular and electronic structure

When radiation passes through a substance, the associated energy may be absorbed or transmitted, depending on its frequency and the structure of the compound. The electrons of an atom or molecule cannot possess any arbitrary energy value but are limited to a set of discrete energy levels. In a molecule's ground state the electrons occupy the lowest possible energy level. For an electron to move from the ground state to a higher energy orbital, or excited state, it must absorb a discrete quantum of energy (in the form of radiation) exactly equivalent to the difference in energy between the two levels. As the energy of a quantum is directly related to its frequency there will only be one point in the electromagnetic spectrum at which quanta posses the right energy to carry out this transition. In addition to raising electrons to higher energy level, the energy absorbed by a molecule may give rise to increased vibration or rotation of its atoms around a bond axis. These too are limited to discrete energy sub-levels.

So a molecule cannot just absorb all electromagnetic radiation but only that which gives rise to specific, discrete energy changes which are permitted to a molecule of that particular structure. The absorption spectrum of a compound is a plot of the amount of radiation absorbed at each wavelength. The different types of spectra arise from interaction of molecules with different regions of the spectrum and require different types of instrumentation.

8.2.3 Instrumentation and definition of terms

Spectrophotometers for use in the UV, visible and infrared region of the spectrum are all based on a common design - this is illustrated in Figure 8.3.In these radiation is passed through a devise which enables us to select a particular wavelength. This radiation is then split into two beams; one passes through the sample being tested/measured while the other passes through a reference cell (often a solvent blank). The light emerging from the sample and reference cells is detected and the difference in the amount of light absorbed by the sample and reference cells is recorded in some way.

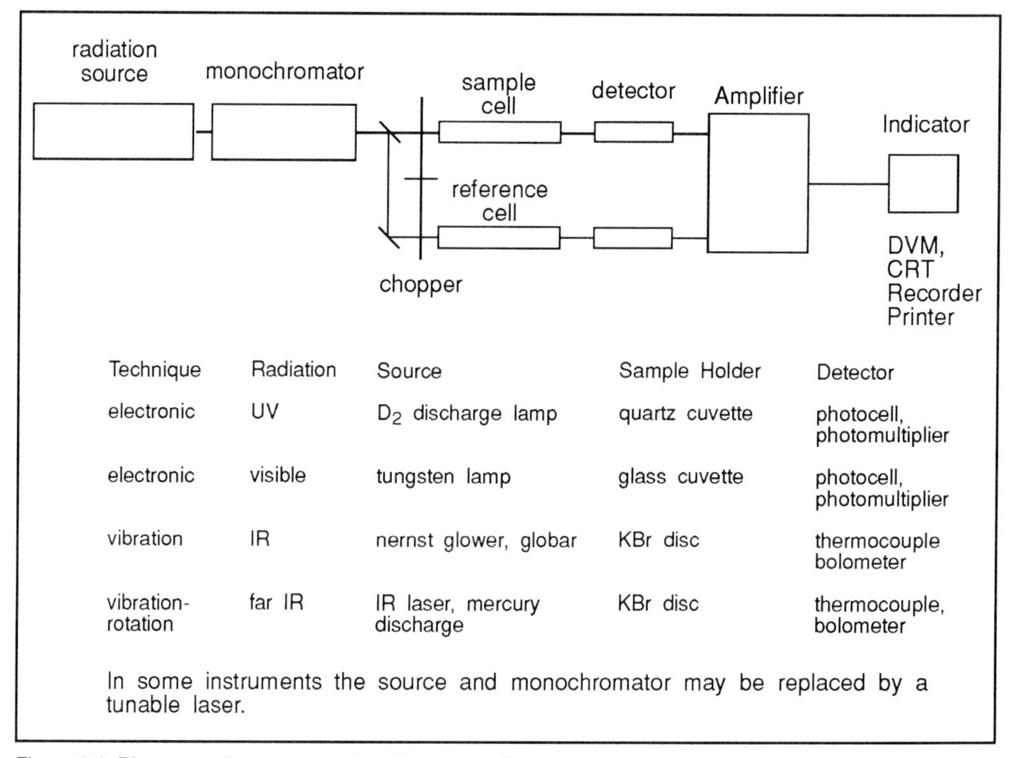

Technique	Radiation	Source	Sample Holder	Detector
electronic	UV	D_2 discharge lamp	quartz cuvette	photocell, photomultiplier
electronic	visible	tungsten lamp	glass cuvette	photocell, photomultiplier
vibration	IR	nernst glower, globar	KBr disc	thermocouple bolometer
vibration-rotation	far IR	IR laser, mercury discharge	KBr disc	thermocouple, bolometer

In some instruments the source and monochromator may be replaced by a tunable laser.

Figure 8.3 Diagrammatic representation of a spectrophotometer

absorbance

If radiation of intensity I_o is incident on the sample and I is the intensity of radiation passed through the sample then the absorbance, A, which is a measure of the amount of radiation absorbed by the sample is given by:

$$A = \log(I_o/I) \qquad \text{(E - 8.6)}$$

Spectrophotometers used in the UV and visible region normally give a record or trace of absorbance at varying wavelengths.

transmittance

The **transmittance** T, is defined as the fraction of radiation passing through the sample.

$$T = I/I_o \qquad \text{(E - 8.7)}$$

Infrared spectra are recorded as percentage transmittance at various wavenumbers.

Π Can you see the relationship between A and T?

If you combine equations 8.6 and 8.7 you obtain the relationship

$$A = \log(1/T) = -\log T \qquad \text{(E - 8.8)}$$

Thus you can convert absorbance values into transmittance values or vice versa.

Π Calculate the absorbance of a sample for which the light intensity is reduced to 16% of its initial value.

In this case $T = I/I_o = 0.16$

Thus $A = -\log 0.16 = 0.80$

The Beer-Lambert law relates the measured absorbance to the concentration and the path length of the sample:

$$A = \varepsilon c l \tag{E - 8.9}$$

extinction coefficient

where c is the concentration expressed as mol m^{-3}, l the path length in m and ε the molar absorption coefficient (formerly called the extinction coefficient) in units of m^2 mol^{-1}. Note that literature values of the extinction coefficient are often in units of dm^3 mol^{-1} cm^{-1} or l mol^{-1} cm^{-1} - the conversion factor to the new SI unit is 10^{-1}. Remember that dm^3 ≡ litre.

The Beer-Lambert law is valid only over a limited range of concentrations and so to use the relationship to determine concentrations in solution it is essential to plot a calibration curve for known concentrations of the compound.

To see if you understand the definitions and principles involved try the following SAQs.

SAQ 8.2

At 525 nm a solution containing 7.5×10^{-5} mol dm^{-3} of $KMnO_4$, contained in a cell of path length 2.0 cm has an absorbance of 0.88. Which one of the following is the molar absorptivity, ε(m^2 mol^{-1}), for $KMnO_4$ at 525nm? Its molar absorptivity is:

1) 5.86 ;

2) 11.73;

3) 58.6;

4) 586.

SAQ 8.3

The following data were obtained for the absorption of radiation by a sample of bromine dissolved in tetrachloromethane using a 2 mm path length cell. Calculate the extinction coefficient of the solution at the wavelength employed.

$[Br_2]$(mol dm^{-3})	0.001	0.005	0.010	0.015	0.020
Transmittance/ %	81.4	35.6	12.7	4.5	1.6

8.3 Ultraviolet UV and visible spectroscopy

8.3.1 Electronic spectra

electronic states

Many molecules absorb radiation in the visible and ultraviolet regions. Absorption in this region arises from changes in electronic states - the transition of electrons from a lower to a higher energy level.

In a visible/ultraviolet spectrophotometer the light beam is split - half passes through a transparent cell (Figure 8.3) containing a solution of the sample whilst the other half passes through a reference cell containing only the solvent in which the sample is dissolved. If the sample absorbs light at a particular wavelength, the intensity of the beam at that wavelength will be less than that passing through the reference cell, at the same wavelength. The instrument plots absorption against wavelength.

Each electronic energy level has associated with it a number of vibrational and rotational energy levels which means that most ultraviolet/visible spectra are not composed of a number of sharp peaks but a few broad absorption bands.

There are of course, different types of electrons in a molecule, σ electrons are held tightly and a relatively large amount of energy is required to excite them - usually energy associated with the extreme end of the ultraviolet region and so these electrons are generally unaffected by the ultraviolet/visible radiation employed in most spectrophotometers.

Π What kinds of electrons might be more loosely held and thus more likely to be excited by ultraviolet radiation than σ-'bonding electrons?

Delocalised electrons of compounds containing conjugated double bonds, ie π-electrons, are more loosely held, as are the non-bonding (*n*) electrons of lone pairs such as are associated with oxygen and nitrogen.

When a molecule absorbs radiation an electron is excited from the highest occupied molecular orbital to the lowest unoccupied molecular orbital at the next highest energy level. The wavelength at which the compound (or a part of the compound) absorbs most energy (absorption maximum) is determined by the difference in energy of these two levels. Conjugation lowers the gap between these two levels meaning that radiation of a lower energy (longer wavelength) can be absorbed. In general, the greater the number of conjugated double bonds a compound contains, the longer the wavelength at which a compound absorbs radiation; this is known as the bathochromic shift. If a compound contains enough conjugated double bonds it will absorb radiation in the visible region and the compound will be coloured. For example, β-carotene is a yellow pigment widely found in nature - this contains 11 conjugated C=C double bonds and has an absorbance maximum at 451 nm. This applies also to cyclic conjugated systems; thus, benzene has an absorption maximum of 254 nm whereas naphthalene (bicyclic ring) has an absorption maximum of 275 nm.

bathchromic shift

chromophores

Individual molecular species or parts of a molecule, such as an aromatic ring or a carbonyl group, which give rise to distinct absorbance bands, are know as chromophores. A list of the absorptions of some typical, isolated chromophoric groups is given in Table 8.1.

Chromophore	Example	Solvent	λ_{max}(nm)	E_{max}(m^2 mol^{-1})
>C=C<	ethene (ethylene)	(vapour)	170	15800
	benzene	n-hexane	204	7900
			256	200
—C≡C—	1-octyne	n-heptane	185	2000
>C=O	ethanal (acetaldehyde)	n-heptane	390	16
	propanone (acetone)	n-hexane	186 280	1000 16
HO(>)C=O	ethanoic acid (acetic acid)	ethanol	208	32

Table 8.1 Characteristic absorption for selected chromophoric groups

8.3.2 Applications

Visible/ultraviolet spectra are not generally used to determine the absolute structure of a compound - the peaks are too broad for that (see Figure 8.4), but the presence of particular peaks can indicate the presence of certain groups or sub-structures and hence may give an idea of the structure of a molecule.

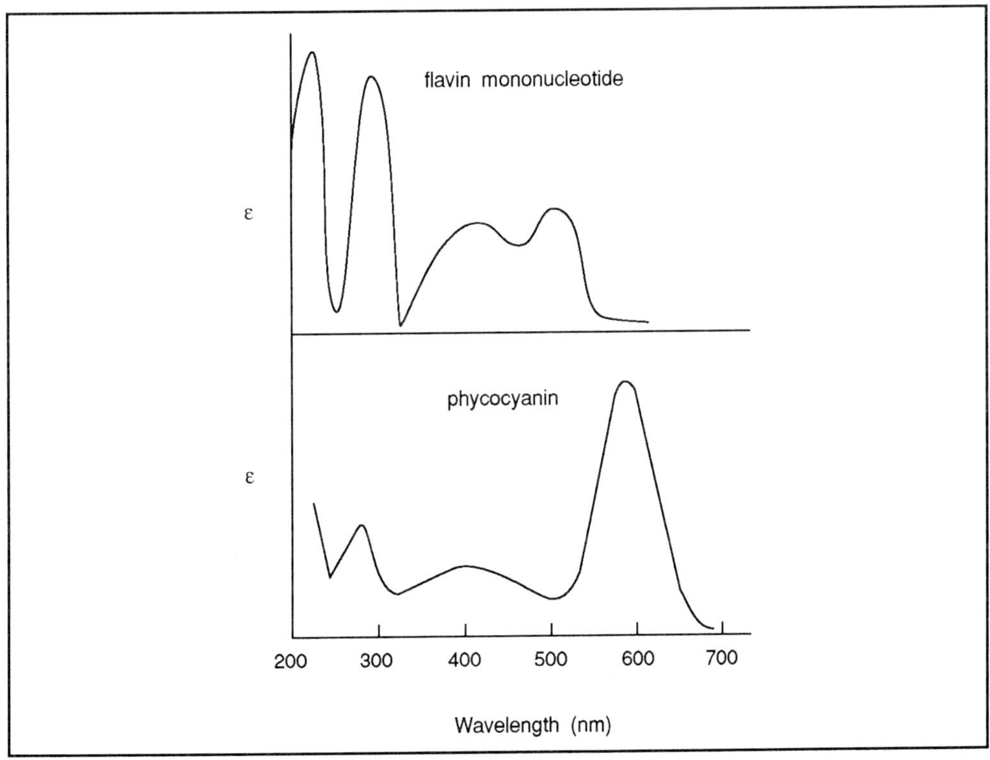

Figure 8.4 Spectra of two biological molecules, flavin mononucleotide and phycocyanin, indicating that spectra can be very different, which often allows us to identify compounds from their spectra.

The spectra may be used in this way to identify in biological mixtures, the presence of classes of compounds such as nucleic acids, proteins or various pigments. So, visible/ultraviolet spectroscopy can be used to help identify compounds, or classes of compound in both the pure form or in mixtures but for absolute structural determinations, other methods are generally employed.

Protein structural studies: Peptide bonds absorb at about 190 nm (a wavelength below which it is impractical to work except in a vacuum, as any oxygen present will absorb strongly in this region). However, proteins also have an absorption maximum at 280 nm.

Π What part of the protein molecule do you think gives rise to an absorption peak at 280 nm?

It is the amino acid side chains which give rise to this absorption maximum, or to be more precise, the three amino acids with side chains containing aromatic (and hence conjugated) moieties - phenylalanine, tryptophan and tyrosine.

protein

Although visible/ultraviolet spectroscopy cannot be used to determine the structure of a protein it can sometimes provide information about structure and conformational changes. Tyrosine (Figure 8.5) is particularly useful in this respect as it possesses an acid-base function in its side chain so that the spectrum of the undissociated form is different from that of the dissociated form.

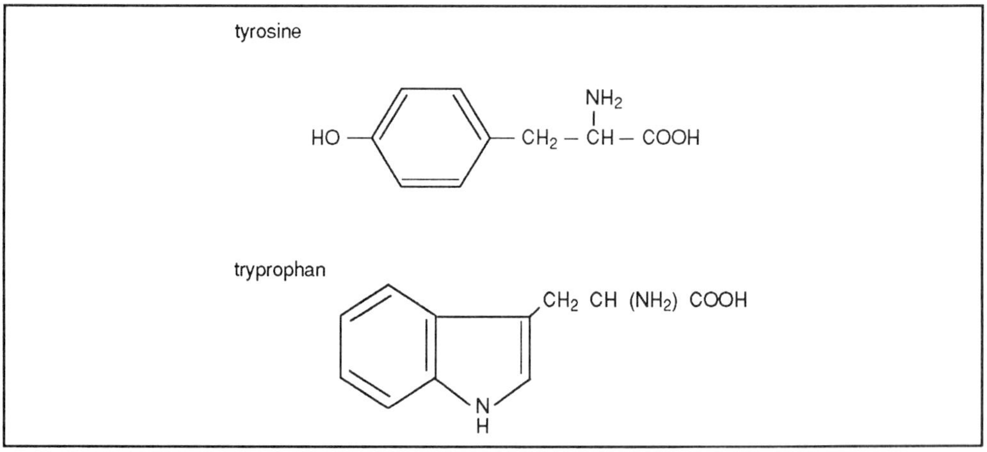

tyrosine

HO — C_6H_4 — CH_2 – CH(NH_2) – COOH

tryprophan

CH_2 CH (NH_2) COOH

N
H

Figure 8.5 Structures of amino acids responsible for absorption at 280 nm.

The environment (eg pH, solvent polarity, temperature) of a chromophore influences its spectrum. Therefore, if the pH of a solvent is altered, the spectrum of the amino acid chromophore at the surface of a protein may change, without a change in the conformation of the protein. If however, the chromophore is inaccessible to the solvent, ie in the interior of the protein, there will be no change in the spectrum. This technique of solvent perturbation has been useful in providing information on the positions of certain amino acids within a protein. Similarly, if a protein is denatured it may expose amino acid chromophores which can be detected in the same way and so the effects of changes in pH, temperature, etc on protein unfolding, can be studied.

nucleic acids

Nucleic acids: It is also possible to study structural changes in nucleic acids. When double helical DNA is denatured, by heating for example, the two DNA strands separate into single stranded coils - a process which is reversed by cooling. This DNA melting is associated with a considerable increase in absorption at 260 nm and so by following the absorption at this wavelength, the effects on DNA secondary structure of a range of parameters can be studied. G-C (guanine-cytosine) pairs are bound by three hydrogen bonds and so have a greater stability to denaturation than A-T (adenosine-thymine) pairs. Therefore, the temperature required to melt the DNA increases with increasing G-C content. By following the absorption at 260 nm as the temperature is increased, one can estimate DNA base composition.

SAQ 8.4

The ultraviolet absorption spectra of molecules are due to which one of the following::

a) rotation of molecules;

b) vibration of atoms within molecules;

c) electronic transitions from the ground state, or from low energy levels to high energy levels;

d) electronic transitions from high energy levels to low energy levels or to the ground state.

SAQ 8.5

The absorption of a sample is given by which one of the following expressions?

(a) I_o/I; (b) $\log (I_o/I)$, (c) I/I_o ; (d) $\log (I/I_o)$

As will become apparent, measurements of absorption in both the UV and visible regions of the spectrum are used in quantitative studies, eg determination of the concentration of proteins in solution, the assay of enzymic activity. (See Chapter 6).

8.4 Infrared spectroscopy

8.4.1 Vibration-rotation spectra

When molecules absorb radiation in the visible/ultraviolet region, they absorb sufficient energy to excite electrons into higher energy orbitals. When they absorb radiation of a longer wavelength (lower energy) such as infrared radiation, the energy absorbed is not sufficient to cause excitation of electrons but it can cause atoms to vibrate about the covalent bonds that bind them. Like the absorption of other forms of electromagnetic radiation, the molecule can only absorb infrared radiation at certain energy levels, ie certain frequencies. A particular region of the infrared spectrum can be referred to by its wavelength but it is more normal to refer to it by its frequency or more specifically, by its wavenumber (cm^{-1}) which is the number of waves per centimetre (8.2.1).

bending and stretching modes

Like weights joined by a spring, molecules can vibrate in a number of ways, the two basic forms being bending and stretching (Figure 8.6).

Two atoms can stretch back and forth (Figure 8.6a), three atoms can stretch either symmetrically or asymmetrically (Figure 8.6b) and three atoms can bend - this essentially involves a change in bond angle, either in or out of the plane of the bond (Figure 8.6c).

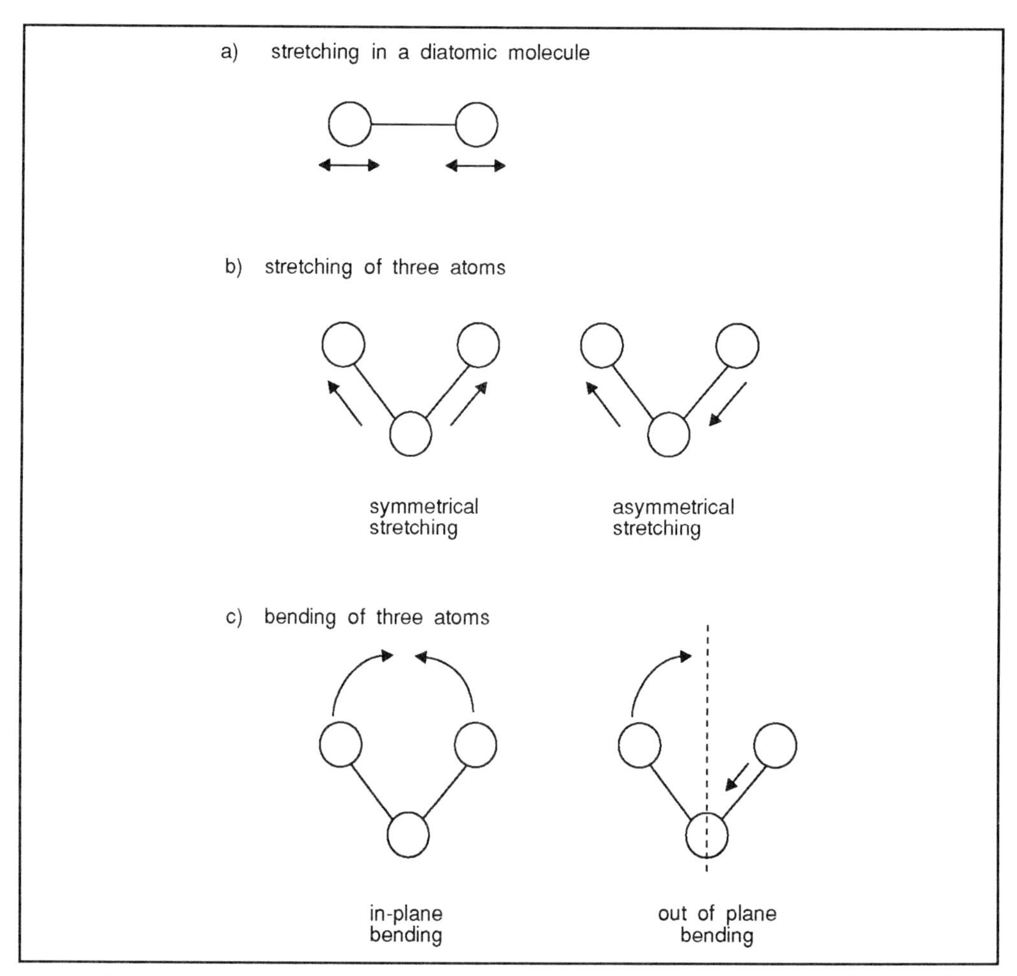

Figure 8.6 Stretching and bending modes.

Not all molecular vibrations result in the absorption of energy. For this to happen, the dipole moment of the molecule must change as the vibration occurs and the intensity of the infrared absorption is proportional to the amount by which the dipole moment changes. Thus, highly polar groups tend to result in very intense infrared absorptions whereas symmetrical vibrations such as that about the C=C bond of ethene do not give infrared absorptions.

Π What factors might determine the frequency at which a bond would stretch?

The two main factors which determine the frequency of a given vibration are the stiffness of the bond and the masses of the bonded atoms. Therefore, double bonds vibrate at higher frequencies than single bonds and heavy atoms vibrate at lower frequencies than lighter atoms.

Particular groups of atoms (also referred to as functional groups) absorb radiation at approximately the same frequency, whatever compound they are part of but as might be expected they do not always absorb at exactly the same frequency - the molecular environment does have some influence. Peaks can be shifted to some extent by hydrogen bonding, or the degree of conjugation, or the degree of electron withdrawal by a neighbouring group, etc.

For any molecule of n atoms there are a total of $3n$ degrees of freedom (ie translational, rotational and vibrational modes). Of these, 3 are for translation (for movement in the three directions mutually at right angles) and for a diatomic or linear polyatomic molecule there are 2 degrees of freedom of rotation (ie rotation about two axes mutually at right angles - the third direction, spin rotation about the bond does not count) and for a non-linear polyatomic molecule there are 3 degrees of freedom of rotation. Thus the total number of fundamental vibrations for a diatomic or linear polyatomic molecule is $(3n-5)$ and for a non-linear polyatomic molecule, $(3n-6)$. To convince yourself of the number of degrees of freedom make some molecular models. However, as we have seen, not all of these vibrations result in an absorption peak in the infrared spectrum.

Π What other reasons apart from vibrations which do not involve a change in the dipole moment of the molecule, might there be for a compound to give a spectrum of less than $(3n-6)$ peaks?

Different bonds may vibrate at the same, or effectively the same, wavenumber so that only one peak is observed for two or more vibrations. Also, absorption may occur outside the region measured by a particular infrared spectrophotometer.

On the other hand, extra absorption peaks can also appear in the spectrum, for a variety of reasons, so the situation can get quite complicated. The complexities of the infrared spectrum can however, in the hands of an expert, provide further information about the structure of a molecule.

Practical aspects: In essence, an infrared spectrophotometer works in the same way as a visible/ultraviolet spectrophotometer (Figure 8.3) although the two instruments may look very different. It is not possible to study samples in aqueous solution as water absorbs strongly in this region, as do organic solvents. Samples are often prepared therefore as a thin film in nujol between two discs of potassium bromide, or the sample is ground with potassium bromide before being compressed into a disc.

8.4.2 Applications

Infrared spectra are very useful in the identification and structural determination of organic molecules. As they contain so many peaks, each relating to some aspect of the arrangement of atoms in the molecule, each spectrum is highly specific for a particular compound. In fact, if two compounds give the same infrared spectrum then it can be assumed that they are the same compound. For this reason, the infrared spectrum of a molecule is known as its fingerprint (Figure 8.7).

fingerprint

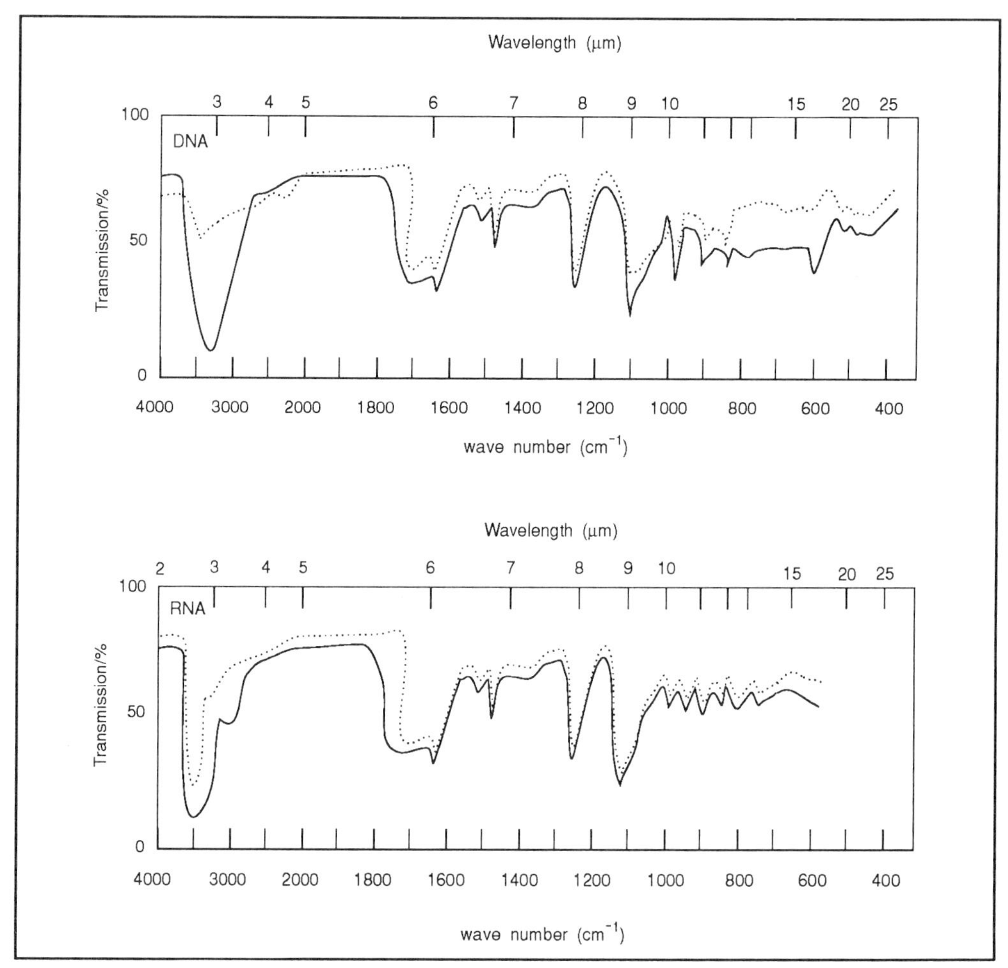

Figure 8.7 Infrared spectra of films of sodium deoxyribonucleate from calf thymus and ribonucleate from rat liver observed at 75% relative humidity (———) and at 0% relative humidatity (---------).

Because of this complexity, infrared spectroscopy is not generally applied to large biological molecules such as proteins or polysaccharides but is used to identify or characterise purified compounds of small or intermediate size such as natural products or drug metabolites. For example, an infrared spectrum might be obtained from a purified, unknown secondary metabolite; it could then be compared with a library of spectra of other secondary metabolites to see if it fits with any of them. Failing that, we know that certain groups give rise to characteristic absorption bands (Table 8.2), so the spectra would help to determine which groups were present in the molecule.

Group	Class of compound	Frequency range (cm^{-1})	Intensity
C - H	alkane	2965-2850 (stretch)	strong
	- CH_3	1450 (bend)	medium
		1380 (bend)	medium
	aldehyde	2900-2820	weak
		2775-2700	weak
C = O*	ketone	1725-1705	strong
	aldehyde	1740-1700	strong
	carboxyl acid	1725-1700	strong
	ester	1750-1730	strong
	amide	1700-1630	strong
C-O	alcohol, ester, carboxylic acid, ether	1300-1000	strong
- O - H	alcohol	3650-3590	variable & sharp
	carboxylic acid H-bonded	3300-2500	variable & broad
- N - H	primary amine & amide	about 3500 (stretch)	medium
- C - X	X = fluorine	1400-1000	strong
	= chlorine	800 - 600	strong
	= bromine	600 - 500	strong

Table 8.2 Characteristic imfrared absorption frequencies of some selected functional groups.
* not conjugated.

Full interpretation of infrared spectra requires a much greater understanding and knowledge than can be developed here but it should be clear that this is a valuable technique and even more so when used in conjunction with other physical techniques such as NMR or mass spectrometry.

8.5 Nuclear magnetic resonance spectroscopy (NMR)

8.5.1 Nuclear spin

Certain types of nuclei such as a hydrogen nucleus (proton) ^{1}H or ^{13}C, ^{31}P, ^{15}N, ^{19}F (usually those containing an odd number of protons) are said to be **spin-active**. These nuclei spin about an axis and since they are also charged they generate around them a tiny magnetic field, or magnetic moment and they can therefore be seen as tiny spinning bar magnets. When these nuclei are placed in a strong external magnetic field they align themselves with respect to the direction of the magnetic field. They can align themselves either parallel with the field - a low energy state, or antiparallel to the field - a high energy state (Figure 8.8).

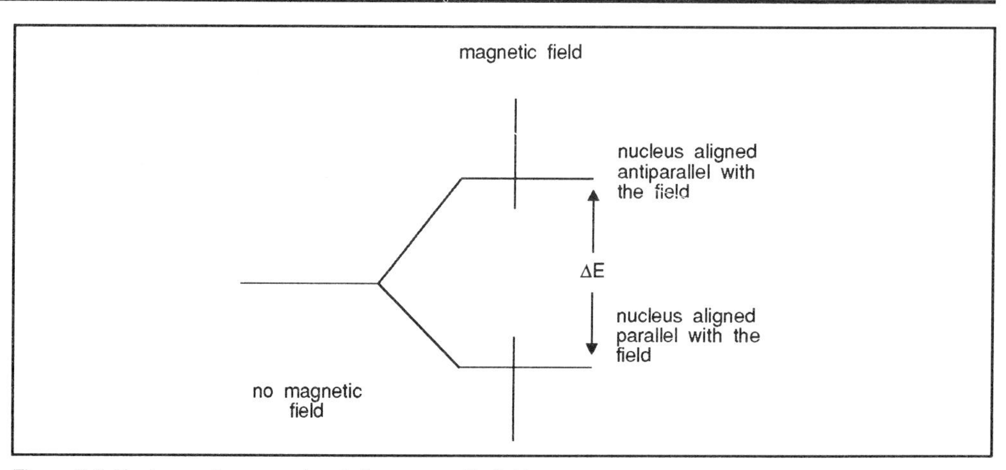

Figure 8.8 Nucleus spin energy levels in a magnetic field.

magnetic resonance

If a proton (or other spin-active nucleus) absorbs the appropriate quantum of energy, it will jump from the lower energy state to the higher energy state. This process is known as **magnetic resonance.** This applies to all spin-active nuclei but for the time being we will concentrate on proton-NMR or ^{1}H NMR, ie NMR involving hydrogen nuclei. The absorption of energy by the nuclei, like all absorption of electromagnetic radiation is quantised. The quantum of energy required to flip the proton from the low to the high energy state is proportional to the strength of the magnetic field. At any given magnetic field strength then, absorption takes place at a specific frequency of electromagnetic radiation. NMR spectrometers are constructed such that the magnetic field strength is very high - thousands of times greater than the earth's magnetic field. At this strength, radiation required to flip the proton from the low to the high energy states, is in the radiowave region. The sample is scanned with radiation over a certain frequency range and when the frequency is at precisely the 'right' strength, the proton flips from one state to the other: magnetic resonance occurs. This causes an electrical current to flow in a coil around the sample and this is amplified and displayed as a peak, or signal on the recorder.

In fact, most modern NMR spectrometers operate in the Fourier transform (FT) mode whereby the sample is pulsed with radiation extending over a wide range of frequencies so that all the nuclei resonate at the same time. The instrument then measures the decay of magnetisation, as the nuclei fall back to the lower energy levels, over time and the complex mathematical process of Fourier transformation converts this time related spectrum into the frequency related spectrum which we can then interpret.

If all hydrogen nuclei absorbed energy at exactly the same frequency, NMR spectrometers would not give us very much information about a compound. Fortunately, the field strength at which absorption occurs is dependent on the magnetic environment of each hydrogen nucleus. This environment depends on two major factors:

- magnetic fields generated by circulating electrons;
- magnetic fields generated by other magnetic nuclei (usually hydrogen nuclei).

8.5.2 Chemical shifts

As we said above, different protons in a molecule absorb energy at different magnetic field strengths. This is because the electrons in bonds and atoms adjacent to the proton, influence the magnetic field around the hydrogen nucleus by generating their own tiny magnetic field. If this secondary magnetic field opposes the applied field, a slightly higher magnetic field strength is required to make the proton resonate. The proton is said to be shielded. If the induced field reinforces the applied field, the proton is said to be deshielded and a smaller applied field will bring about resonance. The extent of the shielding depends on the electron density around the proton; the presence of electronegative groups, ie electron-withdrawing groups, decreases the magnitude of the shielding. Therefore, the different hydrogen nuclei of a molecule will be shielded to different extents depending on the molecular structure and so will resonate at different field strengths.

shielding

The local molecular environment then, causes the different sets of protons in a molecule to be shifted from the position in the magnetic field that a proton free of all electrons would absorb. These sets of altered positions in the spectrum are characteristic of a proton in a given microenvironment (group) and are known as **chemical shifts**.

It is not convenient to measure the actual frequency or magnetic field strength at which absorptions occur so instead, the differences between chemical shifts are measured with reference to a standard compound, usually tetramethylsilane (TMS). With respect to this standard, the signal of which is taken as zero, the magnitude of the shifts are generally in the order of a few parts per million (δ ppm) of the operating frequency.

Let us now look at some simple (diagrammatic) NMR spectra:

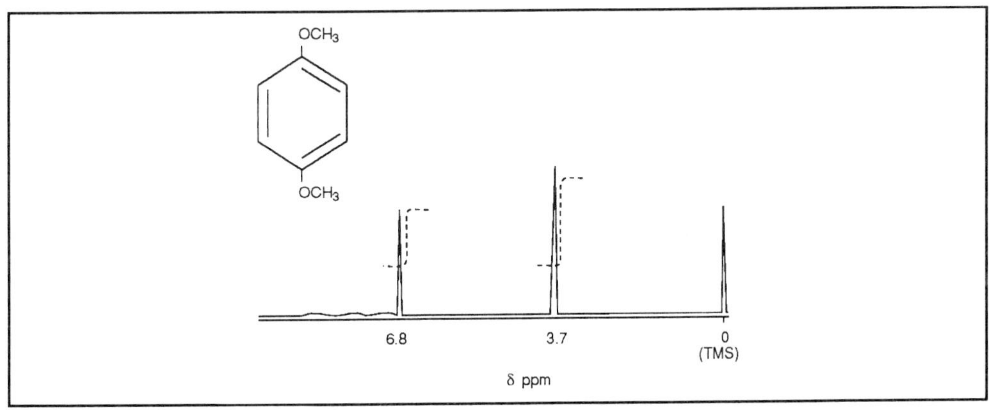

Figure 8.9 NMR spectrum of 1,4-dimethoxybenzene (broken lines are the integrals of the area under the peak).

The spectrum of 1,4-dimethoxybenzene (Figure 8.9) consists of two signals and the fact that there are two signals tells us that there are two types of hydrogen atom in the molecule, with chemical shifts of 6.8 and 3.7 ppm.

Π Try to divide the hydrogens of the molecule into two different types.

There are 6 chemically equivalent (ie in the same magnetic environment) hydrogens on the two methoxy groups and 4 equivalent hydrogens on the aromatic ring. These two types therefore give rise to two signals.

The occurrence of particular peaks with particular chemical shifts in a spectrum, indicates the presence of protons in various functional groups. Some typical chemical shifts are given in Table 8.3

Proton	δ(ppm*)
$C\underline{H}_3(CH_2)_n$	0.9
$CH_3(C\underline{H}_2)_nCH_3$	1.25
$C\underline{H}_3Cl$	3.05
$RC\underline{H}_2Cl$	3.4
$R_2C\underline{H}Cl$	4.0
Aromatic Ar$\underline{H}$	7.2
$RC\underline{H}O$	9.5
$RC\underline{H}_2CHO$	2.2
$C\underline{H}_3OH$	3.4

Table 8.3 Some typical proton chemical shifts (R = alkyl). *The value of δ varies slightly depending on the solvent, temperature and concentration. The chemical shift refers to the proton underlined.

Another major piece of information that we can read from the NMR spectrum is given by the relative magnitude of the peaks - or to be precise - the area underneath the peaks. This corresponds to the relative number of hydrogen atoms giving rise to each signal. For instance, when the area of the dimethoxybenzene signals (above), are measured, they are found to be in the ratio 3:2, or 6:4. This of course is the ratio of the two types of hydrogen atoms - 6 methoxy hydrogens and 4 aromatic ring hydrogens. Most NMR spectrometers will calculate the integral of (area underneath) each peak and these are usually plotted onto the spectrum as a series of integration curves or 'step' curves where the height of each step is proportional to the integral of the peak; these are the broken lines on the spectrum in Figure 8.9.

8.5.3 Spin-spin coupling

Most NMR spectra, however, are not as simple as those shown in Figure 8.9. Their complexity is increased by the phenomenon of signal splitting, which is demonstrated in the spectrum of 1,1,2-trichloroethane (Figure 8.10). From what we have said so far, we might classify the hydrogens of the molecule into two types - the one on the 1C and the two on the 2C.

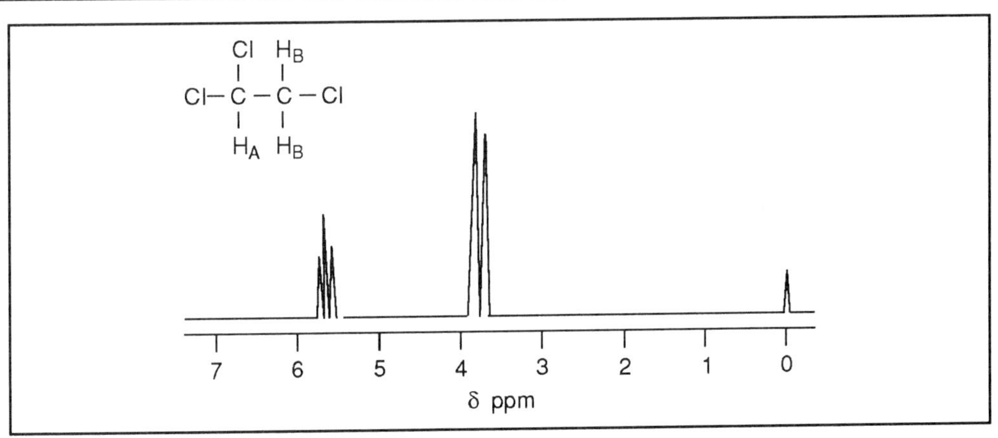

Figure 8.10 Proton NMR spectrum of 1, 1,2-trichloroethane.

Why then are there more than two signals? This signal splitting is due to the magnetic influences of hydrogens on atoms adjacent to those giving rise to the signal.

As we have said, protons can act as tiny magnets and can therefore be aligned either with the applied magnetic field or against it. The magnetic moment of a proton will affect the magnetic environment of a proton on a nearby atom. This is known as spin-spin coupling.

To go back to the example of 1,1,2-trichloroethane, the proton H_A can be aligned with the magnetic field or against it. This in turn means that the protons H_B will have their local magnetic field either increased or decreased, depending on the spin state of H_A. This means that the signal from H_B will be split into two signals - a doublet, one slightly upfield from its original position and one slightly downfield. There will be approximately equal numbers of H_A protons in each spin state so the two signals from H_B will be approximately the same height and half that of the original signal (Figure 8.11).

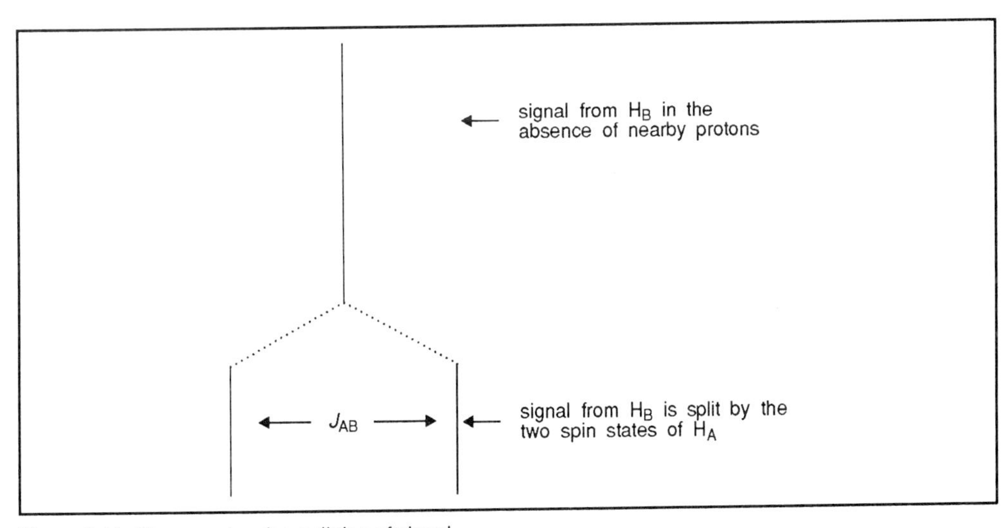

Figure 8.11 Diagram showing splitting of signal.

coupling constant

The splitting between the two peaks, in terms of frequency units, is known as the coupling constant, or J_{AB} and is generally measured in hertz (Hz). The coupling constant for a given system is always the same as it is not dependent on the operating frequency or the size of the applied field.

The H_A proton of 1,1,2-trichloroethane has two nearby protons (those on the adjacent carbon) and is therefore subject to a greater variety of magnetic fields. The H_A proton can 'see' the H_B protons in any one of four states. The H_B protons can be aligned both with the magnetic field, or one can be aligned with and one against, or vice-versa, or both can be aligned against the magnetic field (Figure 8.12).

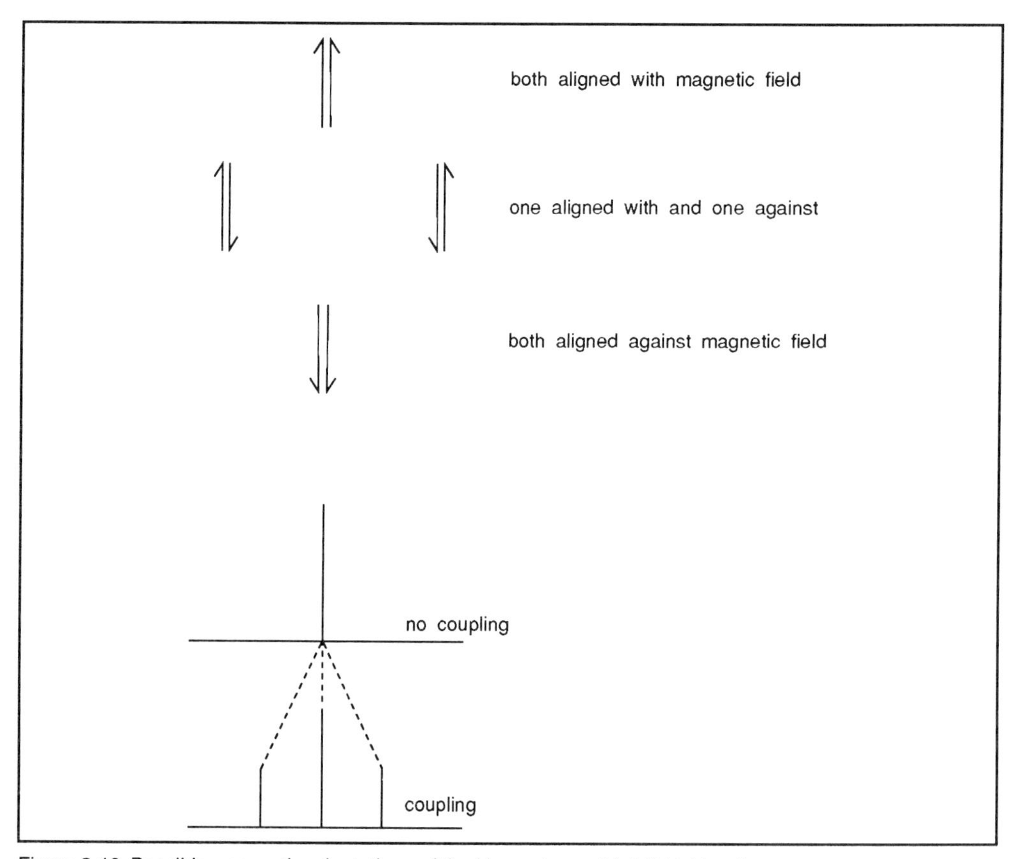

Figure 8.12 Possible magnetic orientations of the H_B protons of 1,1,2-trichlroethane.

Although there are four spin states, two of these are equal in energy - one proton reinforces the field and one opposes it so that they cancel each other out and the signal is not displaced. Of the other two spin states, when both protons are aligned with the field, the signal from H_A is shifted downfield and when both are aligned against the field the signal from H_A is shifted upfield. So the signal from this hydrogen is split into three signals - a triplet, one upfield, one downfield and one undisplaced and of twice the intensity of the other two.

Spin-spin coupling is not observed for protons that are chemically equivalent. Therefore, there is no signal splitting in the signals from the hydrogens of ethane as these protons are all equivalent.

Spin-spin coupling effects are transferred through the bonding electrons and they do not generally occur if the protons are separated by more than three σ-bonds.

By the same reasoning, the signal from a hydrogen which is adjacent to a carbon with three equivalent hydrogens, is split into a quartet and these four singles have intensities in the ratio 1:3:3:1.

Pascal's triangle

This pattern of splitting can be generalised. When a proton, or a group of protons, is coupled to a group of protons containing n equivalent protons, then its signal will be split into $n+1$ peaks. The relative intensities of these peaks can be calculated as we did above, by going through all the possible permutations of the proton's spin states and they turn out to follow the pattern of Pascal's triangle:

1

1 1

1 2 1

1 3 3 1

1 4 6 4 1

1 5 10 10 5 1

1 6 15 20 15 6 1

Thus, the relative intensities of the signals of a triplet will be 1:2:1, of a quartet 1:3:3:1 and so on.

These characteristic signal splitting patterns are extremely useful in providing information about the numbers of adjacent protons, while chemical shifts give information about the chemical type of the proton.

SAQ 8.6

Explain how the peaks on the spectrum (Figure 8.13) correspond to the protons of diethylether. (CH_3CH_2-O-CH_2CH_3).

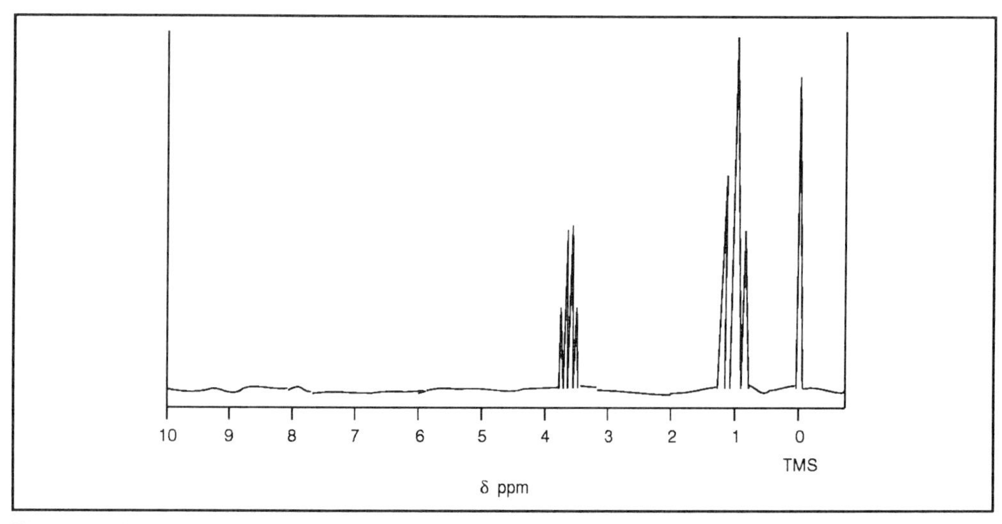

Figure 8.13 The NMR spectrum of diethylethe.

SAQ 8.7

The NMR spectrum of acetaldehyde (CH_3CHO) shows two regions of resonance, a quartet for the aldehydic proton and a doublet for the methyl group. Which one of the following represents the the relative intensity of the components?

a) quartet 3:1:1:3, doublet 3:1;

b) quartet 1:3:3:1, doublet 1:2;

c) quartet 1:1:1:1, doublet 1:2;

d) quartet 1:3:3:1, doublet 1:1.

Π Bearing in mind what you know about spin-active nuclei, can you think of any requirement of a potential NMR solvent?

An NMR solvent must itself be free of ^{1}H nuclei and for this reason NMR is carried out in either aprotic (eg $CC1_4$) or deuterated solvents.

NMR spectroscopy differs from other forms of spectroscopy in that every peak in the spectrum 'means' something and in order to correctly and fully interpret an nmr spectrum every peak must be accounted for.

8.5.4 Carbon-13 NMR spectroscopy

Do not forget that hydrogen atoms are not the only nuclei which have a magnetic spin. The carbon-13 (^{13}C) isotope can also give rise to NMR signals in just the same way. As the ^{13}C isotope has a natural abundance of only about 1.1%, sensitive spectrometers must be used to measure ^{13}C spectra. However, ^{13}C spectroscopy has the advantage that the carbon atoms of a molecule can be observed directly, rather than indirectly through the protons of attached hydrogens. This is obviously an advantage in observing carbon atoms which have no attached hydrogen atoms. Another advantage of ^{13}C NMR is that ^{13}C nuclei absorb over a much wider range of chemical shifts than protons and so there is less chance of signals overlapping.

Π Explain how the relatively low abundance of ^{13}C makes the reading of ^{13}C NMR spectra simpler than it might otherwise be?

As only 1.1% of carbon atoms exist as this isotope, it is very unlikely that a molecule will have two adjacent ^{13}C atoms and consequently ^{13}C- spectra do not show spin-spin couplings between carbon atoms.

broad band decoupling and off resonance

Coupling does occur however, between protons and ^{13}C. It is possible to obtain the spectrum in a form in which all the coupling between the ^{13}C and protons has been eliminated and the signal from each carbon appears as a singlet. This is termed broad band decoupling and is used to obtain precise chemical shifts. Similarly, it is possible to eliminate all couplings except those between a carbon and the protons directly attached to it. This process, known as off resonance decoupling, is helpful because it tells us precisely how many hydrogen atoms are attached to each carbon. For example, if a signal appears as a quartet it must represent a methyl group.

Unlike in ^{1}H NMR, it is very difficult to obtain reliable integration curves for ^{13}C NMR spectra, for technical reasons.

8.5.5 Applications

NMR spectroscopy is of enormous value in the elucidation of chemical structures, particularly of pure, relatively small organic molecules. In terms of biotechnology this means for example, that it is widely employed to determine the structure of secondary metabolites isolated from micro-organisms in the search for potential new drugs, or to confirm the results of a chemical synthesis.

NMR can also be used in biosynthetic studies - by feeding an organism a compound labelled with a stable isotope such as ^{13}C, NMR enables one to see if and where, in a particular compound, the labelled precursor ends up. By this and other means, one can build up a stepwise picture of how a natural product such as an antibiotic, is synthesized by the producing organism.

NMR can be used to study biological macromolecules, although in determining the structure of a molecule such as a protein, one is faced with an extremely complex spectrum which requires computer analysis as well as very high resolution instruments. Even the NMR spectrum of a relatively simple molecule such as lysine (Figure 8.14) is quite complex.

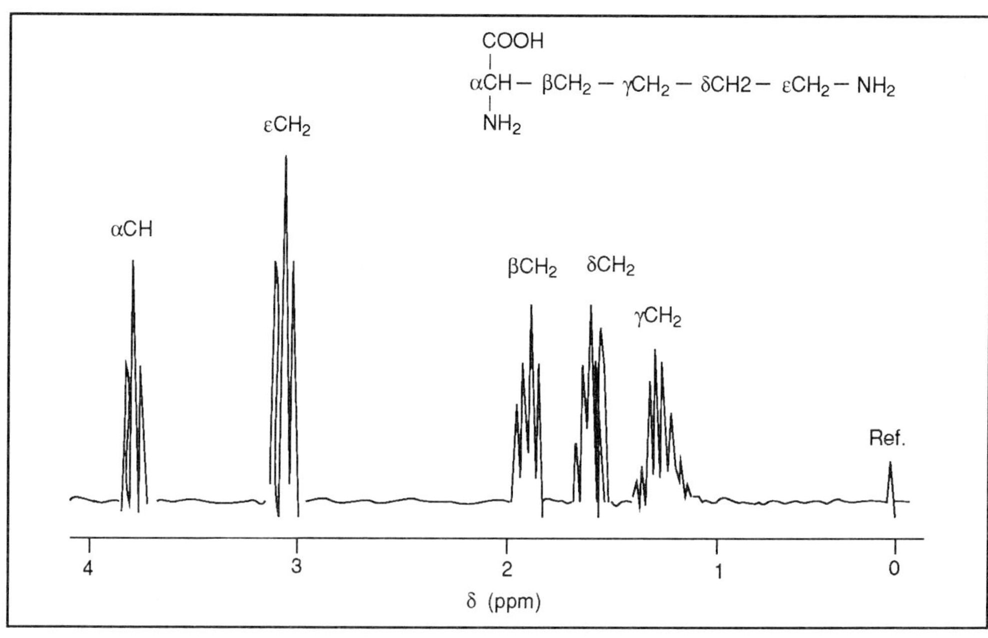

Figure 8.14 Proton spectrum of lysine at 220 MHz

NMR is much more likely then to be used to detect small conformational changes in a protein such as in binding studies of an active site of an enzyme. More recent NMR technology even allows the non-invasive study of metabolic events in whole animals.

SAQ 8.8

Indicate whether the following statements about NMR are true or false:

a) all nuclei can undergo NMR;

b) the chemical shift (δ ppm) of a given proton will increase with increase in operating frequency;

c) spin-spin couplings are due to the interactions of protons with the different spin states of all the other protons of the molecule;

d) a proton on a C-atom coupled to a methyl group will give rise to a signal split into a quartet.

8.6 Mass spectrometry

8.6.1 Basic principles

Mass spectrometry is a technique in which the molecule of interest is vaporized in a vacuum and bombarded with a beam of electrons. This bombardment can knock off an electron causing the molecule to become ionised and forming what is known as the **molecular ion**. The bombardment also imparts to the ion a considerable amount of energy which can cause it to break up, or fragment, into a series of **fragment ions**. So, our molecule is volatilised, ionized and broken into a molecular ion and a series of smaller fragment ions.

mass/charge ration

Mass spectrometry is based upon the fact that we can now sort these ions by employing the rule that a moving ion is deflected by a magnetic field to an extent that is dependent on its mass/charge ratio (*m*/*e*). As most ions have a charge of +1, this effectively means that the ion is deflected to an extent depended on its mass. So, by accelerating the ions to constant velocity and then deflecting them in a magnetic field, we can obtain molecular masses of the molecular ion (and so the molecular mass of the compound) and of the fragment ions. This information can tell us a lot - perhaps more than we might at first imagine - about the structure of a molecule.

Let us take a brief look at the instrument which carries out this process - the mass spectrometer.

Π You might already see how mass spectrometry differs fundamentally from other forms of spectroscopy. What is this difference?

All forms of spectroscopy described in this chapter involve the interaction of electromagnetic radiation with matter. Mass spectrometry however, involves physical bombardment of a compound with electrons (or atoms). It also differs in that it is, by its very nature, a destructive technique but with the very small amounts required for analysis, this is hardly a disadvantage.

8.6.2 The mass spectrometer

The basic layout of a mass spectrometer is shown in Figure 8.15

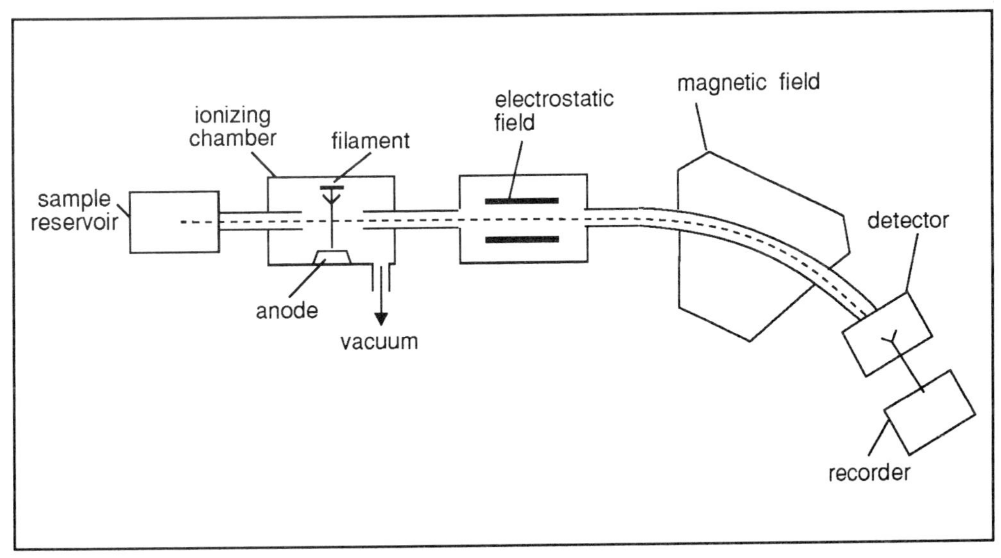

Figure 8.15 Basic layout of a mass spectrometer.

Π Why do you think that it is important to carry out the process in a vacuum?

If other molecules are present the sample ions will collide with them and be prevented from travelling further, or be transformed.

The sample molecules are vaporized in the sample reservoir and a stream of sample vapour is drawn in to the ionisation chamber. Here it is bombarded with electrons produced by a tungsten filament and the ions are then accelerated in an electrostatic field through a series of slits. The path of ions is then bent through a curved tube by means of a magnetic field. As already mentioned, this is the area which differentiates the ions. Ions with the same charge are deflected by the magnetic field to extents dependent on their mass - smaller ions are deflected more and larger ions, less. The magnetic field or the accelerating voltage is varied and this in turn alters the path of the ion beam so that at any one time the ions of a particular mass follow exactly the curve of the tube and pass through a detector slit at the end of the tube and onto the ion detector. This measures the intensity of the ion beam at each m/e value and this gives a measure of the relative abundance of each ion. The recorder then plots the spectrum which is essentially a bar chart showing each ion and its relative abundance.

Π By employing a very narrow detector slit, the resolving power of the mass spectrometer is increased. ie the instrument can distinguish between different ions of very similar masses. However, what disadvantage might there be of narrowing the detector slit?

Reducing the detector slit width means that at any one time there are less ions falling on the detector. This means that the sensitivity of the instrument is correspondingly reduced. Therefore an optimum slit width should be used, balancing sensitivity and resolving power.

8.6.3 Ionization by electron impact (EI)

We will consider the mass spectrum of hexane (Figure 8.16) which has been obtained by the process of electron impact (= bombardment by electrons).

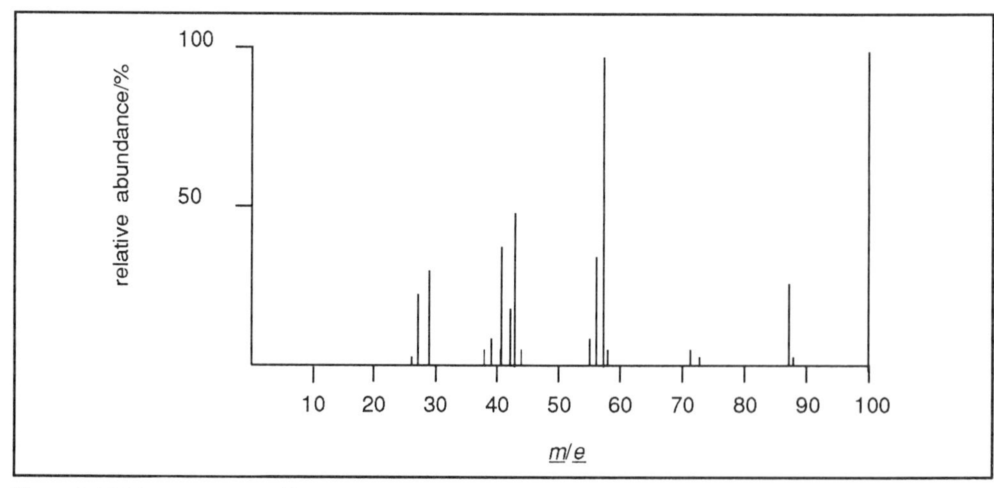

Figure 8.16 Mass spectrum of hexane prouced by electron impact

base peak

The molecular mass of hexane is 86 and at one end of the spectrum there is a peak at *m/e* 86 - the molecular ion (M^+). On any spectrum the peak of the greatest intensity is called the base peak, its intensity is taken as 100 and the intensities of the other peaks which represent the fragment ions, are expressed relative to it. The base peak is sometimes the molecular ion or sometimes a fragment ion. Sometimes there may be no molecular ion peak, if it has been entirely fragmented into smaller ions. Analysis of the fragment ions provides information on the chemical groups present in the molecule and their arrangement.

Π The spectrum of hexane shows the base peak at *m/e* 57. What fragment ion might this represent?

The molecular ion has a mass of 86 and the base peak ion has a mass of 57 - ie hexane minus a fragment of mass 29. From these figures we can surmise that the base peak is produced by cleavage of the molecular ion to produce a fragment of the molecular formula $-C_2H_5$ (molecular mass 29).

$$C_6H_{14} \rightarrow [C_6H_{14}]^{+\bullet} \rightarrow C_4H_9^+ + C_2H_5\bullet$$

The ion of *m/e* 56 corresponds to this same ion less one hydrogen, the ion at *m/e* 71 corresponds to the molecular ion from which $-CH_3$ has been cleaved, and so on.

fragmentation pattern

Each compound has a characteristic fragmentation pattern in much the same way that it has a characteristic infrared spectrum and two compounds with the same mass spectra are almost certainly identical.

Fragmentation patterns are usually very complex but it is often possible to deduce from them a great deal of information about the structure of a molecule, particularly when coupled with data already recorded for the fragmentation of various chemical classes and moieties. Spectra can be further complicated by the presence of peaks arising from rearrangement ions which may be produced by intramolecular atomic rearrangement during fragmentation. However these are often characteristic peaks and can themselves be useful.

If you look closely at the mass spectrum of hexane you will see that there is a small peak at *m/e* 87, ie the molecular mass of hexane +1. This might at first seem illogical - an ion with a mass greater than the molecule being analyzed. It is explained by the fact that not all carbon atoms have a mass of 12. About 1.1% of carbon is naturally present in the form of the isotope ^{13}C, so that molecules containing this isotope will give an isotopic peak at *m/e* 87. Most of the elements found commonly in biological molecules exist as several isotopes. Table 8.4 shows some of these commonly encountered isotopes and their abundance relative to the commonest form of the element.

Isotope	Relative Abundance
^{2}H	0.015%
^{13}C	1.10%
^{15}N	0.37%
^{18}O	0.20%
^{34}S	4.21%
^{37}Cl	24.23%

Table 8.4 Relative abundances of some common isotopes

These isotopic peaks may appear to be irrelevant but they are in fact very useful. By knowing the relative natural abundances of each isotope and relating this to the relative intensities of the isotopic peaks, it is possible to determine, or at least partially determine, the molecular formula of the ions.

SAQ 8.9

From the figures of relative isotopic abundance in Table 8.4 calculate the intensity of the M+1 peak of hexane, relative to the M^+ peak.

Π What might you deduce about a molecule (molecular mass 100-200), if its spectrum gives an M+2 peak of about 0.4% intensity relative to the M^+ peak?

The M+2 ion obviously contains an isotope of 2 mass units greater than the most common form or it contains two isotopes of 1 mass unit greater than the commonest form. As the molecular weight is fairly small, the proportion of molecules containing 2 x ^{2}H or 2 x ^{13}C or one of each, is much smaller than 0.4%, so the M+2 ion is probably due largely to a +2 isotope such as ^{18}O, ^{37}Cl or ^{34}S. Looking at Table 8.4 we can see that the ion cannot contain S or Cl as these would give a much higher abundance of the M+2 peak. But, as ^{18}O has a relative abundance of 0.2%, the intensity of this peak probably indicates that the compound contains two oxygen atoms.

A rule useful in the deduction of the molecular formula is the 'nitrogen rule' which says that a molecule of an even-numbered mass must contain an even number or no nitrogens and an odd-numbered mass must contain an odd number of nitrogens.

Using these and similar lines of reasoning it is often possible to build up the entire molecular formula of a compound.

This process is greatly facilitated by modern high resolution mass spectrometers which can resolve ions differing in molecular mass by only tiny fractions of a mass unit. This is important as nuclides do not have masses which are integers. By definition, carbon-12 has a mass of exactly 12 but ^{1}H for instance, has a mass of 1.0079, ^{16}O mass 15.9949, ^{14}N mass 14.0031, ^{35}Cl mass 34.9689, and so on.

Two compounds with the molecular formulae: C_6H_{14} and $C_4H_6O_2$ respectively, might appear at first glance to both have a molecular mass of 86 - and for most intents and purposes, they do. But if we calculate their respective masses more accurately:

$$C_6H_{14} = (6 \times 12) + (14 \times 1.0079) = 86.1106$$

$$C_4H_6O_2 = (4 \times 12) + (6 \times 1.0079) + (2 \times 15.9949) = 86.0372$$

Mass spectrometers can resolve these two compounds and because the mass of an ion can be determined so precisely, it can only have one possible molecular formula, so that effectively the molecular formula can be determined at the same time.

8.6.4 Other ionization techniques

Apart from bombardment by electrons (electron impact ionization) which gives rise to a range of positively charged ions, there are other ways of ionizing molecules for analysis by mass spectrometry.

chemical ionisation

In chemical ionisation (CI) a reagent gas such as methane or isobutane is introduced into the ionization chamber and bombarded with electrons to give a series of ions which, at the low pressures involved, undergo a series of secondary reactions to give ions such as CH_5^+ and $C_4H_{11}^+$ When these ions come into contact with the volatilised sample molecule, proton transfer readily takes place to give a positively charged M+1 ion. This ion may fragment further but fragmentation is less common by this ionization method and so the predominant ion of the spectrum is often the molecular ion.

field ionization field desorption

In field ionization (FI) and field desorption (FD) the sample ions are formed in the strong electrostatic field established at the tip of a wire electrode to which a high voltage is applied. This too, results in a high proportion of molecular ions and has the advantage that it can be used to study relatively non-volatile compounds such as carbohydrates, which are difficult to study by other ionization methods.

fast atom bombardment

Fast atom bombardment (FAB), employs not electrons but a stream of fast neutral atoms such as xenon or argon, to bombard the sample to produce sample ions. It is a 'soft' technique which results in relatively little destruction of the ionized molecules. Figure 8.17 shows spectra of dihydrocortisone obtained using different ionisation techniques.

8.6.5 Applications

Mass spectrometry is a very powerful technique, widely used for structural determinations of biological molecules. One of its great advantages is that it requires only very small amounts of sample (sub-microgram levels) and although it generally requires that the material be pure, it can be used for identifying the presence of particular molecules in mixtures.

The usefulness of mass spectrometry can be enhanced by coupling it to separative techniques such as chromatography so that as a mixture is separated on the chromatographic system the individual components are analyzed immediately by mass spectrometry. The most common form of this coupling technique is gas chromatography with mass spectrometry (GC-MS). Gas chromatography is particularly suitable for this process as the material which elutes from the chromatography column is already gaseous, but liquid chromatographic systems can also be used (LC-MS). The ions produced in the mass spectrometer can even be fed directly into a second mass spectrometer (MS-MS) to further increase the information that can be obtained about a sample.

Mass spectrometry can be used to help determine the structure of relatively large molecules such as peptides or biopolymers.

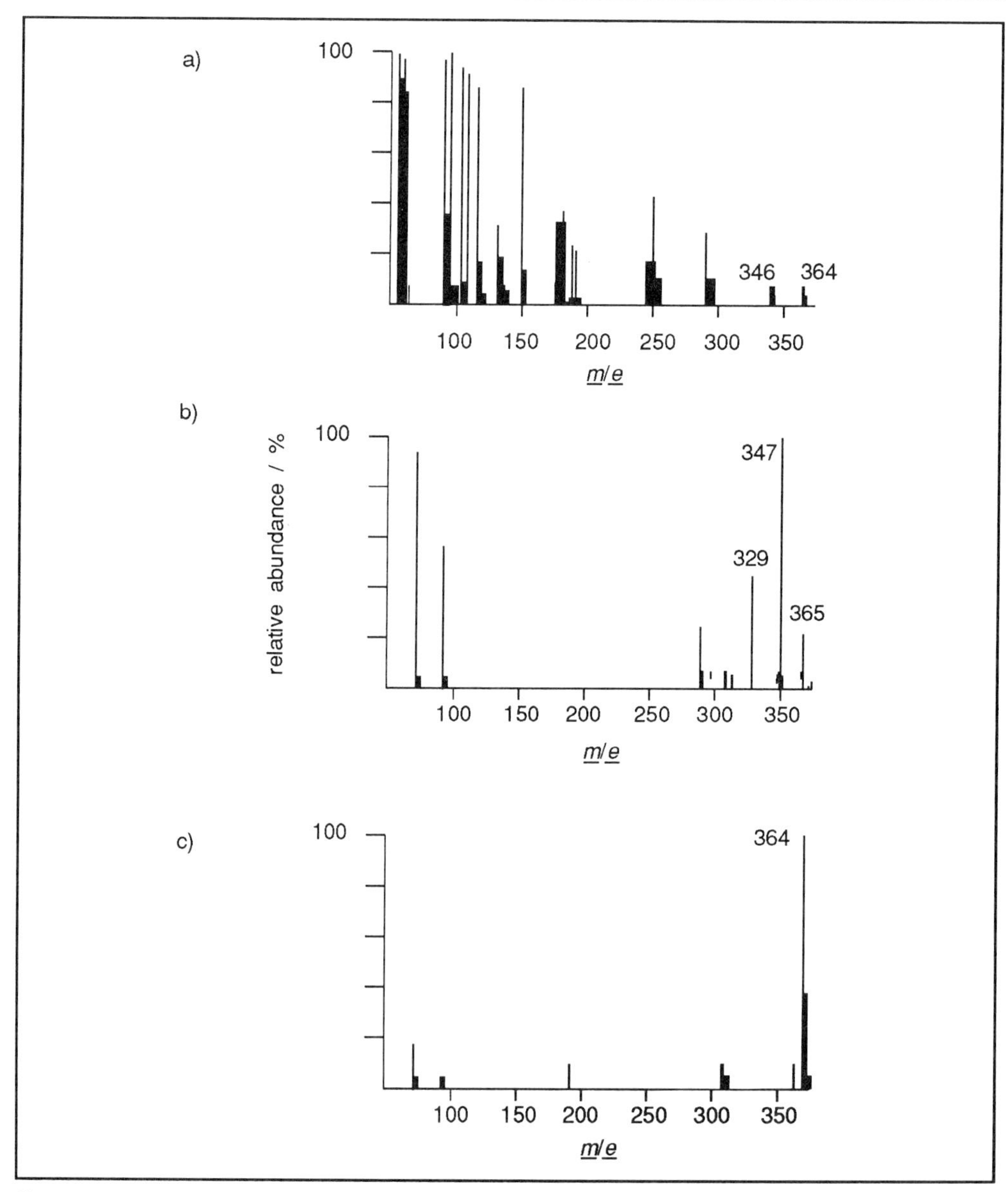

Figure 8.17 Mass spectra of dihydrocortisone obtained by a) EI; b) CI and c) FD ionisation. ($M_r = 364$).

8.7 X-ray Crystallography

When an X-ray beam is aimed at a crystal, some of the X-rays pass right through but some are scattered (diffracted) by the electrons of the atoms in the crystal lattice. These diffraction beams can be detected by X-ray film to produce a diffraction pattern. Because the atoms in the crystal are ranged regularly, the diffraction pattern consists of a regularly repeating pattern of spots with certain specific distances between them and each with a characteristic relative intensity. The pattern is entirely dependent on the three-dimensional arrangement of the atoms in the crystal and from this pattern it is possible to build up a map of the electron densities at different points in the crystal and

diffraction pattern

ultimately to construct a very precise picture of the three dimensional structure of the crystalline molecule.

This technique has been used notably, to obtain three-dimensional structures of proteins, with precise measurements of atomic arrangement, bond lengths and bond angles. This in turn has resulted in a great deal of knowledge about the way enzymes function, how proteins bind other molecules, how they fold and so on. It has also been used to help obtain a complete three-dimensional structure of t-RNA, a histone-DNA complex and even the structure of some viruses.

X-ray diffraction interpretation is however, extremely skilled and painstaking work and has the further requirement of pure samples which are crystallisable - many proteins are not. For these reasons, this is not a technique which is carried out widely on biological molecules.

SAQ 8.10

Complete the following sentences by adding the appropriate techniques:

1) ....is a form of vibrational spectroscopy;

2) ....involves analysis of ionised molecules and fragments;

3) ....involves interaction of electromagnetic radiation with crystalline materials;

4) ....is a form of electronic spectroscopy.

SAQ 8.11

Indicate whether the following statements are true and which are false:

1) mass spectrometry does not involve interaction of a molecule with electromagnetic radiation;

2) infrared spectroscopy involves absorption of electromagnetic radiation of higher frequency than visible/ultraviolet spectroscopy;

3) ultraviolet spectroscopy is capable of determining the molecular formula of a compound;

4) different ^{13}C nuclei resonate at different frequencies.

SAQ 8.12

You have isolated an unknown biological molecule. Which of the techniques described in this chapter might you use to:

1) determine its molecular mass;

2) obtain a detailed three-dimensional picture of the molecule, including bond lengths;

3) determine whether the compound possesses a particular functional group such as a carbonyl group;

4) determine whether a small molecule such as acetate, is one of the building blocks which the organism incorporates into the molecular matrix during synthesis?

8.7 Final remarks

In this chapter we have examined some of the techniques for examinig the structures of biomolecules. We have not attempted to be encyclopaedic and cover all of the techniques that are available. The techniques covered are those which find widespread application. As your knowledge of biochemical analysis increases, you will become aware of other important techniques.For example, polarimetry (which measures the rotation of plane polarised light (light waves vibrating in a single plane) has important applications in investigating the stereo configuration of molecules. Similarly a technique called circular dichroism has important applications in analysing the three dimensional structure of macromolecules (especially proteins). You may encounter these, and other techniques, in other texts of the BIOTOL series. Nevertheless, the knowledge you have gained from this chapter, will have provided you with a sound start to understanding how the structures of biological compounds can be determined.

Summary and objectives

This has been a long chapter in which we have examined important techniques used in the analyses of the structure of biomolecules. Many of these techniques are dependent upon the interaction of such molecules with electromagnetic radiation. We provided you with an explanation of how radiation interacts with molecules and described how these interactions can be used to elucidate chemical structures. We also examined the techniques of nuclear resonance spectroscopy, mass spectroscopy and x-ray crystalography.

Now that you have completed this chapter you should be able to;

- understand and use the relationship between energy, frequency and wavelength of electromagnetic radiation;
- explain the underlying principles of spectroscopy and give a general description of the design of spectrophotometers;
- explain the reason for electronic spectra being in the UV region and vibrational spectra in the IR region;
- state and apply the Beer-Lambert relationship and explain its use in quantitative measurements;
- describe the principles of magnetic resonance and appreciate how the chemical shift of a nucleus is governed by its micro-environment;
- interprete supplied spectral data;
- explain the principles of a mass spectrometry and describe the main forms of ionisation techniques used in mass spectrometry;
- obtain information from a mass spectrum;

The use of radioactive isotopes

The use of radioactive isotopes

9.1 Introduction: commonly used radioactive isotopes

Often in biochemical analyses very small amounts of material need to be quantified. For this purpose we can use spectrophotometric and fluorometric methods to follow enzyme catalysed conversions enabling us to determine small quantities of compounds in a specific way. When still smaller quantities have to be determined, the use of radioactive tracer may be advantageous. In these methods the compound to be determined is labelled by incorporating one or more isotopes that can be recognised by their characteristic radioactive decay. Of the many radioactive elements known only a few are used in the biochemical laboratory. Mostly we use isotopes that emit beta - or gamma radiation. Elements emitting alpha radiation are not commonly used in this field of research.

9.1.1 Radioactive elements establish a characteristic radioactive decay.

Radioactive decay is the phenomenon of spontaneous disintegration of an unstable atomic nucleus

Radioactive decay is the phenomenon of spontaneous disintegration of an unstable atomic nucleus. During decay the atom is transformed into a new state, that can be stable or unstable. In the latter case further decay will take place until a stable end product is formed. Examples of radioactive atoms that decay with emission of electrons (= beta emission) are ^{14}C, ^{3}H, ^{32}P and ^{35}S. In these isotopes, a neutron in the nucleus is split into a proton plus an electron; the latter leaves the atom with high energy and is then called a beta particle. The newly formed atom has the same atomic mass, but the atomic number is increased by 1.

Π Why is the atomic number increased by 1 after emission of a beta particle?

The answer is, of course, the nucleus contains an additional proton.

For the atoms mentioned above the decay reactions are:

$^{14}C \rightarrow {}^{14}N$ + beta particle (β)

$^{3}H \rightarrow {}^{3}He$ + beta particle (β)

$^{32}P \rightarrow {}^{32}S$ + beta particle (β)

$^{35}S \rightarrow {}^{35}Cl$ + beta particle (β)

Each beta emission produces beta particles with a whole spectrum of different energies varying between zero and a specific maximal value (E_{max}) that is characteristic for that isotope.

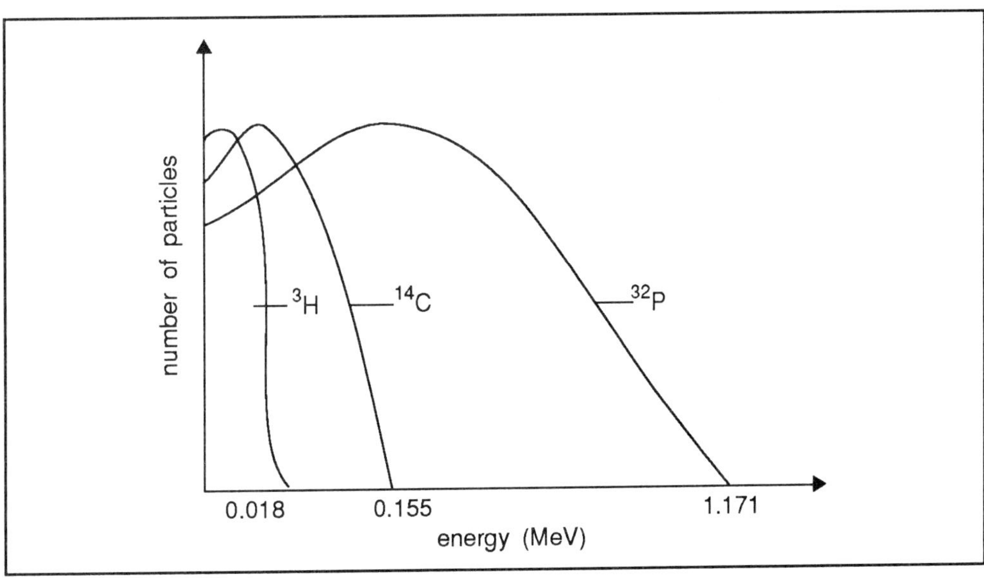

Figure 9.1 The number of particles liberated during radioactive decay as a function of the energy of those particles.

Figure 9.1 shows the spectrum of a few isotopes that are often used to study biological systems. In Figure 9.1, the number of beta particles with a definite energy is plotted against energy. We see that E_{max} for ^{32}P is higher than that for ^{14}C and this is again higher than that of ^{3}H. ^{32}P is said to emit harder radiation than ^{14}C . The hardness of a beta emission is usually indicated by quoting E_{max}, although the number of particles with an energy close to E_{max} is only very small. Probably the average energy would be a better standard, because a larger fraction of the particles will have an energy close to this average value. The average energy is roughly one third of E_{max} (although this depends on the energy spectrum of the isotope).

When beta particles move through matter, their energy is absorbed by atoms they collide with which become 'excited' or ionised. This property is used for the detection of beta particles.

Gamma radiation

Gamma radiation is electromagnetic radiation with a very short wavelength, still shorter than that of X-rays (see Table 8.2 in the previous chapter). Gamma radiation arises when the nucleus falls back to a lower energy level. Very often gamma radiation arises as a secondary effect during emission of a beta particle. Gamma radiation does not cause direct ionisation of atoms. However, during interaction of gamma radiation with the nucleus or with an electron of an atom, secondary electrons can arise which in turn can ionise other atoms; because of this, gamma radiation can be detected in the same way as beta radiation.

9.2 Radioactive decay

Decay of radioactive isotopes is a spontaneous process determined by statistical probability. The number of disintegrations per unit of time is proportional to the number of radioactive atoms present:

$$\frac{dN}{dt} = \lambda . N \qquad \text{(E-9.1)}$$

in which

N = number of atoms of the radioactive isotope

t = time

λ = decay constant

Integration yields the number of radioactive atoms at t = 0. It is thus possible to calculate the number of radioactive atoms remaining at each moment:

$$N = N_o \cdot e^{\lambda t} \quad \text{(E-9.2)}$$

half life

This exponential equation states that during a fixed time interval radioactivity will always decrease by the same fraction. An easy fraction to work with is 1/2. The time interval during which the number of radioactive atoms decreases to half the number present at the beginning of that time interval is called the half life ($t_{\frac{1}{2}}$) of a radioactive isotope. From equation 9.2 we can deduce that:

$$t_{\frac{1}{2}} = -\frac{\ln 0.5}{\lambda} = \frac{0.693}{\lambda} \quad \text{(E-9.3)}$$

Table 9.1 give the $t_{\frac{1}{2}}$ values for some commonly used isotopes.

isotope	emission	E_{max} (MeV)	half life
^{3}H	β	0.018	12.3 years
^{14}C	β	0.155	5568 years
^{32}P	β	1.71	14.2 days
^{35}S	β	0.167	87 days
^{125}I	γ	0.35	60 days

Table 9.1 Properties of some commonly used isotopes.

The half life is important for several reasons. In the first place the half life value sets a limit to the specific (radio)activity, this is the number of atoms decaying per unit mass per unit of time. Moreover, for atoms with a small half life value decay of radioactivity during the experiment has to be taken into account.

Radioactivity is expressed in the unit Curie (Ci) or Bequerel (Bq).

The Curie is a traditional term which is progressively being replaced by Bequerel. Bequerel is, as you will see, much easier to work with.

cpm, dpm

The Curie is defined as the number of disintegrations per second of a gram of radioactive sodium and is equivalent to $3.70 \cdot 10^{10}$ disintegrations per second. A Bequerel is equivalent to 1 disintegration per second; thus 1 Ci is equivalent to $3.70.10^{10}$ Bq.

It is important to realise that the Curie or Bequerel refer to the number of integrations really occurring in a sample. Most detection methods, however, do not detect (count) all disintegrations. We can express radioactivity as the number of disintegrations measured per unit time; thus expression is often in counts per second (cps) or counters per minute (cpm). For quantitative comparison cps or cpm values must be converts to the actual number of disintegrations per unit of time (dps or dpm). In order to be able to do this the efficiency of the counting apparatus must be known.

carrier free, specific activity, radiochemically pure

Radio isotopes are rarely pure; mostly they are diluted with chemically identical non-radioactive isotopes. A chemically pure isotope is called carrier free. The relative amount of a radioactive isotope is represented by the specific activity, this is the number of disintegrations per unit of time per unit of mass of a specific element or compound. Radioactive chemicals are mostly assumed to be radiochemically pure, which means that only one radioactive isotope is present. In practise this is not always true; especially when a compound has been stored for some time undesired side products may have arisen because of decay reactions. Therefore, radioactive compounds must be carefully stored and, if needed, be purified before use.

SAQ 9.1

A sample contains 10^{-3} Curie ^{32}P.

1) How many disintegrations per second will take place?

2) Approximately how many disintegrations per second will be registrated 8 weeks later? (Use data from Table 9.1).

9.3 Methods of detection

When we want to detect radioactive decay we must find a way to measure the interaction of the emitted radiation with the environment. Mostly this is done in one of the following ways:

- ionisation of a gas;
- excitation of a liquid or a solid;
- activation of a photographic emulsion;

When using radioactive materials for studying biological systems the second and third method are most often used.

9.3.1 Gas ionisation counting

A beta particle moving through a gas can 'push' an electron out of a molecule upon collision. As a result of this two charged particles arise in the gas: a positively changed ion and an electron. Beta particles usually have sufficient energy to ionise many molecules. When this occurs in a closed vessel in which two electrodes are placed the ions are attracted by the electrodes where they are discharged. A beta particle passing through the gas will thus be accompanied by a small current between the electrodes. Each beta particle can give rise to many hundreds of ions, but these do not all reach the electrodes. Very often recombination of the positive ion and the negatively charged electron will occur immediately after passage of the beta particle. Obviously the degree

generation of ions by beta particles

to which the ions reach the electrodes depends on the potential (voltage) difference between the electrodes: the greater the potential difference the greater the chance that the ions will reach the electrodes before recombination with electrons occurs.

influence of voltage on ionisation

Figure 9.2 illustrates how the number of ions collected on the electrodes depends on the voltage between the electrodes. We can discriminate several voltage regimes. At low voltages most ions recombine before they reach one of the electrodes. With increasing voltage more and more ions reach the electrodes until saturation is reached. In this voltage region all ions formed by the beta particle in the gas reach the electrodes (primary ionisation region). When the voltage increases further the electrons liberated from the gas will be accelerated in the electric field to such an extent that they in turn will ionise gas molecules; a process of amplification thus occurs. Over a certain voltage range the number of secondary formed ions received by the electrodes is proportional to the number generated directly in the gas by the beta particle. This voltage region is called the proportional region. Finally a second saturation region occurs in which both the primary and the secondary ions are all collected on the electrodes. This region is called the Geiger Muller region. If the voltage becomes so high that gas ionisation occurs in the absence of beta particles, we are in the region of continuous discharge.

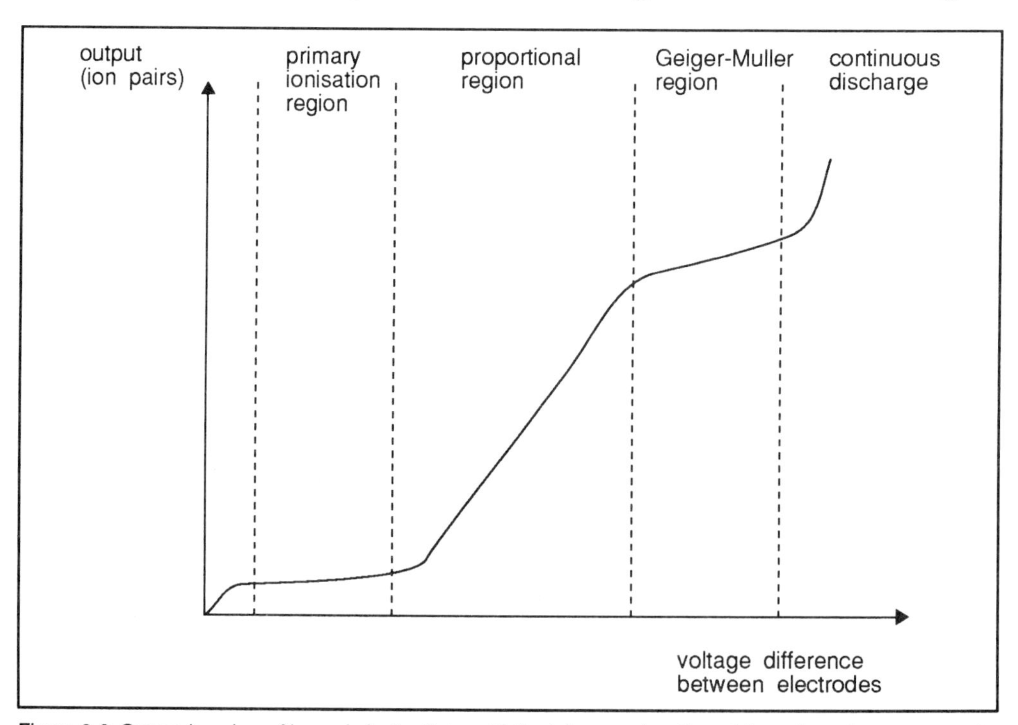

Figure 9.2 Output (number of ion pairs) of a Geiger Muller tube as a function of the voltage between anode and cathode.

In detectors based on gas ionisation a voltage in the proportional region or in the Geiger Muller region is used (gas ionisation counters).The advantage of proportional counting is that particles with different energies can be discriminated by analysis of the pulse height. In this region the size of the pulse, caused by one beta particle is proportional to the energy of the particle. In principle accurate analysis of the energy spectrum also allows for analysis of a mixture of two radioactive isotopes. A problem with proportional counters is that they require an extremely stable voltage source, because small changes in voltage on the electrodes cause large changes in the current. For this reason proportional counters are not frequently used for accurate counting. They have been largely replaced by liquid scintillation counters (counters in which scintillations in a liquid are measured; see following section). In the Geiger-Muller region the voltage on the electrodes is so high that maximal amplification is reached; under these conditions, voltage fluctuations will not cause a large effect.

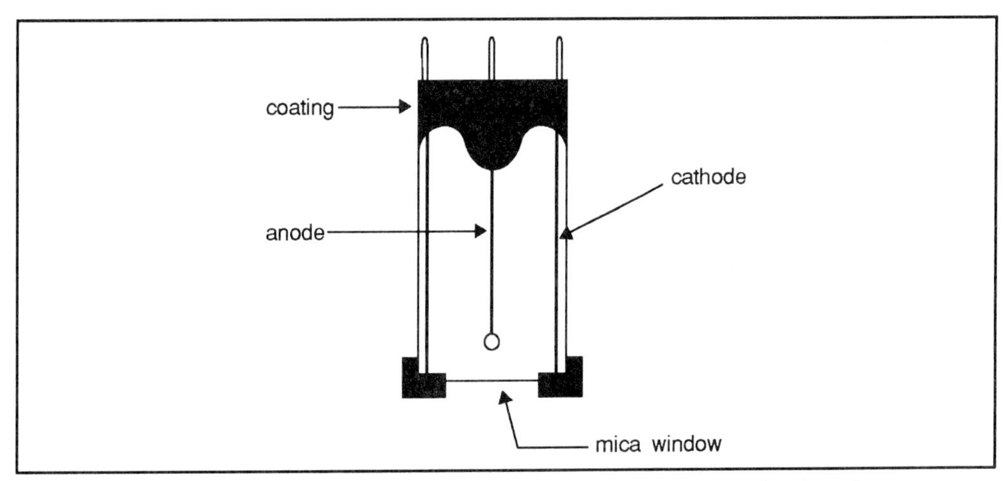

Figure 9.3 Geiger-Muller counter. (Note the coating of the tube has been removed to show the arrangement of anode and cathode - see text for details).

Geiger-Muller counters

Geiger-Muller counters are based on this principle. They are stable, sensitive and relatively cheap counting apparatus. Figure 9.3 shows an often used form of a Geiger-Muller counter. The detector consists of a cylinder with a metal inner surface, serving as the cathode. The wire in the middle serves as the anode. The entrance of the tube consists of a mica or plastic window. The beta particle enters the gas chamber through this window. The gas used mostly is an inert noble gas like helium, argon or neon. The voltage on the electrodes is adjusted to the Geiger-Muller region (about 1000V).

dead time

The pulses between the electrodes can be measured and each pulse corresponds with the passage of a beta particle. The limit to the number of particles to be measured is set by the dead time of the counter. When two beta particles enter the gas chambers very shortly after each other they will be registered as one pulse. With most Geiger-Muller counters it is possible to detect 10^5 counts per minute.

9.3.2 Scintillation counting

Although the Geiger-Muller counter is an efficient apparatus, there are some important limitations to its use. In the first place because of the geometry of the tube not all beta particles will give rise to ionisations.

Even if the sample is placed on the window half the number of the beta particles will move away from the tube. In addition a fraction of the beta particles may enter the tube through the window, but then have insufficient energy to cause ionisations.

SAQ 9.2

We have two samples, one containing 10^{-6} Curie ^{32}P and one 10^{-6} Curie ^{3}H. Which of the two samples will give the greatest number of counts per minute when using a Geiger-Muller counter?

These problems are largely overcome in liquid scintillation counting. In this method the radioactive preparation is added to a solution or suspension containing one or more fluorescing compounds. The particles arising during radioactive decay cause light pulses that will be registered and counted by a photomultiplier. In most solvents the energy of a beta particle is absorbed in such a way that heat is generated or ionisation of the solvent molecules occurs. The solvent molecules can also be brought into an excited state by the beta particles. In principle the excited solvent molecules will fall back to the ground state with the emission of a photon.

use of fluors

Usually the wavelength of the emitted photons is too short to be detected by the photodetectors commonly used. In those situations a small amount of a fluorescing compound (a fluor) is added to the solvent. This compound will absorb the photons emitted by the solvent molecules and will in turn emit photons of a longer wavelength. These photons can be registrated and translated into a pulse by the photodetector. In practice often a second step is added to the process: the photons from the first fluor will be absorbed by a second fluor and thus be translated into emissions of even a longer wavelength. This is to ensure that the wavelength of the emitted lights matches the absorption maximum of the photomultiplier tube. Sometimes the terms primary scintillant and secondary scintillant are used to describe the two fluors used in the 'cocktail' (solvent plus fluors) used in scintillation counting.

background noise

Each beta particle give rise to a light flash, but because this has a low energy only a small electron pulse will arise in the photomultipliers. Therefore, in liquid scintillation counting background noise in an important problem. The photomultiplier noise is the result of heat fluctuations. Photomultiplier tubes emit small pulses of electrons as a result of thermal processes and these represent a 'background noise'. To reduce this problem as far as possible liquid scintillation counters are often cooled to about 5°C. A second way to suppress background noise is to use a coincidence circuit. This consists of two photomultipliers each detecting light flashes. However, pulses are only counted when both photomultipliers detect a pulse at the same time. The reasoning is that pulses caused by noise will only be generated in one photomultiplier tube at a time. Thus these pulses are not counted. Flashes of light caused by a beta particle will give rise to many photons and these photons will enter both photomultiplier tubes giving rise to coincident pulses and these are counted.

^{40}K in glass

Other causes of background noise are usually less important. Normal glass contains for example a rather large amount of ^{40}K (an unstable, radioactive isotope) that gives rise to a certain amount of background. Therefore, in experiments requiring a very low level of background noise counting vessels with a low K content are used. One of the most important advantages of scintillation counting is the fact that the size of the signal is proportional to the energy of the particle causing the light flash. This in principle enables simultaneous counting of more than one isotope in the sample. For that purpose electronics are used that determine the size of each flash and subsequently sort the flashes into energy intervals. This yields an energy spectrum like that shown in Figure 9.4.

gates

If for instance we have a high energy emission, we will get a strong flash of light. This will contain many photons which in the photomultiplier tube will be converted to a large pulse of electrons. For a low energy emission only a small pulse of electrons will be generated. We can electronically distinguish such pulses. We do this by setting "gates" which only allow pulses of a particular size. In the situation depicted in Figure 9.4 we have set three gates (L1, L2 and L3).

L1 is a lower gate. Pulses arising from very low energy emissions are small and cannot pass the lower gate. Pulses arising from emissions of intermediate energy levels are large enough to pass gate L1 but not large enough to pass gate L2. These are counted in channel A in Figure 9.4. Only high energy emissions produce pulses which can pass gate L2 and these are counted in channel B. Thus in channel B we will only count some of the ^{14}C emissions. In channel A we can measure ^{3}H and some ^{14}C emission.

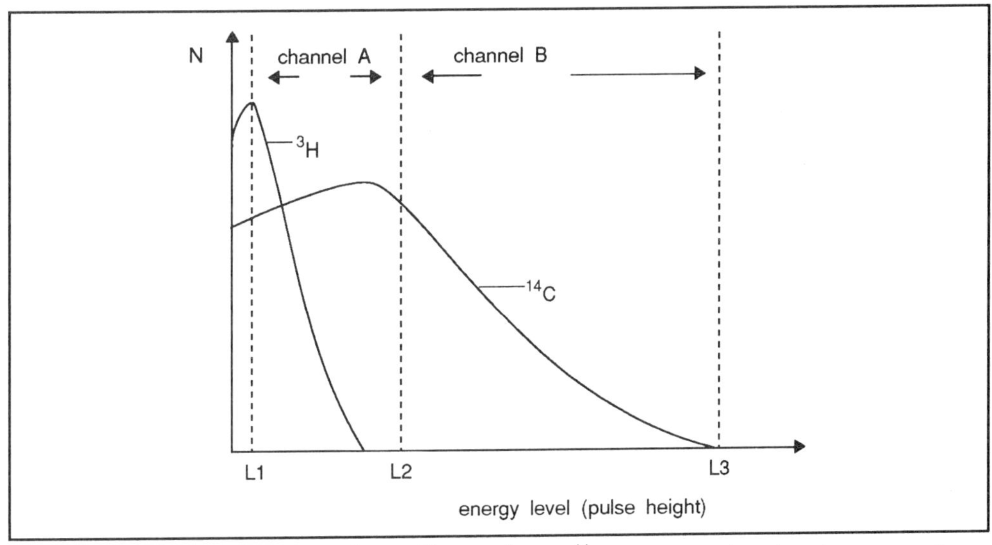

Figure 9.4 Simultaneous counting of ^{3}H and ^{14}C. The ^{3}H and ^{14}C spectra are plotted on the same scale. L1 is the lower limit of most counters, below this limit the thermal noise is appreciable; L2 and L3 are the 'gates' (voltages) used in the example described in the text. The vertical axis is the number of registrations counted at a certain energy level (see text for further description).

In one channel only pulses between L1 and L2 are counted and in the other channel only pulses between L2 and L3. As we have said, in channel B (L2-L3) only ^{14}C disintegrations are counted. Because of the cutting off at the low energy side, however, not all ^{14}C disintegrations are registered. The efficiency of counting therefore is lower than 100%. On the other hand in channel L1-L2 most of the ^{3}H and part of the ^{14}C disintegrations are counted. If we know the shape of the ^{14}C energy spectrum we can calculate the number of counts in channel A caused by ^{14}C and we then know how many ^{3}H registrations there must have been. In practice usually a ^{14}C standard is used to establish what percentage of the ^{14}C disintegrations will be detected in the low energy channel. A skilful channel adjustment will even allow for simultaneous counting of three isotopes whose energies do not fall too close together.

SAQ 9.3

If we want to simultaneously count ^{14}C and ^{3}H in a sample using liquid scintillation counting, in what ratio should we use the two isotopes?

One of the largest problems in liquid scintillation counting is the phenomenon of quenching. We mean by this all those processes resulting in a decrease in the efficiency of energy transfer in the steps leading from the beta particle to a signal in the photomultipliers. There are four important kinds of quenching.

- chemical quenching: this type of quenching is caused by compounds interfering with the transfer of energy between solvent and fluor. Important chemical quenches are oxygen and polar compounds like water and salts;

- dilution quenching: the concentration of the fluor in the solvent determines the probability of energy transfer; dilution of the solution will result in a lower light yield;

- colour quenching: in a coloured sample some of the photons will be absorbed before reaching the photomultipliers;

- optical quenching: all factors decreasing the optical transparency of the sample, like precipitates or phase separations, decrease the light yield.

quench correction spiking

There are several methods available for correcting for quenching. This can be done by adding a known amount of isotope to the sample and recounting the sample, a process known as spiking. This enables us to work out the efficiency of counting of our sample. The other ways involve making use of the fact that the presence of a quenching agent causes the energy spectrum of an isotope seen by a scintillation counter to drift to a lower level. We can represent this by

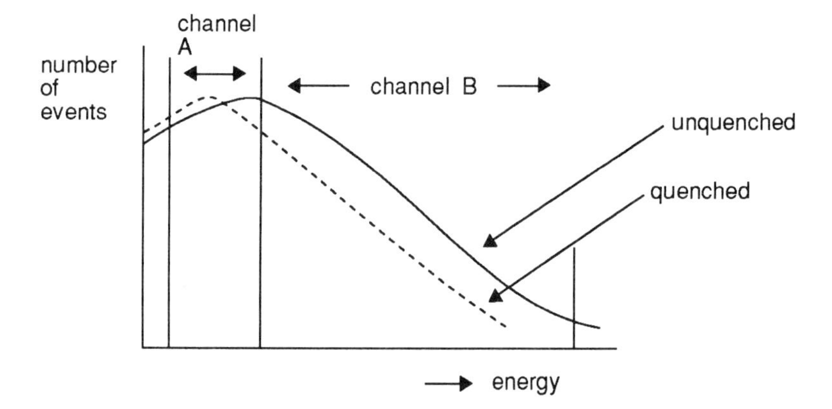

channels ratio

Can you see that the ratio of the numbers of disintegrations detected in the two channels (A and B) changes in the sample when it is quenched? In principle, therefore, we can make use of this information to allow for the effects of quenching. Details of the methods for doing this are given in the appropriate texts listed in the suggestion for further reading given at the end of this text.

NaI crystals

Detecting gamma radiation. In the case of radioactive isotopes emitting gamma radiation we nowadays often use a solid external detector. The sample in a glass or plastic tube is placed in a well in a large NaI crystal. This NaI crystal has been activated by the addition of thallium. Because of their high energy, the gamma rays can leave the sample tube and detach electrons from the atoms of the crystal. These electrons in turn excite neighbouring parts of the crystal which results in fluorescence. This fluorescence can be detected by a photomultiplier. Just as in liquid scintillation counting light pulses are translated into voltage pulses that are registered. In contrast with liquid scintillation, counting of gamma radiation with the crystal method does not need an electric circuit to eliminate background noise. The gamma radiation has so much energy that the smallest pulses can be filtered off without presenting difficulties.

Just as with liquid scintillation counting it is possible to simultaneously count more than one isotope with the NaI crystal method provided that the energy difference is large enough to see a difference in the adjustable channels.

In all counting methods we must take into account two kinds of inaccuracy:

- a statistical error in counting. Because radioactive decay is a statistical process, a certain fluctuation in the number of disintegrations per unit of time will occur. The more disintegrations are collected, the smaller the error. Mostly we allow for an error of plus or minus the square root of the number of registered counts. If we have detected 100 counts, the statistical error will be 10 counts (= 10%). Note that this is only an approximation but it is quite a useful rule of thumb. In the calculation of the statistical error we use the total amount of counts; the background noise is thus given a similar statistical effect as the fluctuation of the radioactive decay;

- the background noise. This is usually determined by counting a sample to which no radioactivity has been added. The value of this counting (often in the order of 10-100 counts per minute) is subtracted from the values found for the other samples.

SAQ 9.4

A liquid scintillation counter measures ^{14}C containing samples with an efficiency of 50% and a background of 100 counts per minute. How many Bequerels of ^{14}C radioactivity must be used per sample to obtain an accuracy of 1% for the radioactivity determination in less then 10 minutes?

9.3.3 Autoradiography

In autoradiography, radioactive materials are placed on a photographic emulsion. The radiation emitted during decay of the radioactive isotope leads to activation of individual silver halide grains in the emulsion and, therefore, these will precipitate as silver grains upon development.

The pattern of precipitated silver in the developed photographic emulsion can be examined, possibly with a microscope. The precipitated silver tell us something about the site at which the radioactive isotope was present in the sample and about its quantity because this is proportional to the staining intensity of the emulsion.

Mostly isotopes emitting beta radiation are used for autoradiography. The energy of the beta radiation can vary from low (^{3}H) to high (^{32}P). The higher the energy of the beta particle, the longer the trace of silver grains will be in the emulsion.

Π Why will the length of the trace increase with the energy of the beta particle?

Of course more energy available will mean more silver halide grains activated. This influences the accuracy of identifying the exact position of the radiation source. For many purposes, however, this is not of practical importance. When the radioactive preparation is placed on top of a photographic emulsion only half of the disintegrations will contribute to the trace of precipitate, namely only those beta particles that move in the direction of the photographic plate. The efficiency of detection is lowered further by the fact that in thicker preparations part of the radiation is already absorbed in the preparation itself. Of course this is more serious in cases when relative weak radiation is used than in cases when hard radiation when used.

9.4 Examples of the use of radioactive isotopes

9.4.1 Ligand binding

When we want to determine the dissociation constant of a ligand - protein complex it can be easy to use radioactive ligand. This enables us to determine the amount of radioactivity bound, after separation of free and bound ligand.

equilibrium dialysis method

An example of this method is the so called equilibrium dialysis method. In this application a protein is placed in a dialysis tube, together with the radioactively labelled ligand for which we want to determine the dissociation constant. The solution outside the dialysis tube also contains ligand, but no protein (see Figure 9.5).

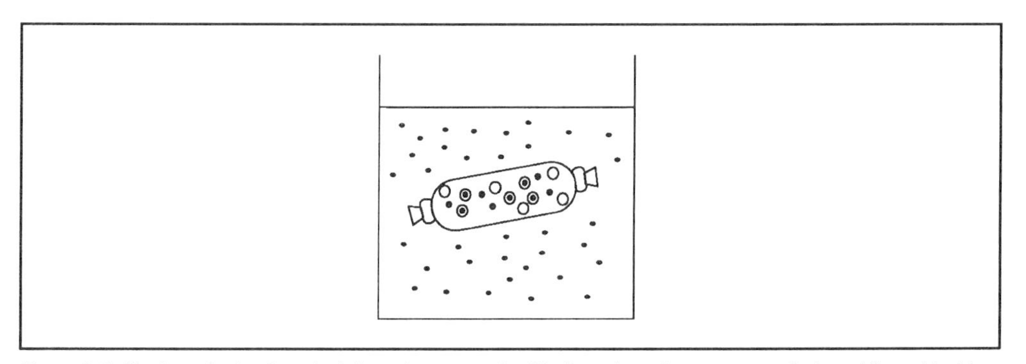

Figure 9.5 Binding of a (radioactive) ligand to a protein. Binding gives rise to accumulation of ligand inside the dialysis tube. The protein is indicated by open circles, the ligand by dots.

After a certain time, equilibrium will be reached in the dialysis system. Under equilibrium, the concentration of free ligand inside and outside the dialysis tube will be equal. However, inside the dialysis tube there will also be bound ligand. Therefore, determining the amounts of ligand inside and outside the tube will enable us to calculate the amount of ligand bound.

In this situation a simple determination of the radioactivity of samples taken inside and outside of the dialysis tube will be sufficient to determine the relative concentrations of the ligand, provided that we know the specific activity of the ligand (that is the amount of radioactivity in a known amount of ligand). By repeating the experiment at different concentrations of the free ligand, we can experimentally determine the dissociation constant.

photoaffinity labelling

A method often used to determine binding sites of compounds to enzymes is photoaffinity labelling. This can be used for the determination of both binding sites of substrates (in the active centre of an enzyme) and of binding sites of activators or inhibitors. The principle is as follows.

An analogue of the compound to be tested is made which contains a photolabile group, this is a group that decomposes under the influence of light while leaving behind a reactive group. An example is 8-azido-ATP. Under the influence of (UV) radiation this compound decomposes into N_2 and a very reactive radical: nitreno-ATP. The nitrogen radical, having a free electron, reacts with a neighbouring functional group with the formation of a covalent bond. Reaction can occur with C-, N- or O atoms for example in a polypeptide chain or protein. If we incubate 8-azido-ATP with a protein that has a specific binding site for ATP, a large part of the 8-azido-ATP will be bound. Exposure to UV radiation will give rise to formation of the nitreno radical and this will subsequently react with neighbouring groups in the polypeptide chain.

locating binding

If such an experiment is carried out with a radioactively labelled 8-azido-ATP containing for example ^{14}C or ^{3}H then we can conclude from the amount of radioactivity incorporated how much ATP is bound. However, we still can go one step further. Hydrolysis of the protein with protein splitting enzymes will enable us to find out the localisation of the binding site in the polypeptide chain.

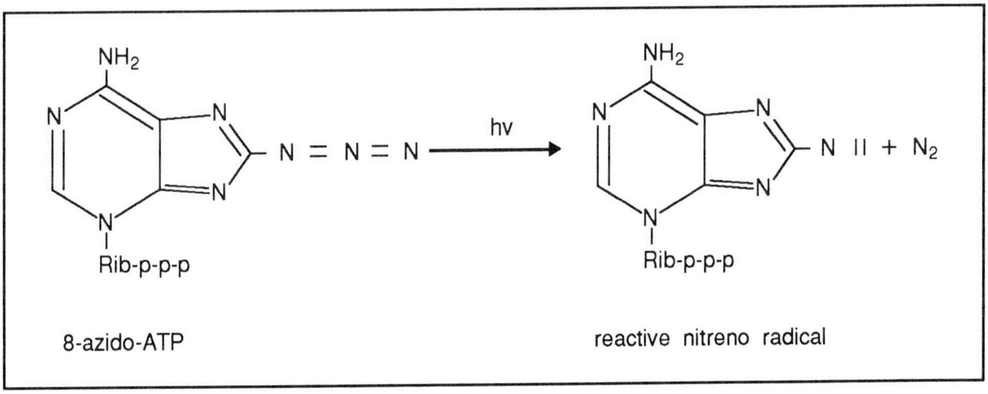

Figure 9.6 8-azido-ATP and its conversion to the active nitreno radical.

9.4.2 Determination of enzyme activity

exchange reactions

In some cases the use of radioactive isotopes is the only way to determine the activity of an enzyme. This is for instance so in the so-called exchange reactions. An example is the ATP-P_i exchange reaction. Such a reaction is catalysed by different (combinations of) enzymes. Enzymes involved in the reaction.

$$\text{ATP} + \text{Enzyme} \rightleftarrows \text{Enzyme-ADP} + P_i$$

are capable of bringing about ATP-P_i exchange. We measure this by determining the incorporation of ^{32}P from inorganic phosphate into ATP in a medium in which cold (= unlabelled) ATP and labelled $^{32}P_i$ are present.

^{32}P labelled ATP

In the reaction from the left to the right ATP is split and P_i is liberated. The P_i liberated will mix with $^{32}P_i$ present in the medium. In the back reaction from the right to the left ^{32}P will be incorporated into ATP. Inorganic and organic phosphate can be separated by extraction of a complex of phosphate and molybdate in an organic solvent. By subsequently measuring the radioactivity in the aqueous and the organic phases we can determine how much radioactively labelled phosphorous has been incorporated into ATP. Enzymatic reactions of the type described in the equation above are also used to label the terminal phosphorous group in ATP with ^{32}P. In this case carrier free $^{32}P_i$ is used to obtain the highest possible specific activity of ^{32}P labelled ATP.

9.4.3 Transport in cells or particles

Radioactively labelled compounds are often used to measure the uptake of compounds in cells. Strictly speaking this can only be done if the compound taken up is not metabolised in the cell. For if the compound is metabolised part of the radioactivity is incorporated into other compounds (examples of deliberate use of this possibility will be dealt with in a later section).

In transport studies cells are incubated with the compound to be taken up and at different times samples are taken and subsequently cells and medium are separated. This can be done by centrifugation or filtration. After this separation step the uptake of the radioactive compound can be determined by measuring radioactivity in the cells (in the pellet or on the filter) or in the medium (the supernatant or the filtrate). If the efflux of the compound that has been taken up is slow, the cells can be washed to remove the radioactivity of the adherent water. For this the cells are mostly diluted with a large quantity of medium at a low temperature to further decrease the efflux. The cells are subsequently concentrated again, for instance by filtration on centrifugation.

9.4.4 Radio-immunoassays

The radio-immunoassay (RIA) is an immunological determination that allows the highly specific determination of an antigen (often a protein) in a sample. The method is based on the competition between a radioactive and a non-radioactive antigen molecule for binding to an antibody.

In a method often used the antibody is covalently attached to a solid support (for example plastic). Subsequently an excess of a radioactively labelled antigen is added; this will result in binding of the antigen to all binding sites. After incubation, the solid support and the liquid are separated and the amount of radioactivity bound is determined. The experiment is repeated with labelled antigen diluted with different concentrations of unlabelled antigen. The dilution of the radioactive antigen with unlabelled antigen will of course result in a decrease in the amount of radioactivity bound. In this way a standard series is obtained. (see Figure 9.7).

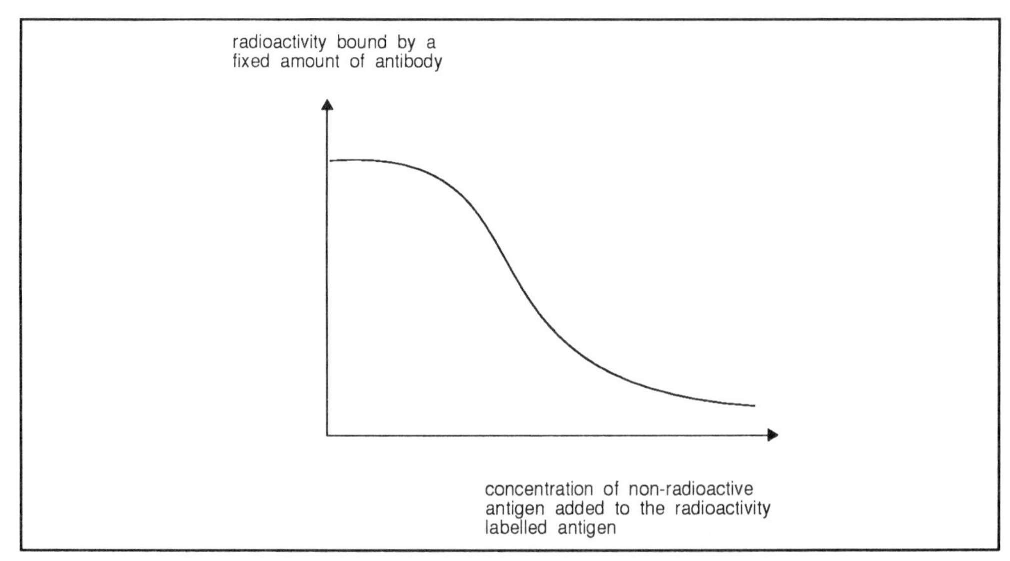

Figure 9.7 Stylised standard curve for a radio-immunoassay (RIA). See text for discussion.

If we now use an unknown sample instead of unlabelled antigen, we can read from the standard curve how much antigen was present in the sample. The RIA- method can be used for all compounds for which specific antibodies are available. Often, but not always, they are proteins. The technique is, however, becoming rapidly replaced by ELISA (see Chapter 7).

Mostly ^{125}I is used for radioactively labelling proteins. This can be done in different ways:

- chemical: in this method Na ^{125}I is oxidised with chloramine-T (the Na salt of N-monochloro-p-toluensulfonamide). The $^{125}I^+$ formed in this reaction reacts with tyrosine and (to a lesser extent) with histidine residues in the protein. After incubation a reducing compound is added and the free ^{125}I is separated from the protein;

- enzymatically: in this method the oxidation of I^- to I^+ is catalysed by the enzyme lactoperoxidase in the presence of H_2O_2. This method is less aggressive from a chemical point of view, but has a lower yield than the chemical method, therefore lower specific activities are produced;

- via conjugates: in this method the proteins are not directly iodinated but iodination takes place via an intermediate. With the help of the so-called Bolton-Hunter reagent (N-succinimidyl-3-(4-hydroxy-5-[^{125}I]) iodophenge-propionate) the NH_2 groups of a protein can be supplied with an iodine containing covalently bound group. The advantage of the latter method is that proteins not containing tyrosine or histidine residues can also be labelled.

9.4.5 Labelling

When we want to follow the fate of a definite metabolite in a cell, we can add the compound to the cell in a radioactive form. From the appearance of radioactivity in different compounds in the cell we can draw conclusions about metabolic pathways. In the past many metabolic pathways have been elucidated by following the fate of added radioactive compounds. Presently the method particularly plays a role in studying the synthesis and degradation of proteins and nucleic acids. An isotope often used for studying protein synthesis is ^{32}S, incorporated into methionine. We can use this amino acid in cell free systems, for example with extracts of reticulocytes or wheatgerm, which both are very active in protein synthesis. In spite of their relatively high activity these systems, only produce small amounts of protein in an absolute sense. Therefore, a very sensitive detection method has to be used. Incorporation of ^{35}S methionine is suitable for this purpose. The radioactive protein products formed can be separated after the reaction using polyacrylamide gel electrophoresis in the presence of dodecyl sulphate, followed by autoradiography. In this way it is possible to determine at once the molecular mass of the products formed.

in vitro protein synthesis detected using ^{35}S methionine

We can also add radioactively labelled compounds to whole cells. Bacteria can take up these directly from the medium and then metabolise them. When the label is present in the medium for a long time (hours) the process is called continuous labelling. Because incubation for a long time will result in a large turnover of the labelled compound, a fairly high concentration of the compound has to be added. Mostly this is done by adding some unlabelled compound. This has the disadvantage, however, that the specific activity of the compound, and therefore also that of the product formed, will be relatively low. After a long time more or less all cell components may become radioactively labelled.

continuous labelling

Another method to bring about incorporation of radioactivity in a cell is called pulse labelling. In this method a compound, for instance an amino acid, with a high specific radioactivity is added to the cells during a short time. Subsequently the reaction is stopped and the different proteins are analysed for incorporation of radioactivity. Most radioactivity will be found in those proteins that were most recently synthesised. This

pulse labelling

was very much the sort of experiment carried out by Calvin and others to determine the metabolic pathway involved in CO_2 fixation during photosynthesis. In this case $^{14}CO_2$ (in the form $H^{14}CO_3$-) was incubated with algae for short periods followed by identification of the early radioactive products. Longer incubations led to the identification of later products in CO_2 fixation.

pulse, chase labelling

The incorporation of a label can also be stopped by a sudden decrease in the specific activity of the radioactive amino acid. This is done by adding a high concentration of compound. This type of experiment is called a pulse chase experiment because the cold (unlabelled) compound chases the labelled compound through its subsequent metabolism. If we use a pulse experiment using an amino acid, we can gain insight into what is happening at a definite moment in protein synthesis. The radioactively labelled proteins are (completely or partly) synthesised in the period between the addition of the label and the subsequent addition of the cold amino acid. The pulse chase method is suitable to follow the fate of a protein, after it has been newly synthesised. It allows, for instance, to follow the modification of a protein after biosynthesis. Such a modification can be chemical, for instance, splitting off a part of the protein or the covalent addition of sugar residues. However, we can also determine by means of cell fractionation if the protein moves to a different cell compartment after biosynthesis simply by following the fate of our radioactive label.

Pulse and pulse chase experiments are also carried out with DNA and RNA. In these cases labelled nucleotides are used. An example is the synthesis of DNA that can be visualised using autoradiography.

9.5 Safety aspects

Working with radioactive compounds is subject to strict safety regulations. The radioactive decay of atoms can have damaging effects on living organisms.

biological consequences of exposure

The biological effects of radioactive radiation are quite complex. In the first place energy rich alpha, beta and gamma radiation can penetrate cells and cause local ionisations, dissociations and excitations. These effects can lead to damage to essential cell components, like proteins or nucleic acids. In the latter case it is even possible that radiation damage may give rise to genetic aberrations.

It is also possible that the radioactive atoms, taken up by the body, are incorporated into cell components. Such an incorporation has as a result that the isotope now causes radiation damage within cells as soon as the atoms decay. Moreover, during radioactive decay a chemical element is converted into another. For instance, ^{32}P will be converted into ^{32}S by emission of a beta particle. As a consequence, phosphorous atoms in, for instance, a phosphate group in DNA will be converted to sulphur atoms. Thus the DNA itself is chemically changed through the radioactive decay.

For the reasons given above, limits have been set to quantity and character of radiation a human being may be exposed to. The higher the energy of the radiation of the radio isotope, the lower the activity that may be used.

Clearly we need to limit exposure to radiation. Exposure is usually measured in Röntgens (traditional) or Sieverts (currently preferred units). Of particular importance is the amount of energy absorbed. These are strictly limited by regulations. Workers and laboratories are categorised into a number of classes depending upon the amount and the nature of the radiation that is used. The regulations provide specifications for organisation and management of laboratories, working protocols, training and expertise of users and the nature of record keeping and health provisions. Before anyone engages in practical work involving isotopes, it is essential that he/she is provided with appropriate safety training. Work with radioactive compounds is only allowed for people with proper training in this field and for individuals working under their supervision. Such training is offered at appropriate laboratories (eg academic institutions, national radiological protection services).

National radiological protection service

Despite such safety issues, radioactive isotopes are extremely useful aids in the analysis of biological systems and you should anticipate that they will be used for a long time to come.

Summary and objectives

In this chapter we have examined the different ways radioactive isotopes are used in studying biological material. We subsequently described the commonly used isotopes and their characteristics, the ways these isotopes can be detected, some application of radioisotopes and briefly considered safety issues for working with radioactive compounds.

Now that you have completed this chapter you should be able to:

- describe the characteristics of radioactive isotopes that emit beta and gamma radiation;
- explain how beta and gamma radiation can be detected;
- give examples of the use of radioactive isotopes in studying biological material;
- understand the safety rules for working with radioactive compounds;
- carry out a variety of calculations using supplied data concerning radioactive isotopes.

Thermal methods of studying biological systems

Thermal methods of studying biological systems

10.1 Introduction

In this Chapter we will consider thermal methods used in the study of biological systems. Your experience will tell you that chemical reactions are accompanied by heat changes. Likewise the chemical changes brought about by living system (metabolism) are also accompanied by changes in heat.

In everyday life this is manifested by the rise in temperature of fermenting liquids, decomposing manure and other organic material, such as straw and cellulose. The heat evolved during the bacterial degradation of organic substances can have disastrous consequences, for example in the spontaneous combustion of refuse heaps as a result of over-heating. We can use the measurement of heat generation as a method of determining metabolic activity.

calorimetry

Calorimetry is a non-specific technique for the direct measurement of metabolic activity. It is a measure of the algebraic sum of the enthalpy changes of all the biological process in the cell, even if the nature of these processes is not fully understood. To increase the specificity and increase the understanding of the complex biological reactions other parameters of the system must be monitored. In some instances the non-specific nature of the method is an advantage in the study of metabolic processes. For example, changes in metabolism as a result of changed environmental conditions, revealed by thermal changes, can often be detected at lower cell populations than is possible by conventional techniques.

The principles of micro-calorimetry are straightforward; we simple measure or monitor a process by determining the amount of heat generated. We focus our attention in this chapter on the practical details of micro-calorimetry and on the application of micro-calorimetry. We have particularly used microbiological examples such as those which are encountered in biotechnological processes.

More details of the theory, practice, applications and reference to the original literature are to be found in 'Biological Microcalorimetry', Ed. A.E Beezer (1980), Academic Press, London and 'Thermal and Energetic Studies of Cellular Biological Systems', Ed. A.M James (1987), Wright, Bristol.

10.2 Measurement of enthalpy changes in biological systems

All heat changes are measured by some form of calorimetry; the exact type of calorimeter used obviously depends on the nature and size of the material under investigation.

The first calorimeter used for biological experiments was described by Lavoiser in 1780. He used an 'ice-calorimeter' to measure the total heat output of small animals, such as mice, and attempted to correlate this heat with their rate of respiration and their body weight.

You are no doubt familiar with the general principles of calorimetry, in which a temperature change, occurring as a result of some reaction or process is measured. Knowing the heat capacity of the calorimeter, eg a Dewar flask, the enthalpy change of the reaction or process can be calculated.

micro-calorimeters

There exists no well defined difference between ordinary 'macro-calorimeters', where temperature changes of the order of 1 - 2 K can easily be measured, and 'microcalorimeters', capable of detecting temperature changes of 10^{-5} to 10^{-6} K. The prefix 'micro' indicates a very sensitive instrument requiring only small sample quantities (0.5 - 2.0 cm^3) with a power sensitivity of 0.1 to 1.0 mW. It is this latter type of microcalorimeter that finds use in the study of biological systems.

Within this group there are several different types including, adiabatic, isoperibolic, isothermal and heat-conduction calorimeters; each has its own advantages and applications. Heat-conduction calorimeters are most commonly used in the study of both cellular and non-cellular systems. Our discussion will be limited to this type.

10.2.1 Principles of heat conduction calorimetry.

In this type of calorimeter, heat released (as a result of reaction) is quantitatively transferred from the reaction vessel to a surrounding heat sink, normally a metal block. The heat flow is recorded by a thermopile between the reaction vessel and the heat sink (Figure 10.1).

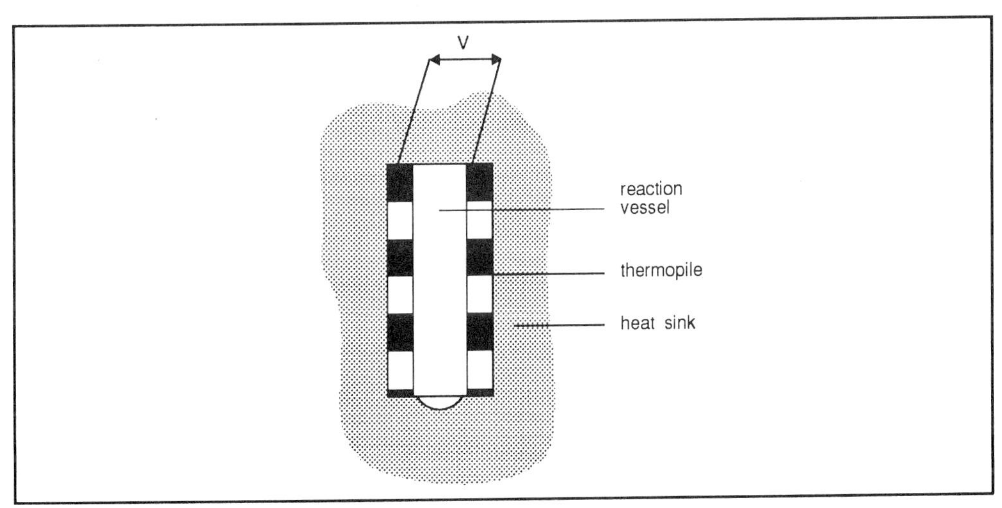

Figure 10.1 Schematic diagram of a section through a thermopile heat conduction calorimeter.

The temperature difference across the thermopile gives rise to a voltage signal, V. If there is a constant heat (q) production in the vessel, this will represent a constant power (p) output. V will reach a steady state value, V_s. The power released in the vessel is quantitatively balanced by the heat leakage to the surroundings:

$$p = \varepsilon V_s \qquad \text{(E - 10.1)}$$

ie the power is directly proportional to the signal voltage, ε is a constant determined by calibration.By measuring V we therefore have a measure of p and thus q.

Typical voltage-time or power -time (p-t) curves for this type of calorimeter are shown in Figure 10.2.

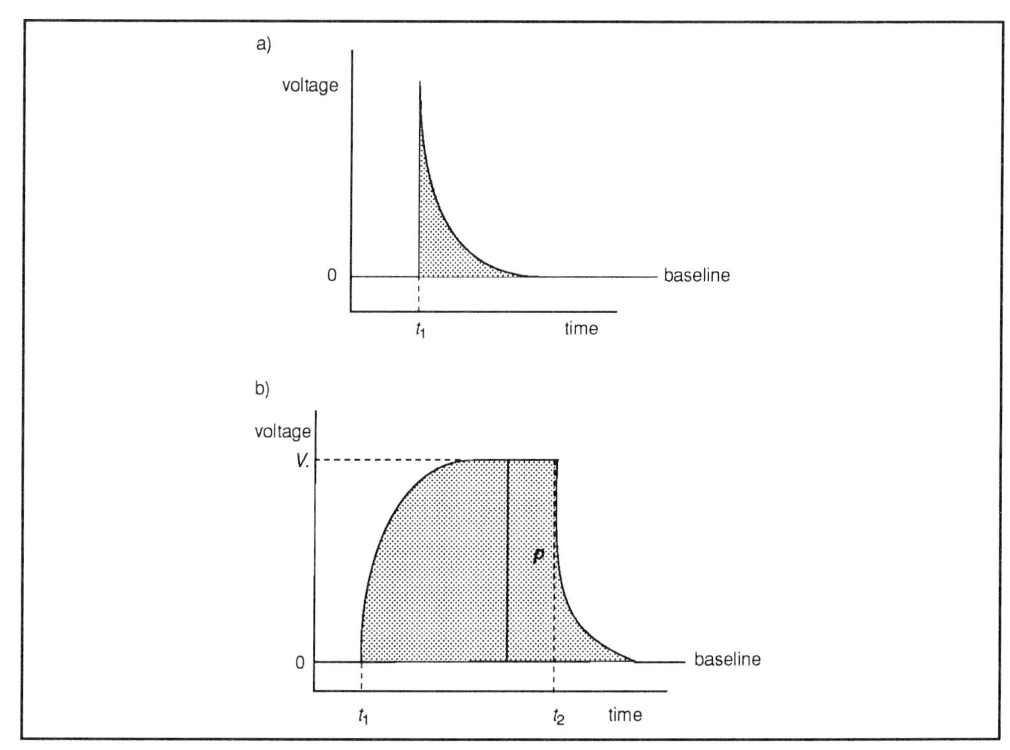

Figure 10.2 Voltage time curves from a thermopile heat conduction calorimeter. a) A short heat pulse released at time t_1. There will be a rapid change of the thermopile voltage. The heat will leak out to the heat sink and the thermopile signal returns asymptotically to its baseline value. b) A constant heat production between t_1. and t_2.. The baseline displacement during steady state conditions, V_s, is directly proportional to the thermal power, p.

Π From a study of the voltage (ie power)time curve (Figure 10.2b), can the heat output over a measured period be determined?

Yes. Power is the rate of work, in this case rate of heat output (J s^{-1}), thus the heat evolved, q, can be obtained by integrating Equation 10.1 between two times, say t_1 and t_2.

$$q = \int_{t_1}^{t_2} p \, dt = \varepsilon \int_{t_1}^{t_2} V \, dt \qquad \text{(E - 10.2)}$$

$$\text{or } q = \varepsilon A \qquad \text{(E - 10.3)}$$

where A is the area under the p-t curve between times t_1 and t_2.

10.2.2 Design of heat conduction microcalorimeters

There are several different designs of heat conduction calorimeters, the most widely used instruments are those manufactured by Seteram (Celuire, France) and LKB Produkter (Bromma, Sweden).

Figure 10.3 shows a schematic diagram of a typical twin heat conduction calorimeter. A large metal (aluminium or steel) block serves as a heat sink for two box-shaped calorimetric vessel holders. Each of these units is surrounded by semi-conducting thermocouple plates interspaced between the unit and the metal blocks forming part of the heat sink. One of these units is used for the sample and the other as a reference; the thermocouples surrounding these units are connected in opposition and so it is the differential signal that is recorded.

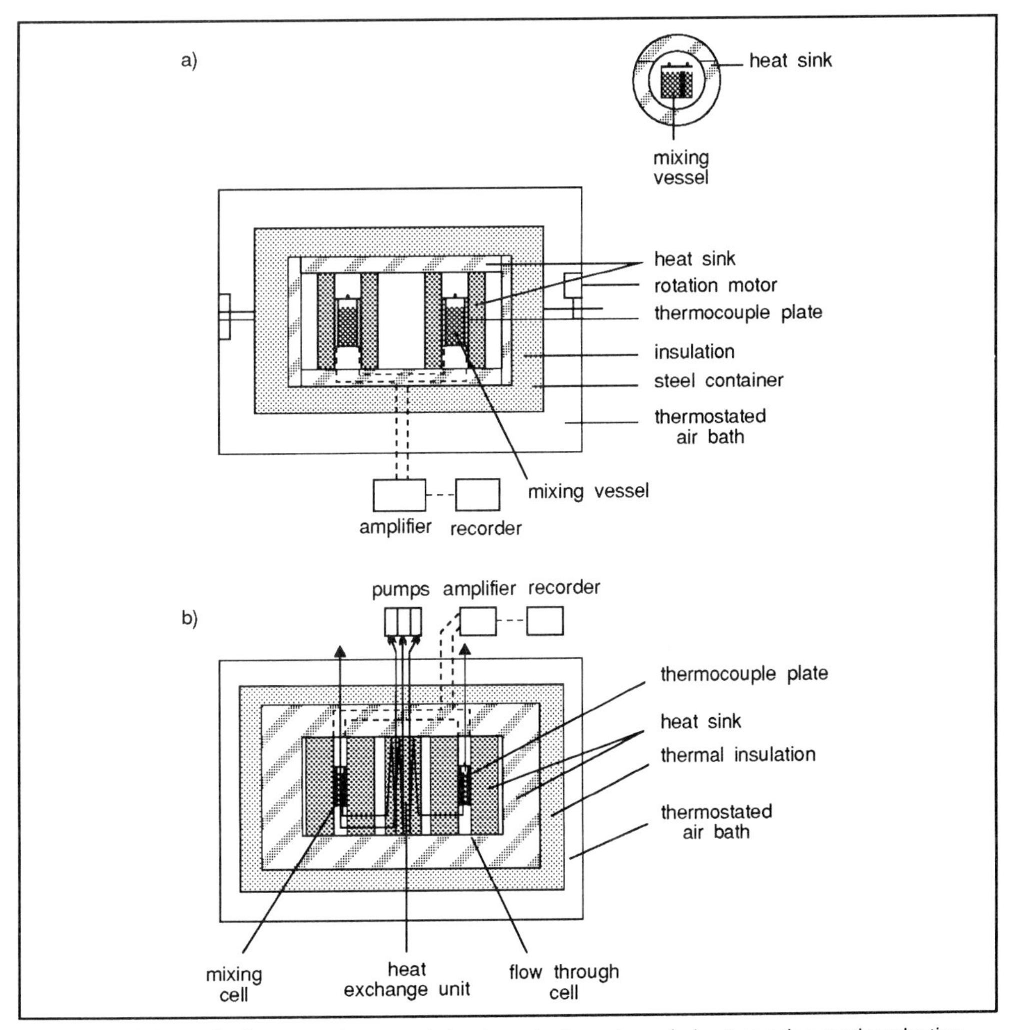

Figure 10.3 Schematic diagrams of some twin heat conduction microcalorimeters using semiconducting thermocouple plates. a) Rotating batch microcalorimeter; b) Flow microcalorimeter.(For details of the calorimeter vessels see Figure 10.4).

Π Can you see any advantage of operating two units as a differential system?

The main advantage with the twin calorimeter arrangement is that any thermal disturbances from the surroundings, which ideally influence both units identically, cancel out. This property is of particular importance in studying small heat effects during long experimental periods, such as the growth of bacteria over a period of several days. Another useful property is that the heat effect produced by one sample can be directly compared with that from a reference sample.

batch calorimeter

The heat sink , when used as a batch calorimeter, is mounted so that it can be rotated to mix the reagents. This is enclosed in a thermostated air-bath, the temperature of which is carefully controlled.

In the batch reaction calorimeter (Figure 10.3a) reagents in the two-compartmented glass reaction vessel (Figure 10.4a) can be mixed by rotation of the calorimetric block. This type of instrumentation is used for binding studies and the measurement of heats of antigen-antibody interaction.

flow calorimeter

The flow calorimeter (Figure 10.3b) is quite similar in design to the batch calorimeter. In a flow experiment one, or two, flows of liquid are pumped through the heat exchange unit and from there to one of the flow vessels. These are designed either as mixing vessels where two reagents are mixed (Figure 10.4c), for example in the study of heats of reaction between acids and bases, or as flow through vessels (Figure 10.4b, d) in which a system liberating (or absorbing) heat is studied. The simple gold tube spiral (Figure 10.4b) can be used for heat measurements on resting cells, while the aerobic steel tube (Figure 10.4d) is widely used for growing cultures where air bubbles from the fermentor, which can give rise to spikes in the *p-t* trace, may enter the calorimeter vessel.

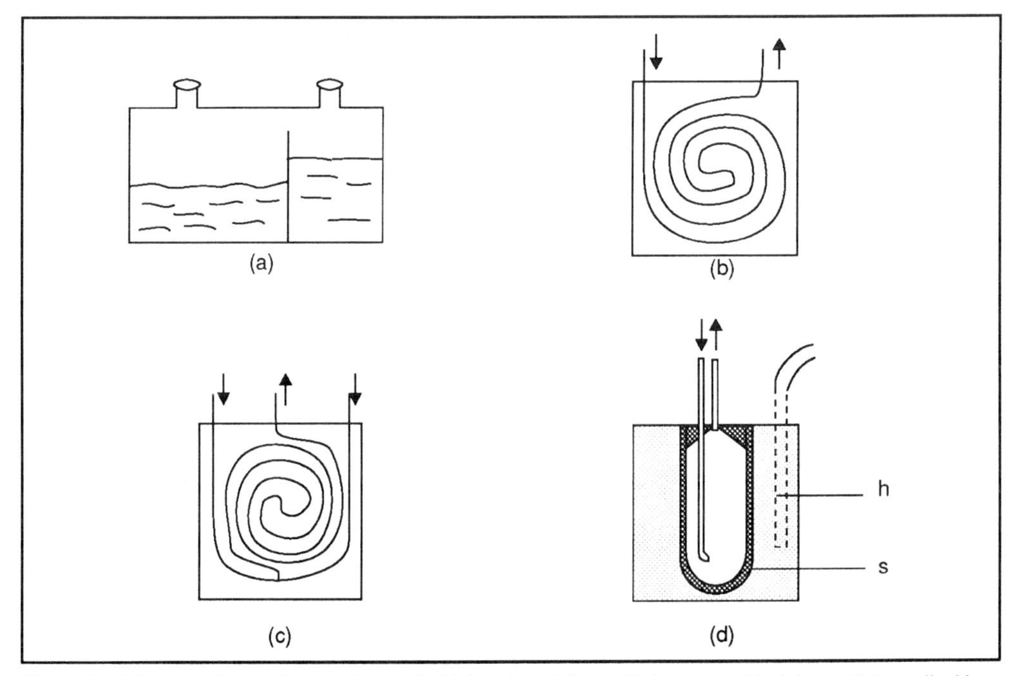

Figure 10.4 Some calorimetric vessels used with batch and flow calorimeters. a) batch or mixing cell with two compartments; b) and c) flow vessels consisting of spiral gold tubes between copper plates; d) flow through vessel for use with mixed gas-liquid flow, consisting of a cylindrical stainless steel vessel(s) inserted into an aluminium block, (h) is the calibration heater.

SAQ 10.1

List the basic components of a batch microcalorimeter and indicate their location on a simple sketch of the instrument.

10.2.3 Precautions in the use of microcalorimeters

In principle the measurement of enthalpy changes in biological systems is simple. However, in addition to the normal precautions of handling biological material, such as sterility, maintenance of the integrity of the material, the viability of the cells and the oxygen tension of their environment, other precautions are essential. The risk of significant systematic errors is naturally much greater in microcalorimetry than in experiments where comparatively large quantities of heat are dealt with. It is important to be aware of possible sources of such errors: mechanical effects (friction etc), evaporation, condensation and adsorption processes; but one must always be on guard for different kinds of artifacts. Fortunately, in practice, many of these errors will cancel out by the procedure used for standardization (calibration) of the technique.

Some of the factors to consider include:

- the careful and precise determination of the calibration constant. This is usually accomplished by applying a measured heat pulse through an electrical resistance located in the reaction vessel holder (Figure 10.4);
- the establishment of thermal equilibrium before making any measurements, by ensuring that the base-line is constant and not liable to drift during the period of the experiment;
- in batch systems, the estimation or elimination of heat changes which may occur as a result of dilution of one or both of the reactants;
- in flow systems, ensuring that the temperature of the inflowing liquid(s) is constant and at that of the block.

10.3 Some applications of microcalorimetry

10.3.1 Heat output of growing bacterial cultures

The shape and position of the power-time curve of growing cells, obtained by flow-calorimetry, depends on such factors as the nature of the organism, the size and age of the inoculum (inocula stored in liquid nitrogen provide a reproducible source), the nature of the growth medium and aeration rate. Most of these factors can be controlled.

SAQ 10.2

List the reasons why batch calorimeters are not usually used to measure the heat output of growing cultures.

The base-line is established by pumping (pump rate usually of the order 90 $cm^3 h^{-1}$) sterile medium from the fermenter, located outside the calorimeter, via polythene tubing (internal diameter = 0.1 cm) to the calorimeter and back to the fermenter (Figure 10.5). All the tubing must be as short as possible and enclosed in a water-jacket. The temperature of the growing culture, the flow lines and the calorimeter block must all be at the same temperature.

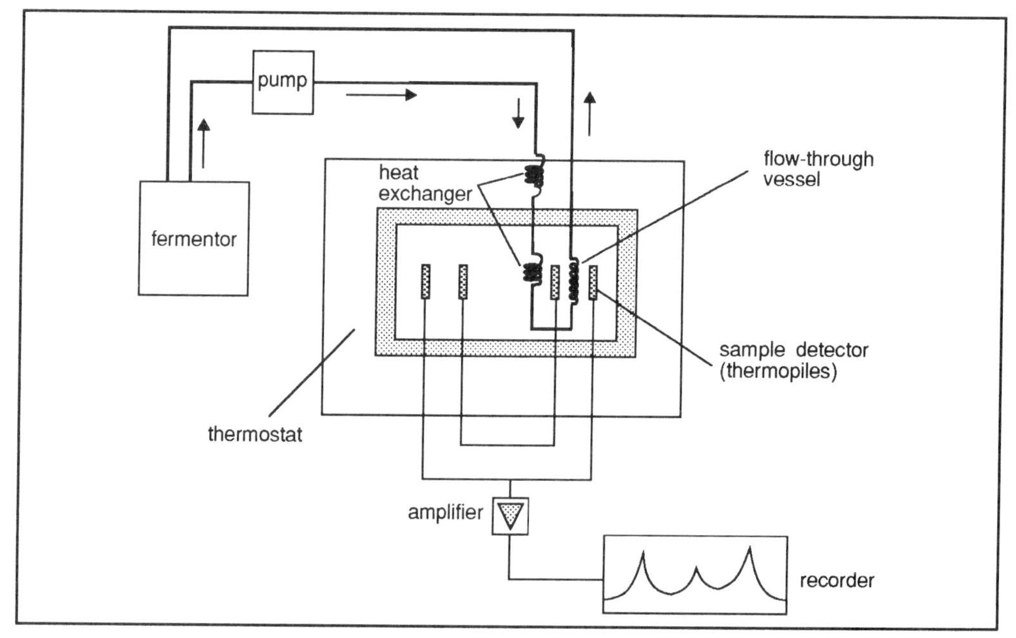

Figure 10.5 Diagrammatic representation of a flow microcalorimeter and its connections to fermentor.

Π Can you suggest reasons why it is necessary to control the temperature of the tubing at the temperature of the block and also to keep the tubing as short as possible.

There are two main reasons, the first is that if the temperature of the inflowing medium or culture differs from that in the calorimeter then the *p-t* curve will be adversely affected making interpretation of the results impossible. Secondly, changes in the oxygen tension of the culture medium in the tubing and calorimeter can result in changes in metabolism; both these factors are minimised by high flow rates and short tubing.

Using a pump rate of 90 cm^3 h^{-1}, the total transit time (ie from fermentor to calorimeter and back to the fermentor) is usually kept down to about 1.3 min and the residence time of a cell in the calorimeter vessel is 0.8 min.

Growth in simple chemically defined medium

In the first instance we will consider energy changes during the growth cycle of a simple organism, *Klebsiella aerogenes*, which grows in an aerated, chemically defined, buffered glucose-limited medium, in which glucose is the sole carbon and energy source.

Π What changes will be observed in the power when the medium is inoculated?

There will be an increase in the power as the organisms start to grow and divide. This increase occurs very rapidly, in fact almost before there is any appreciable increase in the biomass.

The power increases exponentially, with approximately the same rate constant as that for biomass increase, until the glucose is consumed (Figure 10.6). At this point growth ceases and the power falls to zero.

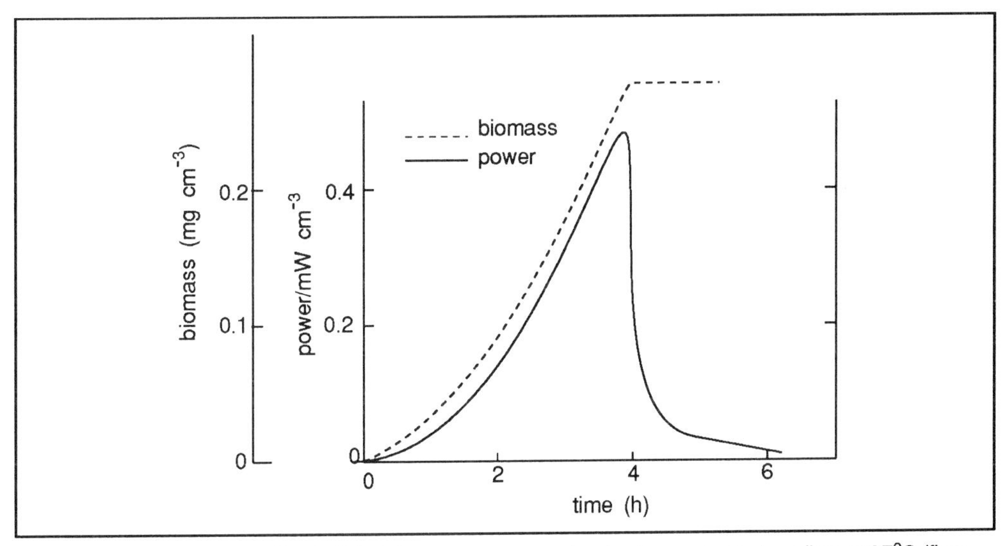

Figure 10.6 Power-time profile of *Klebsiella aerogenes* growing in glucose-limited medium at 37°C (flow rate = 90 $cm^3\ h^{-1}$).

Π Why does the power decrease so suddenly?

Quiet simply, since there is now no carbon source present all metabolic processes cease and there is no further production of heat.

total heat output

The total heat output can be determined by measuring the area under the *p-t* curve from inoculation until the cessation of growth. This can be related to the total biomass formed or to the consumption of glucose. As a typical example for *K.aerogenes*, the heat evolved for the production of 1 g of cell = 14.2 kJ or $\Delta_{sp} H$ = -14.2 kJ (g cell)$^{-1}$ and $\Delta_{met} H$ = -1067 kJ, where $\Delta_{sp} H$ is the enthalpy of production of 1 g of cells measured during the exponential growth phase and $\Delta_{met} H$ is the overall enthalpy change arising from metabolism. This is in quite good agreement with values calculated from theoretical considerations. The 95% confidence limits for these data, measured under standard controlled conditions, are ± 6%.

SAQ 10.3

Explain, giving reasons, how the *p-t* curve would differ if the medium contained excess glucose.

Growth in rich medium

The second example is that of a more nutritionally demanding organism such as *Staphyloccus aureus*; a typical *p-t* curve of this organism, growing in a nutrient broth medium, is shown in Figure 10.7. Metabolism continues long after growth has ceased due to the metabolism or fermentation of residual nutrient in the medium.

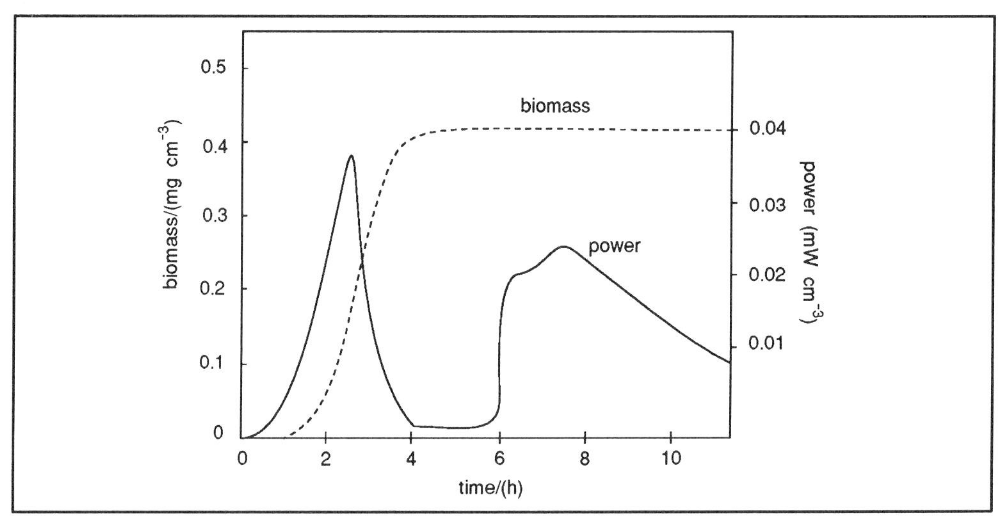

Figure 10.7 Power-time curve during growth of *S.aureus* 13137 in nutrient broth.

Detailed interpretation of the *p-t* curve, along the lines discussed previously, is obviously not possible, but nevertheless if it can be established that the curve is reproducible then it can be used as a reference in the study of metabolic changes brought about by changing growth conditions, eg the effect of antibacterials or the addition of further nutrient (amino acids, nucleotides, vitamins etc).

10.3.2 Determination of mass and energy balances during bacterial growth.

Mass and energy balances are important in biotechnology especially in the design and operation of large scale reactors. It is, for example, essential to know how much heat will be generated in the reactor in order to design suitable heat exchangers. Likewise it is essential to know how much carbon dioxide and other products will be generated. The applicationof mass and energy balances with particular reference to process technology are dealt with in the BIOTOL texts 'Bioprocess Technology: Modelling and Transport Phenomena', 'Operational Modes of Bioreactors' and 'Bioreactor Design and Product Yield'. Here we confine ourself to the application of microcalorimetry to the determination of mass and energy balances in the laboratory.

The determination of mass and energy balances requires detailed monitoring of such parameters as biomass production, carbon content of the medium, carbon dioxide output in addition to the measurement of heat production. Figure 8.8 shows schematically the full assembly of equipment necessary for such studies.

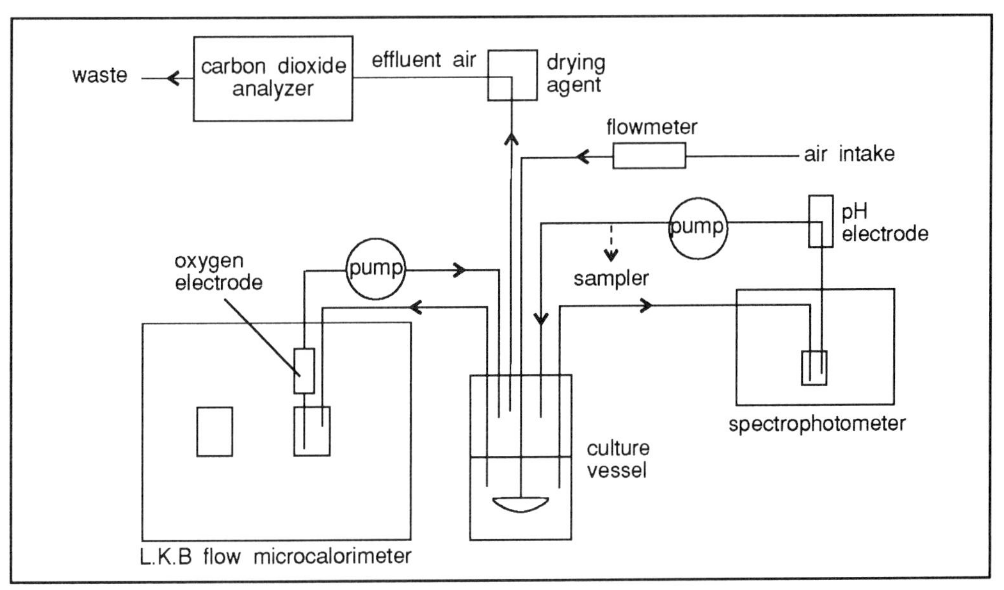

Figure 10.8 Experimental arrangement for monitoring changes of thermal and growth parameters during bacterial growth.

Π Why is it necessary to include an oxygen electrode?

The oxygen electrode is included to ensure that the culture leaving the calorimeter is not depleted in oxygen. Low oxygen tensions can result in changes in the metabolic processes, so that the metabolic processes of cells in the calorimeter may not be the same as those of cells in the fermentor.

The pH electrode is included in the assembly to check that the pH of the culture does not vary by more than 0.1 pH unit. If it is in excess of this figure then it is necessary to take into account heat changes due to the protonation of buffer components.

The total carbon in a growing culture is divided between that in the carbon source in the medium, that in the biomass and that in the carbon dioxide produced. The carbon content of the medium can be determined by the assay of small samples removed from the flow line at various times. The increase in biomass is continuously monitored using a flow cell in a spectrophotometer; a calibration plot is used to convert the absorbance to biomass, expressed as mg cm^3 (or g l^{-1}). The carbon content of the cells is calculated from the elemental analysis of dried cells.

infrared gas analyser

The effluent air from the fermentor passes, via a drying agent, to a differential infrared gas analyser, calibrated using known carbon dioxide/nitrogen mixtures. The air input is split, part going to the fermentor via a flow meter, the remainder passing directly to the gas analyser as the reference. The carbon dioxide-time curve is parallel to the power-time curve, decreasing rapidly on the cessation of growth. From the area under the carbon-dioxide-time plot, the flow rate and the time of flow the total quantity of carbon dioxide produced at any given time can be calculated.

Changes in these parameters during the growth of *K.aerogenes* in glucose-limited medium are shown in Figure 10.9. It is apparent that the rate of production of carbon dioxide and heat run in parallel; detailed analysis shows that the rate constants for both these processes and that for biomass production are identical. When the glucose has been completely consumed there is a marked fall to zero in both outputs. The specific power, that is the power per unit biomass, increases rapidly during the early part of logarithmic growth to a constant value, again decreasing to zero on the cessation of growth.

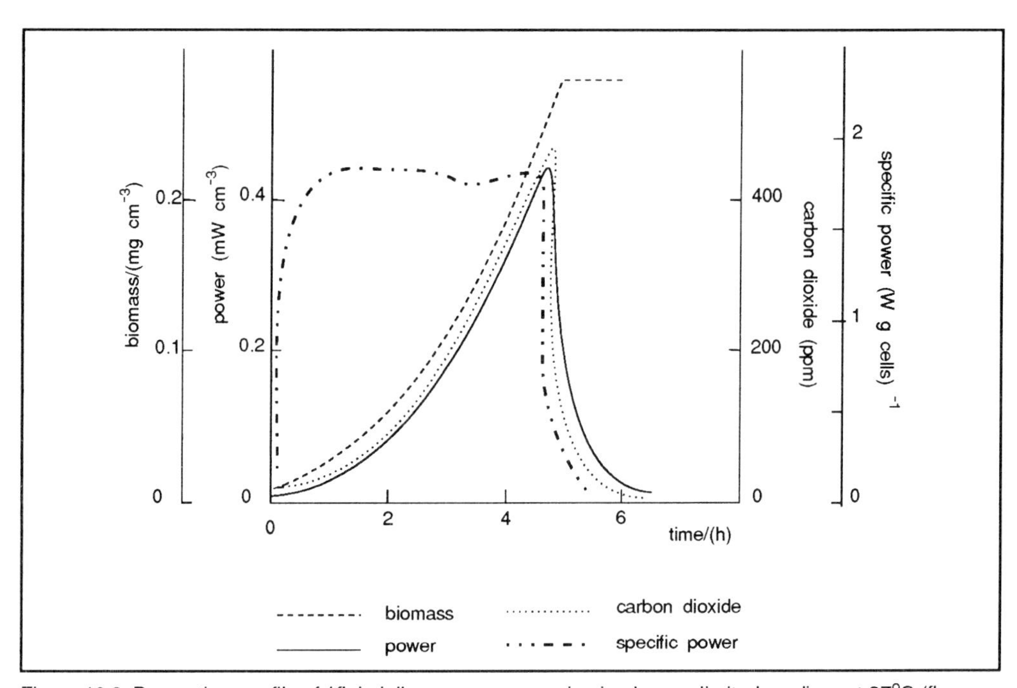

Figure 10.9 Power-time profile of *Klebsiella aerogenes* growing in glucose-limited medium at 37°C (flow rate = 90 cm^3 h^{-1}).

As an exercise in the determination of mass and energy balances try the next two SAQs.

SAQ 10.4

During the exponential growth of *K.aerogenes* in 620 cm^3 of glucose-limited medium (3.3 mmol dm^{-3}) 0.1843 g of biomass (45.1% carbon) are formed and 5.351 x 10^{-3} mol of carbon dioxide are released. Determine the mass balance for this growth and establish whether all the available carbon can be accounted for.

SAQ 10.5

During the exponential growth of *K.aerogenes* in 620 cm^3 of glucose-limited medium (3.3 mmol dm^{-3}) 0.1843 g of biomass (45.1% carbon) are formed with the release of 1.89 kJ of heat. Calculate the total energy initially available and the energy stored in the biomass and attempt to establish and energy balance; account for any deficiency. You will require some information from the results obtained in SAQ 10.4. Note that the complete oxidation of 1 mole glucose to CO_2 yields 28903J.

These calculations have revealed that, in a carbon-limited medium, it is possible to establish a mass balance. The energy balance is not so easy, there is about 11% of the total available energy which cannot be accounted for. Similar values for this deficiency in energy have been reported for a wide range of organisms; it is believed to be that used for biosynthetic and maintenance (repair) processes within the cell.

This method of determining heat and energy balances can be made on any growing culture, provided that the carbon source(s) can be continuously assayed. It is obviously an advantage if the composition of the medium is chemically defined.

10.3.3 Heat output in metabolic inhibitor studies and diagnosis

When a cell suspension in buffered glucose solution flows through the calorimeter there is a steady production of heat.

Clearly microcalorimetry can be used for measuring the heat output from cells. However, we can use changes in heat output to determine perturbations to normal metabolism. This in turn can be used to assay, for example, antimetabolities or can be used in diagnostic tests. We describe two examples below. The addition of the antifungal agent nystatin, which interferes with the metabolism of the cells, causes a decrease in the heat output by yeasts (see Figure 10.11). The extent of the reduction is related to the concentration of the nystatin. The linear relationship between the response (eg the power output after a fixed time of contact) and the logarithm of the dose provides a rapid assay for nystatin and other polyene antibiotics.

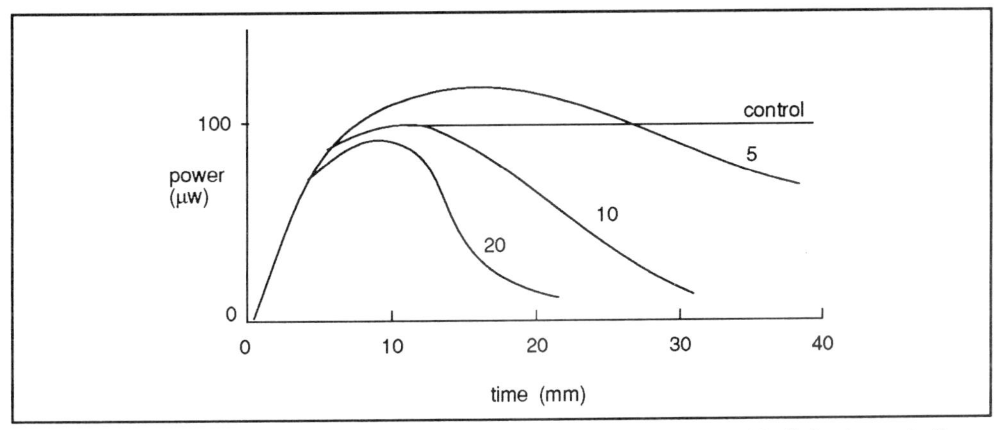

Figure 10.11 Effect of different doses of nystatin on yeast cells incubated anaerobically in glucose buffer solution (numbers are the concentrations of nystatin; units cm^{-3}).

This technique has marked advantages over the conventional agar plate diffusion test in terms of lower detection concentrations, wider range of concentrations detectable and shorter time for assay. Although different shaped *p-t* curves may be obtained for the effect of antibacterial agents on bacterial and yeast strains, nevertheless it is usually possible to establish the linear relationship between the response and log(dose). Another advantage of this technique is that it is capable of being automated.

blood cells

In a similar way blood cells suspended in plasma, serum or buffered glucose solution also exhibit a steady heat output over a period of time. The power output per cell depends on the type of blood cell and in some instances the health of the patient (Table 10.1).

Type of cell		Power output per cell (pW $cell^{-1}$)	
erythrocytes		10×10^{-3}	
platelets		59×10^{-3}	
lymphocytes		2	
granulocytes		1.3	(in phosphate buffer, pH 1.3)
		3.5	(in plasma)
human skin fibroblasts		40	
keratinocytes		40	
adipocytes	(normal)	49	
	(obese)	26	

Table 10.1 Heat production in various blood and tissue cells.

10.3.4 Binding studies using batch calorimetry

Introduction

An important and ubiquitous feature of biological macromolecules is their ability to interact with various small and large molecules with a high degree of specificity and usually with a high affinity.

Π List some interactions which are of importance.

Your list might have included enzyme catalysis, antibody-antigen specificity, hormone-receptor interaction, metal ion binding, together with a host of other protein receptor-mediated processes.

Any molecular description of such processes requires an understanding of the structural and thermodynamic details of the interaction. While the mechanisms of such processes will differ from case to case, determination and interpretation of the thermodynamics are necessary. A knowledge of the variations in the enthalpy and entropy values of the complex-formation enables us to propose models closely describing the chemical interactions on both the macromolecule and the ligand molecules themselves.

The use of the microcalorimetric technique provides a means for the evaluation of the principal thermodynamic quantities (ΔH, ΔG and ΔS) which describe the interacting system.

The binding of a ligand to a macromolecule will often be coupled to other side-reactions and the experimentally observed thermodynamic quantities will contain contributions from these coupled reactions. Means for sorting out these contributions are available but are beyond the scope of this book.

Theoretical considerations

We will focus on the equilibrium binding of a ligand, L, to a macromolecule, M, resulting in the formation of a 1:1 complex. This can be represented by the equation:

$$M + L \rightleftarrows ML$$

for which the association constant K_c can be written:

$$K_c = \frac{[ML]}{[M]\ [L]} \qquad \text{(E - 10.4)}$$

If we start with M_o concentration of macromolecule and we add L_o concentration of ligand, the we can write

$$K_c = \frac{[ML]}{([M_o - [ML])\ ([L_o - [ML])} \qquad \text{(E - 10.5)}$$

or more generally

$$K_c = \frac{[ML]_n}{([M_o]_n - [ML]_n)\ ([L_o]_n - [ML]_n)} \qquad \text{(E - 10.6)}$$

where $[M_o]_n$ denotes the total concentration of macromolecule, namely ([M] + [ML]), following the *n*th addition of ligand solution. $[L_o]_n$ is the sum of the concentrations of bound and unbound ligand following the *n*th addition. $[ML]_n$ is the concentration of the complex following the *n*th addition of ligand. The sum of the heat evolutions following the *n*th addition, Q_n can be expressed as:

$$Q_n = \Delta H\ V_n\ [ML]_n \qquad \text{(E - 10.7)}$$

where V_n is the volume of the reaction solution. Combination of equations 10.6 and 10.7 on rearrangement gives:

$$\frac{1}{K_c} = \frac{[M_o]_n\ [L_o]_n \Delta H\ V_n}{Q_n} - [M_o]_n - [L_o]_n + \frac{Q_n}{\Delta H\ V_n} \qquad \text{(E - 10.8)}$$

Π It would be good practice to derive this equation yourself. Try this before examining our solution.

From Equation 10.6 we can write

$$\frac{1}{K_c} = \frac{([M_o]_n - [ML]_n)([L_o]_n - [ML]_n)}{[ML]_n}$$

$$= \frac{[M_o]_n\ [L_o] - [ML]_n [M_o]_n - [ML]_n [L_o]_n + [ML]_n^2}{[ML]_n}$$

$$= \frac{[M_o]_n\ [L_o]_n}{[ML]_n} - [M_o]_n - [L_o]_n + [ML]_n$$

But from Equation 10.7:

$$[\,ML\,]_n = \frac{Q}{\Delta H\, V_n}$$

Thus:

$$\frac{1}{K_c} = \frac{\Delta H\, V_n\,[\,M_o\,]_n\,[\,L_o\,]_n}{Q} - [\,M_o\,]_n - [\,L_o\,]_n + \frac{Q}{\Delta H\, V_n}$$

Equation 10.8 contains two unknowns K_c and ΔH. From two sets of readings an approximate value for ΔH can be found. If a series of reasonable values for ΔH, around this figure, are inserted into Equation 10.8 the corresponding values for $1/K_c$ can be calculated, and a graph of $1/K_c$ against ΔH constructed (Figure 10.11a). If this is done for all titration steps and if the original assumption about the formation of a 1:1 complex is correct, all the curves will, in the ideal case, intersect at a single point which represents the true value of $1/K_c$ and ΔH. Due to imprecisions in the measurements the intersections may occur over a small area.

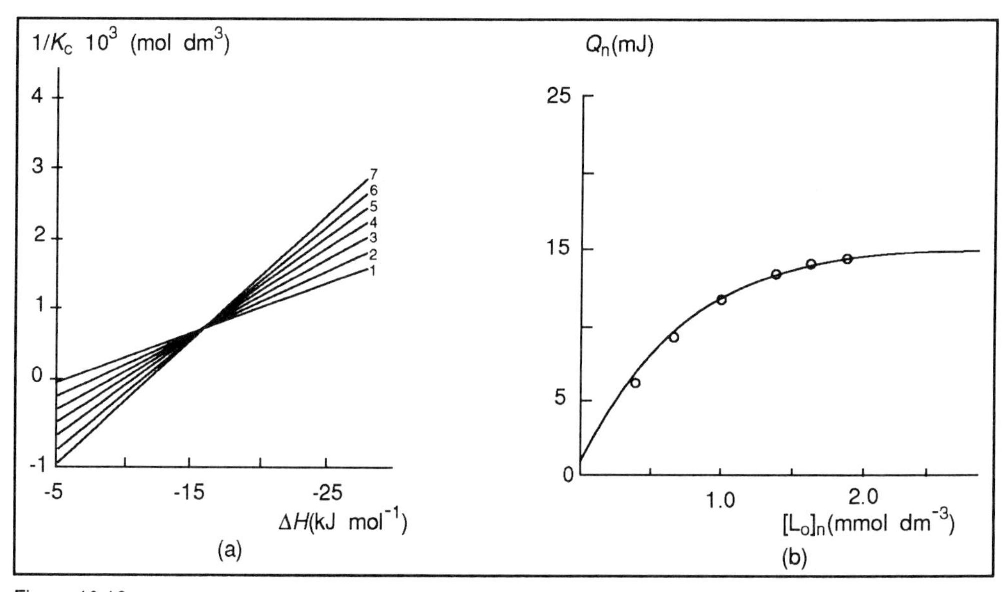

Figure 10.12 a) Evaluation of K_c and ΔH values from results of the calorimetric titration experiments. The numbers in the graph indicate titration steps. b) Experimental heat quantities (Q_n) plotted verses total ligand concentration, $[\,L_o\,]_n$. The curve is constructed from Q_n values calculated from Equation 10.8 using the derived values for K_c and ΔH.

There are other, more complicated, methods which involve the use of iterative procedures. These are described fully in the literature for each particular binding system.

From a knowledge of K_c the value of ΔG° (= $-RT \ln K_c$) and hence ΔS for the binding process can be calculated.

Experimental procedures.

It is, therefore, necessary to measure the heat evolved when varying concentrations of the ligand are brought into contact with the same concentration of macromolecule. Using the two-compartmented cell (Figure 10.4a) a solution of known concentration of the macromolecule is placed in the larger compartment and a solution of the ligand in the other compartment. The reference cell contains the macromolecule solution with water instead of the ligand solution. These are sealed and placed between the thermopiles in the block in the air thermostat; when temperature equilibrium has been established the block is rotated and the solutions mixed. The heat evolved, Q, for this particular addition is then recorded.

The cells are then removed, washed and filled with fresh solutions, this time the concentration of ligand is changed and Q again measured. This procedure is repeated until a sufficient number of additions (usually about 10) have been made to permit the necessary calculations. It is also necessary to measure the heat of dilution of the ligand.

Π What is the problem with this method?

Yes, it is very time consuming, it takes several hours for the block and cells to attain equilibrium after each filling and so the full experiment may well take several days to complete.

titration method

This has been overcome in the so-called 'titration method' using a syringe titration unit attached to the standard batch microcalorimeter. Two motor-driven syringes are attached to the outside of the drum containing the twin cells, within the air thermostat. These are filled with the ligand solution, one is connected to the cell containing the macromolecule solution and the other to the reference cell containing the same volume of water. Additions from the syringes, usually of the order of 10 μl, can be made according to a selected programme. After each small addition the drum is rotated and the heat evolved is measured. Since all the solutions are within the thermostat the whole process of successive additions can be speeded up and a full titration of the reagents achieved with a large saving in time and materials.

Applications and results

The graph of Q_n for the binding of indole-3-propionic acid to α-chymotrypsin is typical of the plots obtained by these methods (Figure 10.12b), the value of ΔH_0 and K_c are given in Table 10.2 which also contains the results of other binding experiments.

Macromolecule	Ligand	ΔH^{o} (kJ mol^{-1})	K_c (dm^3 mol^{-1})
α-chymotrypsin	indole-3-propionic acid	-15.26 (0.2)	1710 (122)
human serum albumin	L-tryptophan	-27.6 (0.8)	14000 (2000)
	benzodiazepin	-41.8 (0.8)	13000 (1000)
HSA/L-tryptophan complex	benzodiazepin	-10.9 (1.2)	1000 (200)
Troponin C	Ca^{2+}	-10.0	1×10^6
Myosin	ADP	-73.1(3.2)	1.2×10^6

Table 10.2 ΔH_0 and K_c values for various binding reactions at 298 K (error limits +/- given in parentheses).

SAQ 10.6

Using the data given in Table 10.2 calculate the values of ΔG^{o} and ΔS^{o} for the binding reactions. Comment on the strengths of the binding in the different reactions. (Remember $\Delta G^{o} = \Delta H^{o} - T\Delta S^{o}$ and $\Delta G^{o} = RT \ln K_c$) Take $R = 8.314$ JK^{1} mol^{-1} and T = 298K.

The technique has also been used in the determination of the heat of antigen-antibody binding, and enthalpy changes in the unfolding of proteins and in the interaction of herbicides with clay and other soils.

10.4 Summary of the applications of flow calorimetry

The following list will give you some idea of the potential of the technique and the type of problems that may be investigated:

- the identification and enumeration of bacteria;
- the evaluation of thermodynamic parameters for microbial growth and metabolism of substrate in batch and continuous cultures;
- the antimicrobial effects of drugs and antibiotics singly and in combiantion, fungicides and slimicides;
- the degradation of solid substrates (eg straw) and the microbial activity of soils and sewage;
- the establishment of assays for antibiotics and phenols;
- the thermal properties of blood cells in a range of diseases, cells grown in tissue culture and the effect of metabolic inhibitors on their heat production;
- ecological studies to provide an insight into how organisms perform and adapt in response to changes in their environment;
- the kinetics of enzyme reactions;
- the binding of ligands to macromolecules.

Summary and objectives

All living organisms produce heat, this can be made use of in the study of their metabolic processes. Calorimetry is a non-specific technique, a fact which can prove advantageous in many situations. The theory of heat conduction calorimetry and the design of microcalorimeters are described briefly. The applications of batch - and flow-calorimetry are discussed. Mass and energy balances can be established by monitoring the heat, biomass and carbon dioxide production and substrate concentration of growing cultures. Changes in these properties, as a result of changed environmental conditions, lead to a better understanding of the metabolic processes involved. Antibiotic assays, based on the heat output of respiring yeast or bacterial cells, may have advantages over conventional methods. Batch calorimetry, used in the study of binding processes, provides thermodynamic data on such interactions.

Now that you have completed this chapter you should be able to:

- give an account of the theory of heat conduction calorimetry and explain how the heat output is computed;
- describe the design and use of microcalorimeters in the batch and flow modes;
- discuss and explain the shape of the power-time curves of growing organisms in substrate-limited and in rich media and also the effect of added antibacterial agents;
- determine the mass and energy balances of growing cells from measurements of the heat, biomass and carbon dioxide production;
- explain how the heat output of respiring cells can be used in antibiotic assays;
- give an account of the principles and practice of the use of batch calorimetry in the determination of thermodynamic parameters of ligand and macromolecular interactions.

Electrometric methods of analysis

Electrometric methods of analysis

11.1 Introduction

Electrometric methods are used primarily to determine the concentration of ionic species, usually in aqueous solution. The method depends on the measurement of the emf of a suitable cell in which one electrode is an indicating electrode (that is its potential is responsive to the concentration of a particular ionic species) and the other a reference electrode. The technique has many applications, particularly with the development of specific electrodes, in biological, clinical and environmental studies.

Electrochemistry can be a baffling subject, largely due to the different sign conventions used. We begin this chapter with a brief account of the basic principles before embarking on the study of the range of electrodes that are now avilable.

11.2 Basic principles of electrochemistry

Π Make a brief sketch of what you think an electrochemical cell looks like. Label, as far as possible, its various parts. (You will be able to check your figure later).

All electrochemical cells consist of a series of conducting phases in contact:

- electrodes, generally metallic, and;
- one or more liquid electrolytes.

At any phase boundary, where two phases of different composition are in contact, there is a difference of potential; the 'phase boundary potential'. The electromotive force (emf) of the cell is the algebraic sum of all these phase boundaries, which can either be electrode/liquid or liquid/liquid boundaries.

11.2.1 Reversible electrodes

Π Define an electrode.

half-cell

An electrode, a term used as a synonym for a half-cell, is an electronic conductor in contact with an electrolyte which acts as a source or sink for electrons. The electrode material may be inert or take part in the electrochemical reaction.

A reversible electrode, or half-cell, is one in which electrochemical equilibrium is established between the electronic conductor and the electrolyte. The following types are recognised;

- **metal electrode**, solid element in contact with a solution of its ions, eg $Cu^{2+}|Cu$;

- **redox system**, inert metal in contact with a solution containing ions in different oxidation states, eg Fe^{3+}, Fe^{2+} | Pt;
- **inert electrode in contact with neutral solutes in different oxidation states**, eg acetaldehyde|ethanol. This type, which often requires the presence of hydrogen ions, is very common in biochemistry;
- **gas electrode, inert electrode with gas in equilibrium with its ions in solution**, eg H^+ | H_2, Pt;
- **metal, insoluble metal salt electrode**, metal coated with an insoluble salt in contact with a solution containing a common ion, eg Ag, AgCl|Cl^-;
- **cationic responsive electrodes**, electrodes whose potential depends on the concentration of a cation, eg glass electrodes.

Electrodes, such as zinc in contact with sulphuric acid, are irreversible because a chemical reaction occurs immediately (ie before connection to another electrode has been made); these electrodes are of no use in analytical electrochemistry.

Galvanic cell

When two electrodes, or half cells, are connected, either directly or through a bridge, if there is any interaction between the two electrolytes, then a Galvanic cell is formed. These are also known as chemical cells in which the free energy change, due to a chemical reaction, gives rise to an electromotive force (emf). Concentration cells (Section 11.3) involve the use of up to four electrodes.

11.2.2 Reversible galvanic, or chemical, cells

In the previous section we mentioned reversible electrodes. In this section we shall see exactly what is meant by the term reversible when applied to electrochemical cells and electrodes.

reversibility

The reversibility of a galvanic cell can be tested by connecting to an external source of emf which is adjusted to balance the emf of the cell exactly (as in a potentiometer circuit); under these conditions there is no current flow and no cell reaction. If the applied external emf is decreased by an infinitesimal amount, current will flow from the cell and a chemical change will occur.

Π What do you think would happen if the external emf is greater than the emf of the cell?

Current will now flow in the opposite direction and the cell reaction will be reversed.

Each electrode in a galvanic cell must be reversible for the cell to be reversible. Even then such a cell will only behave reversibly when the current flowing is infinitesimal and the system is virtually always in equilibrium.

Daniell cell

Consider the simple Daniell cell, which can be written:

$$- \text{Zn} \mid \text{ZnSO}_4 \vdots \text{CuSO}_4 \mid \text{Cu} + \qquad E\,(298\text{K}) = 1.1\ \text{V}$$

The solid lines represent the solid/liquid boundaries and the dotted line the liquid/liquid boundary (such boundaries are usually made in a ceramic disc). The emf is the sum of the three phase boundaries.

Π Can you suggest a reason why a ceramic disc is used to separate two liquids?

Quite simply it is to prevent excessive diffusion of one electrolyte into the other; if diffusion is allowed to continue there will be a continual change, with time, in the emf.

chemical cell

Such a cell is known as a chemical cell because as a result of chemical reactions at the electrodes there is an overall chemical reaction in the cell. The emf of the cell, E, arises from the changes in Gibbs function and is given by:

$$\Delta G = - \; n F E \qquad \text{(E - 11.1)}$$

Where n is the number of electrons involved in the cell reaction and F is the Faraday constant (96 500 C mol^{-1}).

Consider Figure 11.1 the copper electrode on the right hand side is positive and when the outer circuit is completed through a high resistance or digital voltmeter (DVM) positive current flows from left to right in the cell and right to left outside.

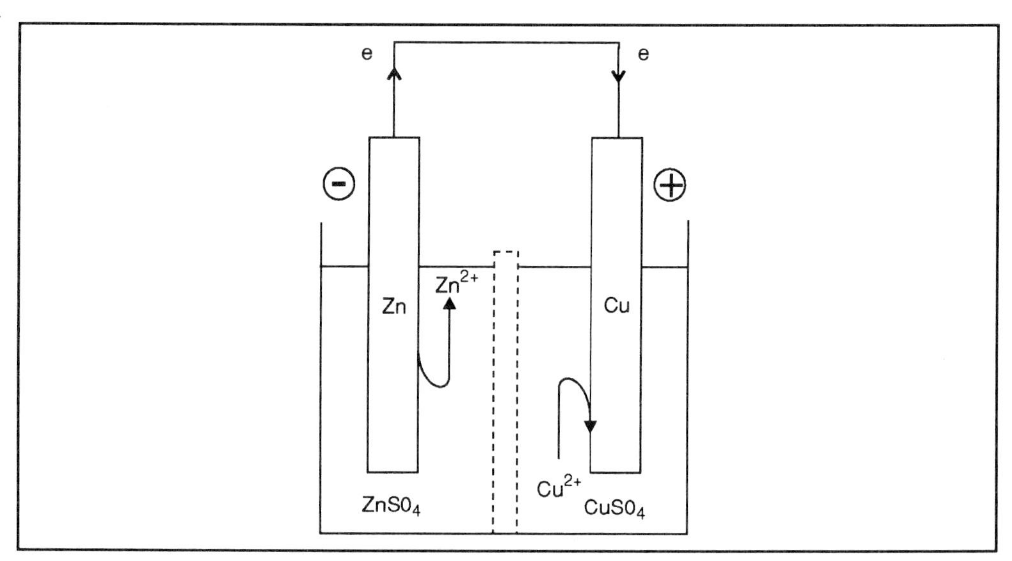

Figure 11.1 The Daniell Cell.

Since only electrons carry the current through the outer circuit the electron flow is from the zinc electrode through the wire to the copper electrode. Inside the cell the current is carried by zinc ions passing into solution, at the boundary the current is carried partly by zinc ions passing from left to right and sulphate ions passing from right to the left. At the copper/copper sulphate electrode the current is carried by copper ions which are deposited.

This can be summarized by considering the electrode reactions:

negatively charged electrode	Zn	→	$Zn^{2+} + 2e$	oxidation
positively charged electrode	$Cu^{2+} + 2e$	→	Cu	reduction
Cell reaction	$Zn + Cu^{2+}$	→	$Zn^{2+} + Cu$	

oxidation
reduction

The reaction involving the loss of electrons is known as oxidation while reduction involves the gain of electrons.

All electrode reactions can be summarized in the general form:

$$\text{oxidized state} + n\,\text{e} \underset{\text{oxidation}}{\overset{\text{reduction}}{\rightleftharpoons}} \text{reduced state}$$

or as we shall use it later:

$$\text{ox} + n\,\text{e} \rightleftarrows \text{red}$$

Thus in every cell, oxidation must occur at one electrode and reduction at the other.

SAQ 11.1

From the information given so far calculate the free energy change of the Daniell cell reaction.

conventions

The following conventions are useful to remember:

in writing cells: reduction occurs at the right hand electrode and oxidation at the left hand electrode; thus the electrode at the right of a chain of conducting phases is positive. If when you measure the emf of a cell you record which is the positive electrode you will be able to write down the cell in the correct manner and hence deduce the electrode and cell reactions;

in writing electrodes: we use the format oxidized:reduced, (for example, $M^+ \mid M$ and AgCl, Ag $\mid Cl^-$), with a full vertical line to indicate the junction of the two phases.

SAQ 11.2

For the following reversible electrodes write down the electrode reaction and give the conventional method of formulating the electrode, ie oxidized state | reduced state:

1) Mg in $MgSO_4$ solution;

2) H_2 in HCl solution using platinum as an inert electrode;

3) Cl_2 in NaCl solution using platinum as an inert electrode;

4) Ag, AgCl in KCl solution;

5) Fe^{3+}, Fe^{2+} in solutionusing platinum as an inert electrode;

6) acetaldehyde, ethanol in solution using platinum as an inert electrode.

11.2.3 Electrode potentials and the Nernst equation

Using the generalised equation for an electrode reaction:

$$\text{ox} + n\,\text{e} \longrightarrow \text{red}$$

the change in Gibbs function (free energy change) is given by:

$$\Delta G = \Delta G^{\circ} + RT \ln (a_{red} / a_{ox}) \qquad \text{(E - 11.2)}$$

substituting $\Delta G = -nFE$ and $\Delta G^{\circ} = -nFE^{\circ}$ and rearranging gives:

$$E\,(\text{O,R}) = E^{\circ}(\text{O,R}) + (RT/nF) \ln (a_{ox} / a_{red}) \qquad \text{(E - 11.3)}$$

reduction potential

This is the Nernst equation for a reduction potential, where E (O,R) is the potential of the electrode when the activities of the oxidized and reduced forms are a_{ox} and a_{red}, and E° (O,R), the standard (reduction) potential is that potential which would be obtained when both the oxidized and reduced forms are present at unit activity. (For the present time we will use activity; no doubt you will recall that activity is a thermodynamic concentration term which allows for deviations from ideal behaviour). Note that the O,R in brackets refers to oxidised, reduced.

For cationic (metal) electrodes the activity of the reduced form (present in its standard state), $a_{\text{red}} = 1$ and so for such a reaction

$$M^{n+} + n\,e \longrightarrow M(s)$$

the electrode potential is written:

$$E\,(M^{n+}, M) = E^{\circ}(M^{n+}, M) + (RT/nF) \ln a_{M^{n+}} \qquad \text{(E - 11.4)}$$

This electrode potential therefore depends on the activity (concentration) of the metal ions in solution.

Π From the equation for the silver, silver chloride electrode [(d) in SAQ 11.2] write down the equation for its electrode potential in the conventional form.

You should have written as:

$$E\,(\text{AgCl,Ag,Cl}^-) = E^{\circ}(\text{AgCl,Ag,Cl}^-) + (RT/F) \ln a_{\text{AgCl}} / (a_{Ag}\; a_{\text{Cl}^-})$$

But this can be simplified, since both AgC1 (oxidized form) and Ag are in their standard states and therefore at unit activity. The simplication gives us:

$$E\,(\text{AgCl,Ag,Cl}^-) = E^{\circ}(\text{AgCl,Ag,Cl}^-) - (RT/F) \ln a_{\text{Cl}^-} \qquad \text{(E - 11.5)}$$

You will notice that by convention we write electrode potentials in the form E(O,R), eg $E\,(M^{n+}, M)$.

So far we have considered the electrode potentials from the more theoretical point of view; but how can we measure an electrode potential? It is not possible to measure the potential of a single electrode; any complete circuit must, of necessity, contain two electrodes. In practice, we measure the potential of all electrodes with reference to a common electrode. By convention, this is chosen to be the standard hydrogen electrode, whose potential at 298K, 1 atmosphere pressure and a hydrogen ion concentration (more accurately activity) of 1 mol dm^{-3} is arbitrarily set to be zero, ie $E^{\circ}(H^+, H_2) = 0$.

standard electrode potential

The standard electrode potential of a half-cell may be determined by combining it with a standard hydrogen electrode preferably in a cell without a liquid junction and measuring the emf. As a simple example for the cell:

Pt, H_2 (g, 1 atmos) | H^+ ($a_{H+} = 1$) ox and red forms | Pt
(standard hydrogen electrode) (system to be measured)

$$E = E(O,R) - E^\circ(H^+, H_2) = E(O,R) \text{ (since, by definition } E^o(H^+,H_2,=0))$$

$$= E^\circ(O,R) + (RT/nF) \ln a_{ox} / a_{red} \quad \text{(E - 11.6)}$$

By carrying out measurements of the emf of the cell, E, at different values of the concentrations of the oxidized and reduced forms and extrapolating to zero ionic strength the value of E °(O,R) can be calculated. Further information on these determinations, including the use of secondary reference electrodes may be found in texts on electrochemistry.

All half-cells can be arranged according to their standard (reduction) potentials. If a cell is constructed with two metals dipping into solutions of their own solutions, with a bridge separating them (for example the Daniell cell), the metal which in has a higher redox potential (lower tendency to give up electrons) will be the positive pole. The metal with a greater tendency to give up electrons (that is the one with a lower redox potential) will be the negative electrode.

Π Given the standard electrode potentials: E° (Cu^{2+}, Cu) = 0.339 V, and E° (Zn^{2+}, Zn) = −0.761 V, calculate the standard emf of the cell obtained by joining these two electrodes.

We hope that you remembered that the emf or standard emf of a cell is the difference between the two reduction potentials. In this case the copper is the positive electrode and so:

$$E^\circ = E^\circ(Cu^{2+}, Cu) - E^\circ(Zn^{2+}, Zn) = 0.339 - (-0.761) = 1.10 \text{ V}$$

Some typical reduction potentials of some biological half-cells are given in Table 11.1.

System	Half-cell reaction	$E^{\circ\prime}$ / V
cyt a^{3+}/cyt a^{2+}	$Fe^{3+} + e \rightarrow Fe^{2+}$	+0.29
cyt c^{3+}/cyt c^{2+}	$Fe^{3+} + e \rightarrow Fe^{2+}$	+0.254
Fe^{3+}/Fe^{2+}, haemoglobin	$Fe^{3+} + e \rightarrow Fe^{2+}$	+0.17
fumarate/succinate	$CHCOO^{-}$ (–$CHCOO^{-}$) $+ 2H^{+} + 2e \rightarrow CH_2COO^{-}$ (–CH_2COO^{-})	+0.031
MB/MBH_2 *	$MB + 2H^{+} + 2e \rightarrow MBH_2$	+0.011
pyruvate/lactate	$CH_3COCOO^{-} + 2H^{+} + 2e \rightarrow CH_3CH(OH)COO^{-}$	-0.185
acetaldehyde/ethanol	$CH_3CHO + 2H^{+} + 2e \rightarrow CH_3CH_2OH$	-0.197
$FAD/FADH_2$	$FAD + 2H^{+} + 2e \rightarrow FADH_2$	-0.219
$NAD^{+}/NADH$	$NAD^{+} + 2H^{+} + 2e \rightarrow NADH + H^{+}$	-0.320
H^{+}/H_2	$H^{+} + e \rightarrow H_2$	-0.421

* The symbols MB and MBH_2 represent the oxidized and reduced forms of methylene blue, which is used as a redox indicator.

Table 11.1 Standard Reduction potentials, $E^{\circ\prime}$, for some biological half-cells at 298 K and at pH 7.0. (NB the values given are for pH 7, in contrast to those for E°, for chemical systems, which are at pH 0).

Π At pH7, what is the emf of the H^{+}/H_2 electrode?

Its value is -0.421V. (See Table 11.1). pH, of course has its main effect on $E^{\circ\prime}$, values of electrodes involving H^{+} ions. Of course, by definition $E^{\circ\prime}$, of the H^{+}/H_2 electrode (at pH0) is = 0V.

SAQ 11.3

For the cell:

$$- \text{Pt, } H_2\ (g, 1\ \text{atmos})\ |\ H^{+},\ Cl^{-}\ |\ \text{AgCl, Ag} +$$

where the activities of H^{+} and Cl^{-} are a_{H+} and $a_{Cl^{-}}$ respectively:

1) write down the electrode and hence the cell reactions;

2) write down the equation for the emf of the cell;

3) given that $E^{o}(AgCl, Ag, Cl^{-}) = + 0.2224$ V calculate the standard emf of the cell.

equilibrium constant

Standard electrode potentials are also of use in calculating the equilibrium constant of a reaction which may be difficult to measure by conventional means. From thermodynamics, we know that $\Delta G^{\circ} = RT \ln Keq$ or, expressed in another way $Keq = \exp(-\Delta G^{\circ}/RT)$. Since we can determine G^{o} from E^{o} values (Equations 11.1, 11.2), thus we can use E^o values to determine K *eq*. If the concentrations of reactants and products are specified the spontaneity or otherwise of a given reaction can be predicted.

You should now have a basic knowledge of the basic principles of electrochemistry to enable you to understand the workings and applications of reference and indicating electrodes.

SAQ 11.4

Using the data in Table 11.1, calculate the value of the equilibrium constant for the reaction:

$$CH_3CHO + NADH + H^+ \rightleftarrows CH_3CH_2OH + NAD^+$$

11.3 Concentration cells

A concentration cell is one in which there is no overall chemical reaction; the reaction occurring at one electrode (or pair of electrodes) is reversed at the other (or other pair). There may, nevertheless, be a net change of Gibbs function (free energy) because of a difference in the concentration of one or other of the reactants concerned at the electrodes. The electrical energy arises from the change in Gibbs function accompanying the transfer of material from one concentration to the other.

There are many different types of concentration cell; the one which is of most concern to us is the concentration cell without transport.These are cells in which there are no liquid junctions. There is no direct transfer of ions or electrolyte, material transport occurs indirectly as a result of chemical reactions. Such cells result when two simple, reversible galvanic cells whose electrodes are reversible with respect to each of the ions constituting the electrolyte are combined in opposition. Consider two cells of the type:

$$Pt, H_2(g) \mid HCl\ (aq) \mid AgCl, Ag$$

with different molalities of HCl, m_1 and m_2, where $m_1 > m_2$. Each of these has the cell reaction:

$$0.5H_2(g) + AgCl\ (s) \rightleftarrows Ag(s) + H^+ + Cl^-$$

If they are now connected in opposition through the silver we have a concentration cell without transport, thus:

$$- Pt, H_2\ (g) \mid HCl\ (m_2) \mid AgCl, Ag --- Ag, AgCl \mid HCl\ (m_1) \mid H_2\ (g), Pt +$$

Π Assuming that the pressure of the hydrogen gas at the two terminal electrodes is the same, can you explain why an emf is generated from such a combined cell?

To do this it is necessary to write down the reactions in the two cells; since the more negative cell (left hand one) drives the more positive cell backwards they are:

More negative cell at m_2 : 0.5 H_2 (g) + AgCl (s) ——> Ag(s) + H^+ + Cl^-

More positive cell at m_1 : Ag (s) + H^+ + Cl^- ——> 0.5H_2(g) + AgCl (s)

From which you can see that for each mol of H^+ and mol of Cl^- that are lost from the more positive cell at molality m_1 one mol of H^+ and one mol of Cl^- are formed in the more negative cell at molality m_2. Thus the net result of taking 1 Faraday of electricity from the cell is the transfer of 1 mol of H^+ and 1 mol of Cl^- from m_1 to m_2 (ie down a concentration gradient); it is this transfer which gives rise to the free energy change and hence the emf:

$$\Delta G = RT \ln (a_{H^+,2}/a_{H^+1}) + RT \ln (a_{Cl^-,2}/a_{Cl^-,1})$$

and

$$E = (RT/F) \ln (a_{H^+,1}/a_{H^+,2}) + (RT/F) \ln (a_{Cl^-,1}/a_{Cl^-,2}) \qquad \text{(E - 11.7)}$$

Where activity $a = m\,\gamma$, (γ is the activity coefficient; you should recall that γ values are always less than 1 and approach 1 at infinite dilution.)

SAQ 11.5

1) Assuming ideal behaviour (ie activity coefficients are unity) rewrite Equation 11.7 and hence calculate the emf of the concentration cell at 298 K when the ratio of $m_1 / m_2 = 10$. Does the emf depend on the actual values of m_1 and m_2? Use R = 8.314 J K^{-1} mol^{-1}.

2) Using the tabulated values of the activity coefficients, calculate the theoretical emf of the cell when (a) $m_1 = 0.1$ and $m_2 = 0.01$ and (b) $m_1 = 0.01$ and $m_2 = 0.001$.

molality of HCl	γ_{H^+}	γ_{Cl^-}
0.1	0.83	0.76
0.01	0.91	0.90
0.001	0.97	0.96

Comment on your results.

11.4 Reference electrodes

A reference electrode is a reversible, non-polarizable half-cell at which the potential remains in equilibrium, against which the potential of a test half-cell can be measured or controlled.

11.4.1 The calomel electrode

calomel electrode

The most common reference electrode is the calomel electrode, which consists of mercury, mercury (I) chloride and potassium chloride of specified concentration, ie Hg_2Cl_2, Hg, Cl^- (Figure 11.2). The electrode potential is given by:

$$E_{cal} = E^{o}_{cal} + (RT/2F)\ \ln\ K_s - (RT/F)\ \ln\ a_{Cl}$$

$$= E'_{cal} - (RT/F)\ \ln\ a_{Cl^-} \qquad (E - 11.8)$$

where $K_s = a_{Hg^{2+}}\, a^2_{Cl^-}$.

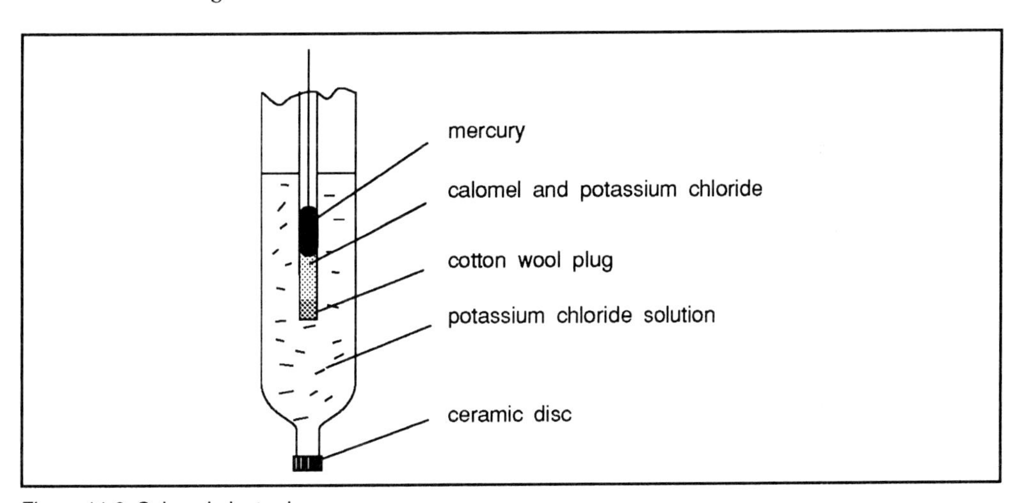

Figure 11.2 Calomel electrode.

The electrode thus behaves as a reversible chlorine electrode, the electrode potential, E_{cal} depends on the concentration of the potassium chloride and the temperature, Table 11.2.

KCl concentration	E_{cal} (298 K) (V)	E_{cal} (V) [as a function of T (K)]
0.1 mol dm^{-3}	0.334	$0.3335 - 7.0 \times 10^{-5}\ (T - 298)$
1.0 mol dm^{-3}	0.281	$0.2810 - 2.4 \times 10^{-4}\ (T - 298)$
saturated	0.242	$0.2420 - 7.6 \times 10^{-4}\ (T - 298)$

Table 11.2

The first electrode listed in Table 11.2 is preferred for accurate work as it has the lowest temperature coefficient; the saturated electrode is the most convenient owing to the ease of replacement of the solution. In commercial electrodes, the liquid junction between the half-cell and the test solution is made by leakage of the potassium chloride through a ceramic disc.

The electrode is the most common reference electrode used in non-aqueous systems.

11.4.2 The silver, silver chloride electrode

This reference electrode can be used in aqueous and non-aqueous solutions. It is prepared by anodizing a freshly plated silver electrode (either a strip or disc) in hydrochloric acid until an even deposit of AgCl is formed; this is unaffected by sunlight. It behaves as a reversible chlorine electrode with a potential given by:

$$E\ (\mathrm{AgCl, Ag, Cl^-}) = E^\circ (\mathrm{AgCl, Ag, Cl^-}) - (RT/F)\ \ln\ a_{\mathrm{Cl}^-} \qquad \text{(E - 11.5)}$$

The standard electrode potential in aqueous solution at T K is given by:

$$E^\circ\ (\mathrm{AgCl, Ag, Cl^-}) = 0.222\,39 - 645.52 \times 10^{-6}\ (T - 298) - 3.284 \times 10^{-6}\ (T - 298)^2$$

11.5 Indicator electrodes

11.5.1 Introduction

Indicator electrodes are those whose electrode potential changes with the concentration of an electroactive species in solution. These electrodes can be used to measure the concentration of ionic species in solution, for example hydrogen ions (pH), anions and cations and, as we shall see later, such substrates as penicillin, urea, glucose and amino acids.

The basic electrode is the hydrogen electrode itself which is also the standard reference electrode for all electrode potentials. Although it is the most accurate, it is not the electrode most commonly used for routine measurements; this role has been taken over by the glass electrode.

11.5.1 The glass electrode

The hydrogen electrode is not suitable for routine use for the rapid measurement of the pH of a solution. It has been replaced by the glass electrode which consists of a bulb of special glass blown on the end of a glass tube, containing hydrochloric acid and a reference silver, silver chloride electrode (Figure 11.3).

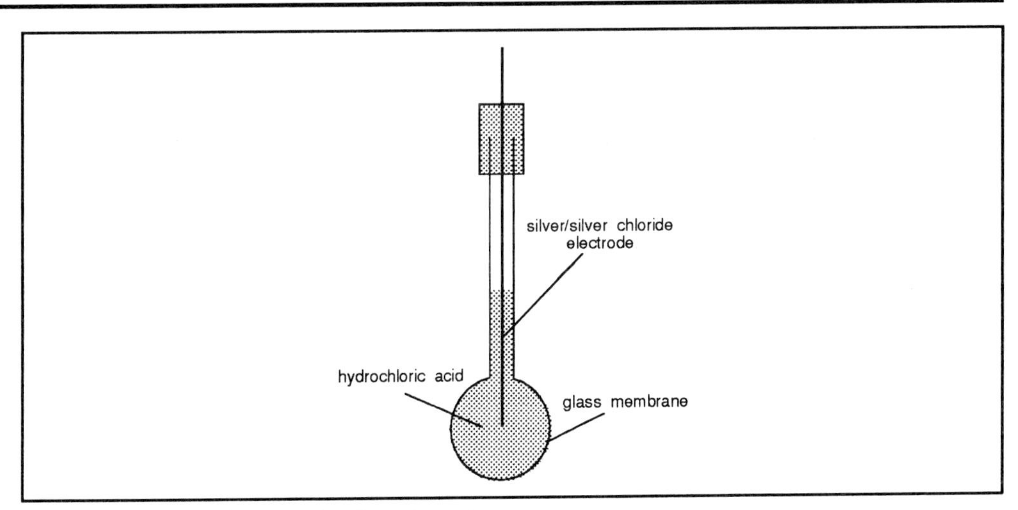

Figure 11.3 Essential features of a glass electrode.

The complete cell for the measurement of the pH of a solution is:

Ag, AgCl | HCl (0.1 mol dm^{-3}) | glass | test solution : KCl (satd.) | Hg_2Cl_2, Hg

<——— glass electrode ———>

The electrode potential of the glass electrode may be represented by the equation:

$$E_G = E'_G + (RT/F) \ln a_{H+} \qquad \text{(E - 11.9)}$$

The surface of the glass electrode must be hydrated for it to function as a hydrogen-indicating electrode; the surface must not be allowed to dry or used in solution of pH > 11.

In use, the glass electrode assembly (ie with the calomel reference electrode) must be calibrated with solutions of known pH values to eliminate asymmetry potentials in the glass. The accuracy of measurement is ± 0.01 pH, or better if extreme care is taken, over the pH range 2 - 10. It reaches equilibrium immediately in any solution, and its response is unaffected by the presence of any gas, oxidizing and reducing agents, and poisons in the generally accepted sense. It has no appreciable salt or protein error. It is available commercially in many forms for the measurement of pH in solutions, emulsions, pastes and on surfaces and for measurement on a micro scale .

The electrode suffers from asymmetry potentials, which necessitates regular standardization. It is very sensitive to previous treatment and should be well washed after use, but not allowed to dry.

11.5.2 Ion-selective electrodes

ion-selective electrode

These are membrane electrodes that respond selectively to one (or several) ionic species in the presence of other ions. The word 'membrane' is used in its widest sense denoting a thin layer of electrically conducting material separating two solutions across which a potential develops. The first ion-selective electrode was the hydrogen ion - responsive glass electrode.

These electrodes find many applications: for example in titrations, water analysis for metal and halide ions,hardness determinations, biosensors (Section 9.6.6), and gas sensing probes.

Ion-selective electrodes, classified according to IUPAC, include the following.

Homogeneous membrane electrode

Homogenous membrane electrodes. A solid-state electrode based on a single crystal or compacted disc of an insoluble salt (eg AgX for X = Cl^-, Br^- and I^-), or a mixture (eg silver iodide/silver sulphide for S^{2-}) sealed on the end of a tube containing a reference solution and electrode, usually a silver chloride electrode (Figure 11.5).

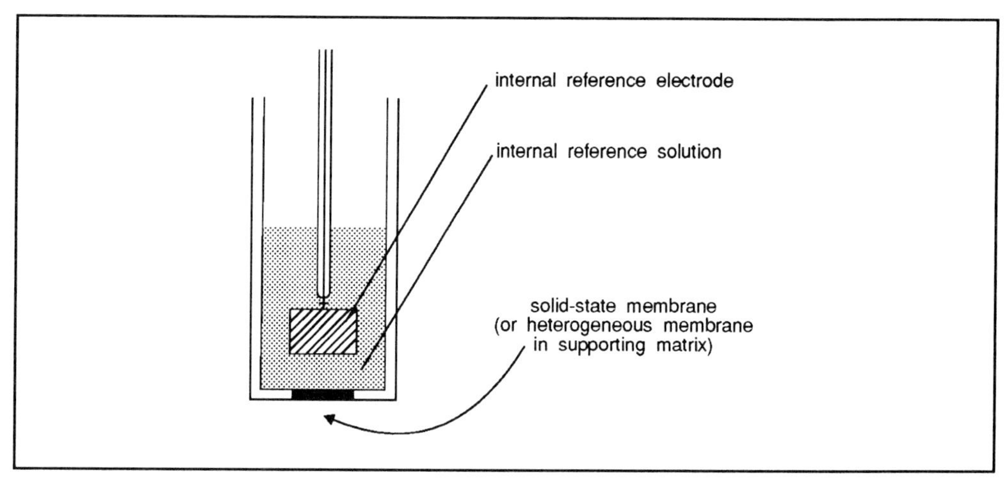

Figure 11.5 Diagrammatic representation of a homogeneous/heterogeneous membrane electrode.

Heterogeneous membrane electrode

Heterogeneous membrane electrodes. In this type of electrode the active material (eg precipitates of insoluble metal salts) is dispersed in an inert matrix material, such as silicone rubber, to give suitable mechanical properties. The prepared membrane is cemented onto the end of a tube containing the reference solution and an electrode (Figure 11.5).

Rigid matrix glass electrodes

Rigid matrix glass electrodes. These electrodes which are responsive to potassium, sodium, ammonium or silver ions are obtained by varying the composition of the glass. Selectivity is not good when other metal ions are present. The principal applications of such cation-responsive electrodes are in water analysis and clinical biochemistry.

Liquid ion-exchanger electrode

Liquid ion-exchanger electrodes. In this type of electrode the ion of interest is incorporated in a large organic molecule with a low water solubility. The organic material, dissolved in an organic solvent, is separated from the aqueous solution under test by a porous membrane (eg cellulose acetate) holding the liquid ion-exchanger (Figure 11.6).

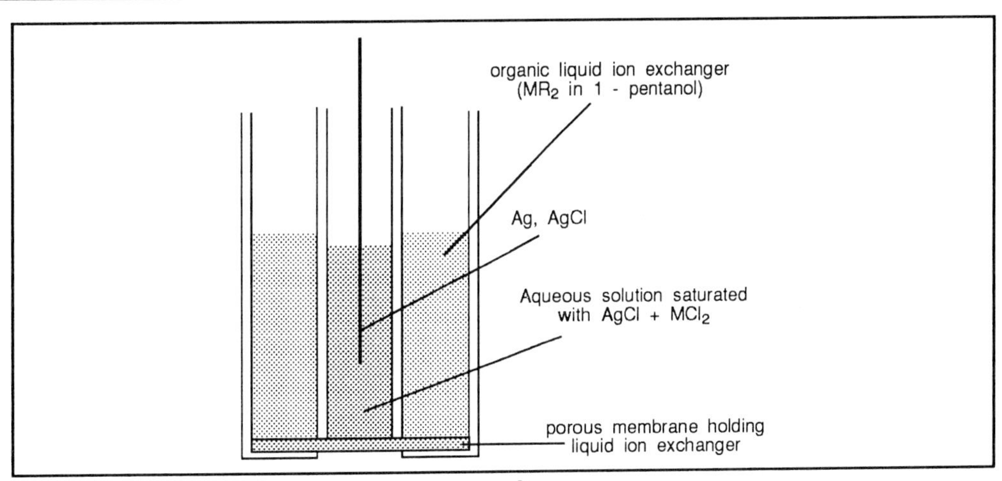

Figure 11.6 Liquid membrane electrode sensitive to M^{2+}.

Anion exchangers are either long-chain alkyl ammonium salts or salts of non-labile metal complexes of the type ML_3X_2, for example L for perchlorate and nitrate electrodes is a substituted 1,10-phenanthroline. Cationic exchangers usually consist of a long-chain alkyl compound with low water solubility which forms a stable compound with the cation under study. The calcium electrode, using calcium (bis-di-n-decyl) phosphate, $[(C_{10}H_{21})_2PO_2]_2Ca$ dissolved in di-n-octylphenylphosphonate, shows Nernstian behaviour (29.4 mV/pCa) over the range 1 - 10^{-5} mol dm^{3} . By Nernstian behaviour we mean that a plot of emf (E) against $\ln[Ca^{2+}]$ is linear with a slope of RT/nF (see section 11.2.3).

PVC-matrix membrane ion-selective electrodes are an extension of the liquid-exchanger electrodes in which the liquid ion exchanger is incorporated in PVC with the aid of a solvent. On evaporation of the solvent, this leaves a flexible membrane with the ion-exchanger trapped in the matrix. The membrane is then cemented onto the end of a tube containing the reference solution and electrode (Figure 11.5). Coated wire electrodes can be prepared by dipping a platinum wire or graphite rod into a solution of PVC/sensor/mediator in a solvent and allowing it to evaporate.

11.5.3 Ion-selective mini-electrodes

These are electrodes with an outside diameter of about 1 mm with medical applications. Commercially available mini-glass electrodes are available for hydrogen, sodium and potassium. The electrodes have been miniaturised to a point where they fit into an injection needle (Figure 11.7).

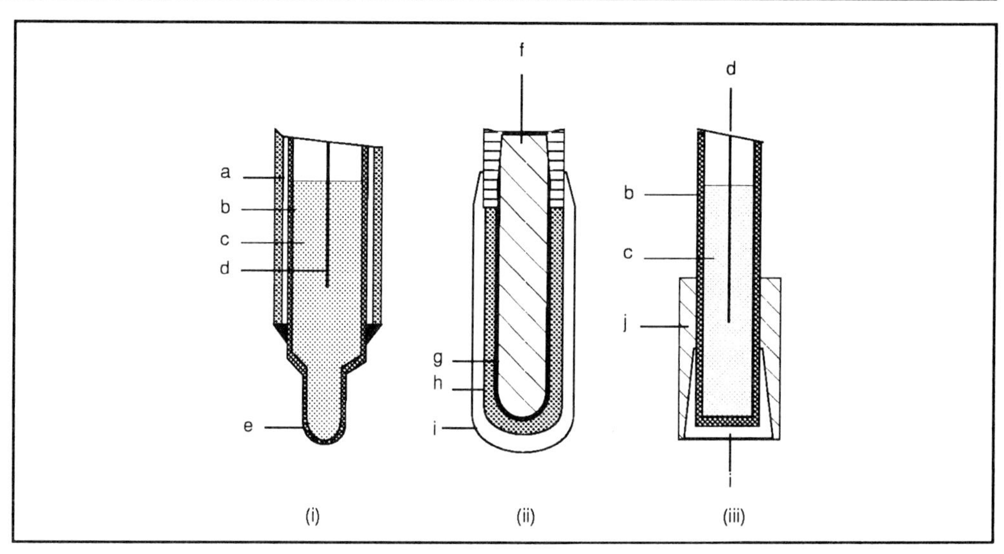

Figure 11.7 Various types of mini-ion-selective electrode. (i) pH electrode for catheter use; (ii) coated wire electrode; (iii) liquid exchanger electrode. a, steel capillary tube (outside diam. 2mm); b, glass sheath; c, internal filling solution; d, reference silver/silver chloride electrode; e, pH glass membrane; f, silver wire, g, silver, silver chloride electrode; h, gelled contact electrolyte; i, PVC membrane; j, polythene tubing. Adapted from Hibbert and James (1984) Dictionary of Electrochemistry Macmillan Reference Books, New York.

Liquid membrane electrodes offer more latitude in the preparation of miniaturised electrodes, for example the coating of thin wires with polymeric membrane material (Figure 11.7 ii). The potential of such electrodes varies with time because of the ill-defined half-cell between the membrane adjacent to the metal and the metal itself.

Continuous monitoring of potassium ions in plasma and whole blood is achieved in catheter tip electrodes, obtained by bonding valinomycin-PVC membranes into PVC tubing.

11.5.4 Ion-selective micro-electrode (ISM)

For an electrode for intracellular measurements, it has to penetrate a cell without causing undue damage; the effective tip must be as small as possible. Several types of ISM are available (Figure 11.8; the liquid-exchanger microelectrodes have a higher selectivity compared to ion-sensitive glass electrodes).

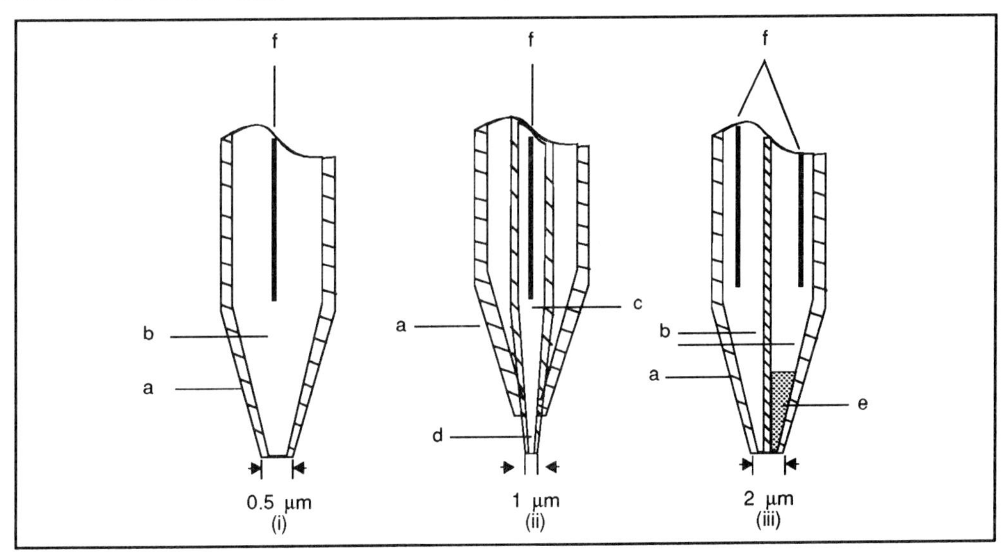

Figure 11.8 Various types of micro-ion-selective electrodes. (i) reference electrode; (ii) pH or cation selective glass electrode; (iii) combined reference/liquid-exchanger electrode (available in the form of a four barrelled electrode for the simultaneous determination of K^+, Na^+ and Ca^{2+}. a, glass capillary; b, internal reference electrolyte; c, internal electrolyte; d, cation-sensitive glass; e, liquid ion-exchanger; f, silver-silver chloride electrode. Adapted from Hibbert and James (1984) Dictionary of Electrochemistry Macmillan Reference Books, New York, London.

Potassium-sensitive microelectrodes have been used in the study of the central nervous system and potassium-, sodium-, calcium- and chloride- selective electrodes have been used for measurements in the brain.

11.5.5 Gas-sensing membrane probe

Gas-sensing membrane probes consist of an ion-selective electrode combined with a suitable reference electrode to form a complete cell.

A typical sensor (Figure 11.9) is based on a glass electrode with a convex, pH sensitive tip, held against a gas-permeable membrane (100 μm thick) to sandwich a layer of internal electrolyte between the glass and the membrane. When the tip is immersed in a sample, the gas diffuses through the membrane until the partial pressure of the gas is the same on both sides of the membrane. This equilibrium partial pressure determines the pH of the medium which is measured by the glass electrode and the reference silver chloride electrode dipping in the internal electrolyte solution. Since the membrane is hydrophobic, ions cannot enter the probe and therefore will not have a direct effect on the measurement.

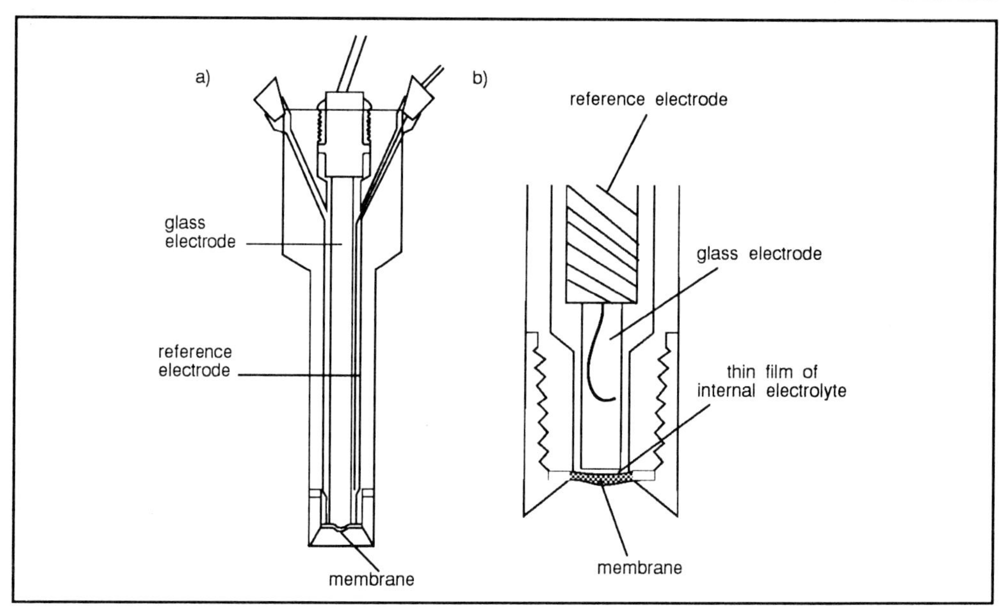

Figure 11.9 Typical gas-sensing probe: a) overall layout; b) cross-section of sensing tip. Modified from Hibbert and James (1984) 'Dictionary of Electrochemistry', Macmillan Reference Books, New York, London.

The dependence of the emf is related to the concentration of the gas, c_x, according to the Nernst equation:

$$E = E' \pm (RT/F) \ln c_x \qquad \text{(E - 11.10)}$$

where the negative sign applies to basic gases (eg ammonia) and the positive sign to acidic gases (eg sulphur dioxide).

The cell must be calibrated at a controlled temperature with solutions of known concentrations of the gas and the limits of Nernstian response established. Gas-sensing probes show outstanding selectivity.

Π Can you think of any potential interfering species?

The only interfering species are volatile substances which can diffuse through the membrane and which have acidic or basic properties comparable to those of the gas under study. The pH of the test solution must be adjusted so that all the material to be measured is present as dissolved gas.

Gas-sensing probes are used extensively in continuous flow analytical systems.

Oxygen Probe

The detection and measurement of oxygen is of great importance. Examples of gas-sensing probes, in which gaseous oxygen diffuses through a membrane so that the partial pressure of the gas is the same on both sides of the membrane. The membrane, polyethylene or Teflon, holds the reference electrolyte in contact with the cathode (Figure 11.10).

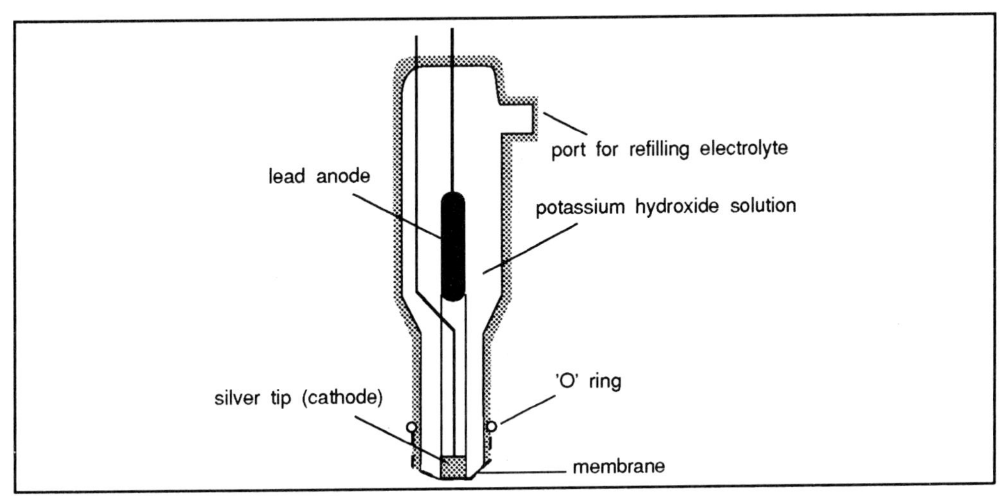

Figure 11.10 Oxygen probe. Adapted from Hibbert and James (1984) 'Dictionary of Electrochemistry' Macmillan Reference Books, New York, London.

Oxygen is reduced at the silver cathode and lead is consumed at the anode (the electrolyte is 1 mol dm^{-3} potassium hydroxide). At the anode:

$$Pb + 4OH^- \longrightarrow PbO_2^{2-} + 2H_2O + 2\ e$$

at the cathode:

$$O_2 + 2H_2O + 4e \longrightarrow 4OH^-$$

giving an overall cell reaction:

$$O_2 + 2\ Pb + 4OH^- \longrightarrow 2PbO_2^{2-} + 2H_2O$$

The current produced is passed through a standard resistance and the emf recorded on a DVM. The oxygen tension is directly proportional to the recorded potential.

When the aqueous solution flows past the electrode, the dissolved oxygen diffuses across the membrane. Minimum flow conditions are critical.

Π Can you explain why the flow conditions are critical?

If the flow rate is too low the sample water around the electrode will be depleted of oxygen, giving rise to a low reading.

The probe has a high temperature coefficient (about 6% per degree change); automatic temperature compensation is provided in modern instruments. The probe must be calibrated using air-equilibrated water (to give 100% saturation). Then a 5% (w/v) sodium sulphite solution is added to give 0% saturation level. A correction for salinity must be made when making determinations in sea water.

The probe is very versatile finding use in monitoring oxygen concentrations in river and ocean water and in sewage and in following the respiration of isolated cell suspensions and, in a miniaturized form, for monitoring oxygen in arteries.

Ammonia-sensing probe

Ammonia-sensing probe. This, the most widely used probe, is employed in the analysis of fresh water, effluent and sewage. The response range is from 1 to 10^{-7} mol dm^{-3}, and although the probe is capable of detecting low levels of ammonia, it cannot detect very small changes in the ammonia concentration at high concentrations. The potential of the probe is very dependent on temperature, hence samples and standards should be at the same temperature (in the range 5 to 40°C).

Π To what pH should the test solution be adjusted?

To ensure that all the ammonia is in the free state in solution the pH should be adjusted to > 12. Any ammonia present as a complex must be released, for example by treatment with EDTA.

Π What are likely to be interfering species?

Volatile, basic species (such as organic amines) will pass through the membrane and interfere with the measurement.

Sulphur dioxide probe

Sulphur dioxide probe. Used in the measurement of the concentrations of sulphite, bisulphite and metabisulphite in solution after acidification.

The probe is insensitive to gradual temperature variations in the range 0 - 40°C. Calibration is necessary; the response is linear in the range 3 - 3000 mg dm^{-3}.

11.6 Biosensors

There are electrodes developed for environmental or clinical analysis and fermentation control. A biosensor comprises a chemically responsive biological material (enzyme, lectin, antibody, microoganism or organelle) immobilized in close contact to a suitable transducing element designed to convert a chemical response into an electrical response. Ion-selective, glass, solid state and gas-sensing electrodes are usually chosen to detect the product of the biological (enzymatic) reaction (Figure 11.9).

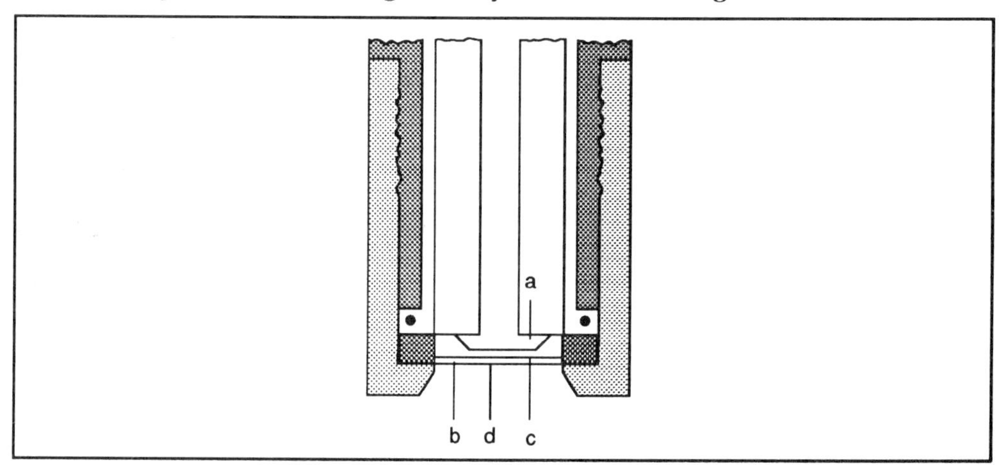

Figure 11.11 Different types of potentionmetric enzyme electrodes: a, Potentiometric sensor; b, trapped or immobilized enzyme; c, nylon spacer; d, dialysis membrane. Redrawn from Hibbert and James (1984) 'Dictionary of Electrochemistry' Macmillan Reference Books, New York, London.

Some examples of biosensors and the species detected are given in Table 11.3.

Detection system	Species detected	Typical substrates
H^+ glass electrode	H^+	penicillin, glucose, acetylcholine, urea
NH_4^+ glass electrode	NH_4^+	urea, amino acids, glutamine
NH_3 gas sensor	NH_3	aparagine, creatinine, 5' AMP, urea, serine*
CO_2 gas sensor	CO_2	urea, uric acid, tyrosine, glutamic acid*
I^- solid-state electrode	I^-	glucose
* intact bacterial cells instead of enzyme		

Table 11.3 Some typical potentimetric sensors for immobilized enzymes.

The key feature of biosensors is the specificity of the bioreagent used. For example let us consider a system in which H^+ ions are detected. If the entrapped-enzyme converts, for example, an aldelyde to an acid by the following type of reaction:

$$R - CHO + \tfrac{1}{2}O_2 \rightarrow RCOOH \rightleftarrows RCOO^- + H^+$$

then by measuring the H^+ ions produced, we have a measure of RCHO. Of course, the selectivity of such an electrode depends upon the specificity of the enzyme. Enzymes are the most commonly employed biological agent used in biosensors.

Biosensors suffer from problems associated with the instability of the bio-reagent. But, because of the specificity of the bioreagent, these types of sensors are potentically extremely useful for detecting and measuring bioproducts. Biosensors are being developed to measure a wide variety of compounds some of which are listed in Table 11.3. A fuller description of biosensors is provided in the BIOTOL text 'Technological Application of Biocatalysts'.

Enzymes may be immobilized by (1) crosslinking to serum albumin, Teflon or nylon using glutaraldehyde as a bifunctional reagent, (2) occlusion in a polymer and (3) liquid trapping. With tissue slices, a cellophane membrane between the slice and the gas-sensing probe protects the probe from lipids and other biological material (Figure 11.10).

SAQ 11.6

Explain the theory underlying the determination of glucose using glucose oxidase with (a) a pH glass electrode and (b) an iodide-selective electrode. Note that the reaction catalysed by glucose oxidase can be written as glucose $+O_2 + H_2O \rightarrow H_2O_2 +$ gluconic acid. Note also that H_2O_2 will oxidise I^-.

Summary and objectives

Electroanalytical galvanic cells consist of an indicator and a reference electrode. The basic principles of electrochemistry are reviewed and the conventions for writing electrodes and cells established. The various types of reversible electrodes are classified. The Nernst equation for electrode potentials is stated and the importance of standard reduction potentials discussed. The theory of concentration cells without transport is considered with special reference to ion-selective electrodes. Detailed consideration is given to reference electrodes, and membrane electrodes including glass, solid state, heterogeneous, liquid-exchanger electrodes and biosensors. The theory and applications of gas-sensitive probes are considered.

You should now be able to:

- describe the construction of galvanic and concentration cells without transport and explain the origin of the emf for both types of cell;
- explain the convention for formulating reversible electrodes and cells;
- state the Nernst electrode potential equation and describe its use in calculating the emf of a simple chemical cell;
- classify electrodes as anionic, cationic, redox, membrane (including glass, solid state, heterogeneous, liquid-exchangers, biosensors) and describe their construction, underlying theory and applications;
- give an account of gas-sensing probes and their applications;
- list examples of biochemicals that have been measured using biosensors.

Responses to SAQs

Responses to Chapter 2

2.1

1) The plot of absorbance against dry weight is obtained as follows.

Dry weight of cells in undiluted culture = (2.8456 - 2.7656) 10 = 8 mg cm^{-3}. Dry weight values are obtained from this concentration by taking into account the dilution.

The absorbance of medium minus cells is subtracted from each of the other absorbances. These values are then plotted against the corresponding dry weight values.

Thus your data should be:

culture density (mg cm^{-3})	corrected absorbance (at 540nm)
0.8	0.990 (from 1.050 - 0.060)
0.4	0.77
0.32	0.660
0.2	0.41
0.11	0.230

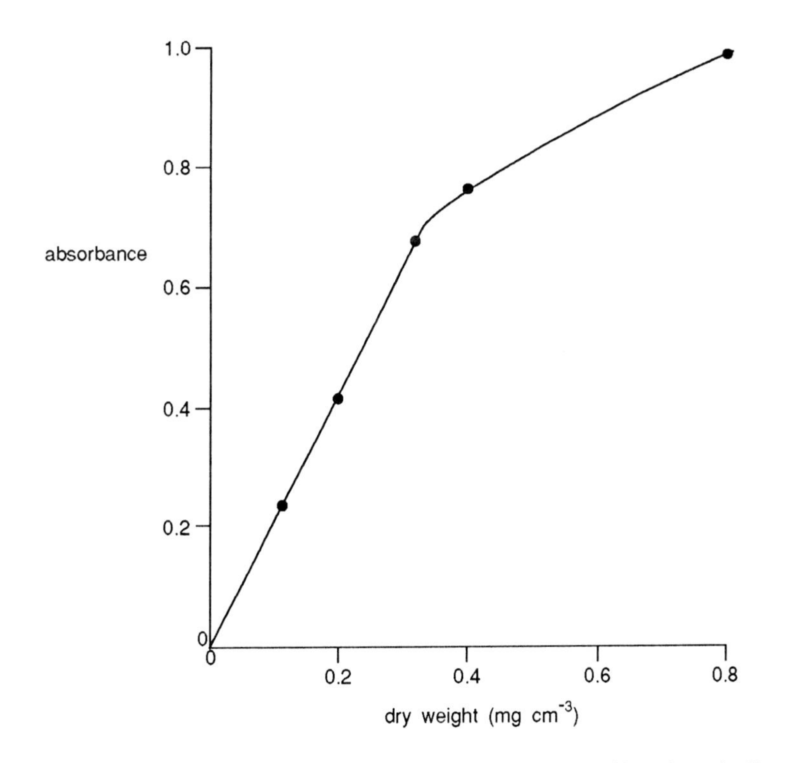

This calibration curve could of course be improved by taking more samples and by taking duplicate samples.

2) Absorbance due to cells = 0.560 - 0.060 = 0.50. We can see from the graph that an absorbance of 0.5 is equivalent to a dry weight of 0.24 mg cm^{-3}. Since the culture was diluted five-fold, the culture density is 0.24 x 5 = 1.2 mg cm^{-3}.

2.2

All the sources of error listed in the question are possible sources of error in using the pour plate technique.

Care must therefore be taken to minimise error from these sources. For example, the carry-over of cells from low to high dilution is minimised by using a fresh pipette for each transfer when creating the dilution series. CFUs cm^{-3}, estimated by pour-plate and spread-plate techniques, is given to only one decimal place to reflect the potentially large error associated with the measurement.

2.3

At the 100-fold dilution we can see that the transfer of 1 cm^3 of the diluted suspension gave rise to 85 colonies per plate, this dilution is selected for the count.

The original sample therefore contains 8.5 x 10^3 CFUs cm^{-3}. In practice we would have carried out the determination at least in duplicate, preferably in triplicate.

You should note that both large and small colonies on the plate are counted. The difference in colony size in the pour-plate technique can be accounted for by most cells being fully entrapped within the agar gel (small colonies) but some cells being able to develop on the surface of the agar (large colonies). The count is expressed to only one decimal place to reflect the accuracy of the count.

2.4

For culture A, the number of CFUs cm^{-3}

$= \frac{126 + 93 + 81}{3} \times 10^6 \times 10 = 1.0 \times 10^9$ CFUs cm^{-3} (x 10^6 is the dilution x 10 because we used 0.1cm^3 of sample).

For culture B, the number of CFUs cm^{-3}

$$= \frac{102 + 91 + 107}{3} \times 10^8 \times 10 = 1.0 \times 10^{11} \text{ CFUs cm}^{-3}$$

2.5

Method of biomass estimation	Factors influencing accuracy
1) protein	complex growth medium
2) ATP	relatively slow sampling speed
3) absorbance	Cell clumping; presence of non-cellular particulate material; cell lysis; changes in biomass morphology
4) wet weight	changes in biomas morphology; cell lysis; presence of non-cellular particulate material
5) DNA	none of the factors listed

2.6

50 mg	-	dry weight
1.0 mg	-	Biuret protein and DNA
0.1 mg	-	Lowry protein and absorbance
0.00001mg	-	cell count

2.7

From the data given we can calculate the OTR = ϕ (C_{in} - C_{out}) = 1 (0.5 mmol O_2 min^{-1})

= 0.5 mmol min^{-1} = 16 mg min^{-1}

(since molecular mass of O_2 is 32)

But OTR = $\frac{\mu X}{Y_o}$

Thus 16 = $\frac{0.2X}{0.5}$ $\therefore$ X = 40mg

2.8

1) This was a fairly open ended question. We have listed 9 factors which influence the reliability of estimates of biomass based on metabolic activities. Hopefully your answer included many of these.

a) Poor accuracy of measurement of nutrient or product.

b) High growth yield coefficient.

c) Variable growth yield coefficient.

d) Substrate also used for product synthesis.

e) Chemically unstable substrate or product.

f) Metabolism by resting cells ie substrate consumption or product formation poorly correlated with growth.

g) Product further metabolised ie not an end-product of metabolism.

h) Growth in complex media in which more than one component serves as carbon source.

i) Some substrate converted to cell storage material.

2) a) Dry weight measurement - iv - the method lacks sensitivity

b) Protein measurement. None of the listed merits and limitations particularly apply to this method.

c) Measurement of O_2 consumption rate - (i, ii, iii, v) This technique is quick, sensitive and measures live (metabolically active) cells. It does, however, require specialist equipment.

2.9

1) A good selection here would be to use absorbance, although a total cell count could be used. To use absorbance however you would need to have a calibration curve relating turbidity to dry weight. Absorbance measurements are certainly less tedious than cell counting. The other methods listed are unsuitable because they would take too long to give a result (eg dry weight, protein and DNA estimation) and/or give the data that was not required (viable cell count, most probable number).

2) Here the anticipated biomass concentration would be low and probably contaminated with dust particles. Thus dry weight, absorbance and O_2 consumption are unlikely to yield good data. Total cell counts and viable cell counts could only be done if the cells were first concentrated (by filtration or centrifugation). The most probable number technique looks most promising but even this has its problems. For example, in selecting the medium to be used, how can we guarantee that all of the cells present in the condensate are capable of growing in the medium selected? In cell probability therefore this method will under estimate the biomass present.

The alternatives of protein and DNA measurement might be used but the biomass in the condensate would first have to be concentrated by centrifugation. The situation presented in this question illustrates that decision making in this area of study is not easy and often we are forced to accept less than perfect compromises.

3) We would recommend protein measurement. This procedure is more straightforward than DNA estimation and is not subject to the distortions caused by differences in cell shape encountered in estimations based on the turbidity of the cultures.

4) The measurement of O_2 consumption would appear to be the best as it gives continuous and instant feedback. It is, however, a costly procedure to set up. An alternative would be to use the turbidity (absorbance) of the culture as the monitoring mechanisms. This is a cheaper procedure to set up and can be built into the bioreactor to give a continual measurement. It is, however, subject to some distortions particularly as a result of the cells becoming attached to the optical faces.

5) Viable cell counting would be the best in this case as the point of interest is the survival (viability) of the biomass, not its total quantity.

6) We would need to use a method which would not be dependent on cell morphology or on having intact cells. The better candidates here are protein or DNA estimations.

In our responses to these questions, you will have seen that often the selection of suitable methods is not straightforward. We can usually exclude some methods as totally inappropriate, the choice between the remainder depends upon such factors as the desired degree of precision and accuracy, the availability of equipment and costs.

Responses to Chapter 3 SAQs

3.1

1) The components of the homogenisation medium are formulated by:

- empirical evaluation. For example, inhibitors of proteases may be added to the homogenisation medium to ensure that the relevant purified protein remains intact;

- rational analysis and copying of physiological solutions. Since the cytoplasmic pH of cells is around 7.5 to 8.0, the homogenisation medium should reflect the physiological solutions of the cell and is therefore weakly buffered to the cytoplasmic pH.

3.2

The principal parameters which must be controlled to maximize efficiency of mixing in an homogeniser are:

- the clearance between the rotating pestle and the walls of the glass vessel. Most Potter-Elvehjem homogenisers have a gap clearance of between 0.05 and 0.50 mm. This value must be carefully chosen so that minimal damage is caused to the released subcellular organelles;

- the number of upward and downward displacements of the pestle within the glass vessel;

- the speed of rotation of the pestle;

- the length of time of homogenisation;

- the ratio of homogenisation medium to biomass concentration.

Responses to Chapter 4 SAQs

4.1

In all numerical problems it is advisable to ensure that all the units are compatible - this is best achieved by use of the SI system.

$$\omega = \frac{\pi \times 60\,000}{60} = 2\,000\,\pi\ \text{s}^{-1}$$

From Equation 4.1

$$\begin{aligned}\text{applied centrifugal force} &= \omega^2 x \\ &= (2000\,\pi\ \text{s}^{-1})^2\,(0.06\ \text{m}) \\ &= \underline{2.37 \times 10^6\ \text{m s}^{-2}}\end{aligned}$$

and from Equation 4.2

$$\text{RCF} = \frac{\left(2.37 \times 10^6\ \text{m s}^{-2}\right)}{\left(9.80\text{m s}^{-2}\right)} = \underline{241\,700}$$

In this question there is a mixed bag of units and it is essential to be very clear about the units to use. We recommend that in entering numbers into any equation you include each unit and ensure that the answer has the correct unit.

4.2 Using Equation 4.3a

$$v = 2\,\omega^2 x r^2 (\rho_p - \rho_m) \setminus 9\eta$$

$$= \frac{2 \times (2\pi \times \frac{50\,400}{60}\, s^{-1})^2 \times (0.05\, m) \times (2.7 \times 10^{-9}\, m)^2 \times \left[(1.330 - 0.998) \times 10^3\, kg\, m^{-3}\right]}{9 \times (0.890 \times 10^{-3})\, kg\, m^{-1}\, s^{-1}}$$

$$= \frac{2 \times 2.79 \times 10^7 \times 0.05 \times 7.29 \times 10^{-18} \times 0.332 \times 10^3}{9 \times 0.890 \times 10^{-3}} \qquad \frac{s^{-2}\, m\, m^2\, kg\, m^{-3}}{kg\, m^{-1}\, s^{-1}}$$

$$= \underline{8.43 \times 10^{-7}\, ms^{-1}}$$

4.3 As the sedimentation rate of the particle is closely associated with the centrifugal field and the size of the particle, it is apparent that the particle will remain stationary when the density of the particle and the suspending medium are equal, (Equation 4.3a).

$v = 0$ when $\rho_P = \rho_M$

We must remember however, that there may be a sharp difference between the observed separation of particles and the expected separation of particles by centrifugation. The sedimentation rate is also dependent upon the shape of the particles as this may also influence ρ_P.

4.4 The main problem associated with this technique is that only an impure preparation or enriched fraction is almost invariably obtained. The initial suspension comprises a population of small and large particles. The applied centrifugal field may result in small particles which were originally suspended near the bottom of the centrifuge tube forming a pellet with larger particles originally suspended near the top of the tube. The resultant pellet will comprise a heterogenous mixture of small and large particles. Other difficulties with this technique are that there are significant differences in size of the same organelle within a single cell type. In addition, different organelles may adhere to each other within the pellet and prove exceedingly difficult to separate even with the repeated washing steps.

4.5 The ideal gradient material should:

- be inert towards the biological material, the centrifuge tubes and rotor;
- not interfere with the monitoring of the sample material; eg it must be relatively easy to monitor the exact concentration of the sample throughout the density gradient by, for example, analysis of the refractive index of the solution;
- be easily separated from the fraction after centrifugation;
- be stable in solution and be readily available in a pure and analytical form;
- exert minimal osmotic pressure.

4.6 There are two main reasons why precise loading is not critical in isopycnic centrifugation:

- many such centrifugation are performed with self-forming gradients where the sample is mixed thoroughly with the salt of the heavy metal or iodinated gradient medium prior to centrifugation;
- the particles will separate according to their buoyant density which is critically dependent upon the difference in density of the particle and the suspending medium.

4.7 $\omega = 50400 \times 2\pi/60 = 5.2779 \times 10^3\, s^{-1}$

$$\omega^2 = 2.7856 \times 10^7 s^2$$

using the formula for s (Equation 4.6)

$$s = \frac{\ln x_2/x_1}{\omega^2(t_2 - t_1)} = \frac{\ln 6.731/5.949}{(2.7856 \times 10^7 s^2)(70 \times 60 \text{ s})}$$

$= 10.55 \times 10^{-13}$ s $=$ 10.55 Svedburg

4.8

$$s = 2r^2(\rho_P - \rho_M)/9\eta$$

$$r = \left\{\frac{9\eta s}{2(\rho_P - \rho_M)}\right\}^{1/2} = \left\{\frac{9 \times (0.890 \times 10^{-3} \times 10.55 \times 10^{-13}) \text{ kg m}^{-1}}{2(0.335 \times 10^3) \text{ kg m}^{-3}}\right\}^{1/2}$$

$= (12.6 \times 10^{-18})^{1/2} = 3.55 \times 10^{-9}$ m

For a spherical molecule (we have assumed that the molecule is spherical)

$$M = \tfrac{4}{3}\pi r^3 \rho_P N_A = \tfrac{4}{3}\pi (3.55 \times 10^{-9} \text{ m})^3 \times (1.333 \times 10^3 \text{ kg m}^{-3})(6.022 \times 10^{23} \text{ mol}^{-1})$$

$= 150.49$ kg mol^{-1}

$= 150\,490$ g mol^{-1}

4.9

Use Equation 4.10 to evaluate M and find f from Equation 4.9 and f_o from the assumption that the molecule is spherical of radius r. From the ratio f/f_o obtain a value for the axial ratio (Table 4.1). From Equation 4.10.

$$M = \frac{RTs}{D(1 - \vartheta\rho_m)} = \frac{(8.314 \text{ J K}^{-1} \text{ mol}^{-1})(298.15 \text{K})(5.04 \times 10^{-13} \text{ s})}{(6.97 \times 10^{-11} \text{ m}^2 \text{ s}^{-1})(1 - 1.0024 \times 0.734)}$$

$= 67.83$ kg mol^{-1} $= 67\,830$ g mol^{-1}

Hence $\underline{M_r = 67\,830}$

From Equation 4.9

$$f = \frac{RT}{N_A D} = \frac{(8.314 \text{ J K}^{-1}\text{mol}^{-1})(298.15 \text{ k})}{(6.022 \times 10^{23} \text{ mol}^{-1})(6.97 \times 10^{-11} \text{ m}^2 \text{ s}^{-1})}$$

$= \underline{5.91 \times 10^{-11} \text{ kg s}^{-1}}$

The volume of a molecule = $\vartheta \times M_r/N_a$

$$\vartheta_M = \frac{(0.734 \times 10^{-3} \text{ m}^3 \text{ kg}^{-1})(67.83 \text{ kg mol}^{-1})}{(6.022 \times 10^{23} \text{ mol}^{-1})(6.022 \times 10^{23} \text{ mol}^{-1})}$$

$= 8.27 \times 10^{-26}$ m^3

From which we can calculate the effective radius:

$$r = 3\left(\frac{3\vartheta_M}{4\pi}\right)^{1/3} = \left(\frac{3 \times 8.27 \times 10^{-26}}{4\pi}\right)^{1/3} = 2.70 \times 10^{-9} \text{ m}$$

$$f_o = 6\pi\eta r = 6\pi(0.890 \times 10^{-3} \text{ kg m}^{-1} \text{ s}^{-1})(2.70 \times 10^{-9} \text{ m})$$

$= 4.53 \times 10^{-11}$ kg s^{-1}

The ratio $f/f_0 = \frac{5.91 \times 10^{-11}}{4.53 \times 10^{-11}} = 1.30$

The axial ratio from Table 4.1 is therefore about 6, ie the molecule is about six times longer than it is wide.

4.10 Using Equation 4.14 gives a direct value for M irrespective of its shape.

$$M = \frac{2RT \ \ln c_2/c_1}{(1 - \vartheta\rho_M)\ \omega^2 (x_2^2 - x_1^2)}$$

$$= \frac{2\,(8.314\,J\ K^{-1}\ mol^{-1})\ (293.15\ k)\ \ln 9.40}{(1 - 0.749 \times 0.9982)\ (2\pi \times 120\ s^{-1})^2\ [\ (0.065^2 - 0.055^2\ )m^2]}$$

$= 63.45$ kg mol^{-1}

$M_v = 63\ 450$

Responses to Chapter 5 SAQs

5.1

1) Ethanol, CH_3CH_2OH, has a hydroxyl group which will readily form hydrogen bonds with hydroxyl groups of other ethanol molecules. There is therefore relatively strong interaction between ethanol molecules and the liquid has a relatively high boiling point.

2) Chloroethane, CH_3CH_2Cl, has no groups which allow it to hydrogen bond to other chloroethane molecules but it does possess an electronegative chlorine atom which confers upon the molecule a permanent dipole. Dipole-dipole interactions between chloroethane molecules will therefore lead them to align themselves in a relatively ordered way (Figure 5.2)

3) and 5)

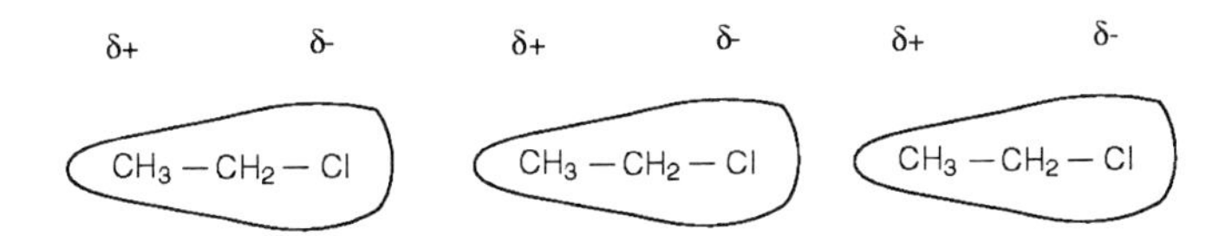

Ethane and hexane are totally non-polar molecules. They possess no hydroxyl groups or similar pairs of atoms that would enable them to hydrogen bond. They also have no electronegative atoms so they have no permanent dipole. To associate with other ethane or hexane molecules, these compounds must rely on van der Waals forces. That is, the electron cloud around the molecule will be continually shifting so that there exist continually changing temporary or induced dipoles which interact with similar temporary dipoles in adjacent molecules.

4) Sodium acetate is an ionic compound existing in the form of sodium and acetate ions. These ions can interact strongly by ionic (coulombic) forces, with ions of the opposite charge in adjacent molecules thus building up a strong, ordered crystal structure with sodium ions surrounded by acetate ions and vice versa.

The order of boiling points are:

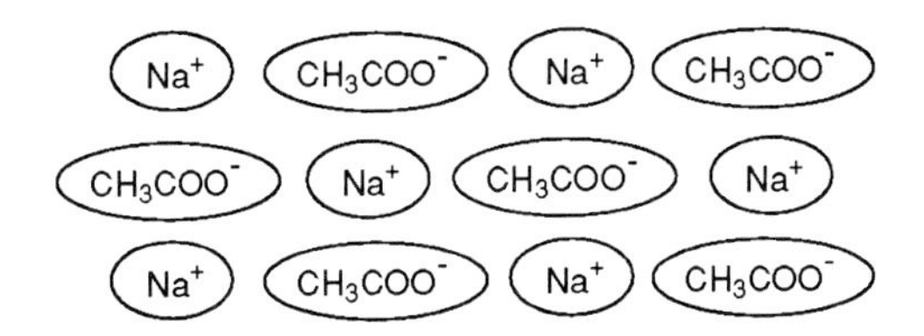

sodium acetate, ethanol, hexane, chloroethane, ethane

The order of hexane and chloroethane may have surprised you. Read on for an explanation.

Sodium acetate would be expected to have the highest boiling point because of the strong lattice structure described above. (In fact, it decomposes before boiling). Ethanol would have the next highest boiling point (78.3 °C) because the liquid molecules are bound together by the relatively strong hydrogen bonds.

The dipole-dipole interactions constituting the major forces between chloroethane molecules, are weaker than either of the above, so that they would be expected to have a lower boiling point (13.1 °C) than either sodium acetate or ethanol.

Ethane and hexane are held together by even weaker forces - the temporary dipole-dipole interactions caused by fluctuation of the electron cloud - van der Waals forces. Correspondingly then, ethane has a much lower boiling point (-88.6 °C) than any of the previously discussed compounds. Hexane on the other hand, has a much higher boiling point (68.9 °C) than might on first thought be expected. Despite being of the same class of compound as ethane and despite the fact that its liquid form is held together by forces weaker than those at work in liquid chloroethane, hexane has a much higher boiling point than both.

This is because the boiling point of a liquid is not dependent solely on the major intermolecular forces but also on its molecular mass. A molecule of higher molecular mass will tend to have a higher boiling point because greater energy is required to release the molecules from their liquid state. In a homologous series of molecules such as alkanes, boiling points will tend to increase with increasing molecular mass.

Intermolecular forces therefore play a large part in determining a compound's physical properties and are also extremely important in determining how molecules function in biological systems, as well being very relevant to the factors involved in their purification.

5.2

a) $K = C_o / C_{aq} = 3$

If x g of A are extracted from the aqueous layer to attain equilibrium

Then $C_o = x/100 \quad C_{aq} = (10 - x)/100$

Thus $K_D = x/(10 - x) = 3$

$x = 7.5$ g

The mass of A remaining in the aqueous layer = 2.5 g

b) If y g of A are extracted from the aqueous layer for the first 25 cm^3 of ether, then:

$C_o = y/25$ and $C_{aq} = (10 - y)/100$

$$K_D = \frac{y/25}{(10-y)/100} = \frac{4y}{10-y} = 3$$

$$y = \frac{30}{7} \text{ g}$$

ie a fraction 3/7 of the original weight of A has been extracted.

For each of the four extractions 3/7 of the amount of A remaining in the aqueous layer after the previous extraction is removed by ether. Therefore, the total fraction of A extracted by 4 successive 25 cm^3 aliquot of ether is:

$$\frac{3}{7} + \frac{3}{7} \text{ of } \frac{4}{7} + \frac{3}{7} \text{ of } \left(\frac{4}{7}\right)^2 + \frac{3}{7} \text{ of } \left(\frac{4}{7}\right)^3 = 0.8933$$

Thus the mass of A extracted = 0.8933 x 10 g = 8.933 g

and the mass remaining in the aqueous phase = 1.066 g

(You may have worked out the mass extracted after each 25 cm^3 addition - you should have obtained the same result).

This result shows that the larger the number of extractions, the more complete is the extraction for a definite total volume of solvent.

5.3

They are all important. If a compound is strongly soluble in the mobile phase and interacts only very weakly with the stationary phase, it will move relatively quickly through the column, ie will have a low retention time. If the mobile phase interacts strongly with the stationary phase the molecule will have to compete with the mobile phase for binding sites on the stationary phase and this will serve to further lower the retention time of the molecule. If on the other hand, the compound is retained very strongly by the stationary phase, it will elute from the column much more slowly.

5.4

Adsorption takes place through the formation of intermolecular interactions such as hydrogen bonding, dipole-dipole interactions and van der Waals forces. Ionic bonding does not generally play a part however, as this involves formation of a bond too strong to be easily broken.

5.5

(a) If the distribution coefficient is 0 then from equation 5.6 V_e must equal V_o. .This means that the solute is eluted in the void volume and is therefore too large to enter any of the pores of the gel. Its molecular mass is therefore above the cut-off point of the gel filtration medium. This is analogous to standard chromatography whereby a K_D of 0 represents a compound which does not adsorb at all to the stationary phase.

b) If the distribution coefficient is 1 then the elution volume (V_e) minus the void volume (V_o.) must be the same as the internal pore volume. This means that the solute has available to it all the space within the beads. This solute is therefore very small and will elute last.

c) All molecules separated by size exclusion chromatography must, by definition, have a K_D value between 0 and 1. A distribution coefficient of greater than 1 indicates that the solute is affected by processes other than size exclusion, ie the solute is adsorbed by the gel.

5.6

The table of results is as follows.

Protein	RMM	log (RMM)	elution volume(cm^3)
cytochrome c	12 400	4.09	205
chymotrypsinogen	25 000	4.40	188
ovalbumin	45 000	4.65	170
transferrin	66 000	4.82	150
lactate dehydrogenase	135 000	5.13	130
catalase	240 000	5.38	105
β-galactosidase	520 000	5.72	90
protein X			140

By plotting the data (log RMM against elution volume) it can be seen that the elution volume is approximately directly proportional to lg (RMM). By simply reading the graph then, we can see that an elution volume of 140 cm^3 corresponds to lg (RMM) of 5.0, so we can estimate that X has a RMM of approximately 100 000.

5.7

1) Amino acids can be well separated by ion exchange chromatography or zone electrophoresis but nowadays it is common to separate them by reverse phase HPLC.

2) D- and L-valine are of course, optical isomers and as such will behave identically on all chromatography systems except those which involve interaction with a chiral compound. Therefore they can be separated on an HPLC system with a chiral stationary phase. Alternatively, the amino acids can be derivatized with an optically pure chiral reagent to give stereoisomers which can then be resolved by reverse phase HPLC.

3) An antibody by its very nature has an antigen to which it binds specifically. This interaction makes the antigen ideal for use as a stationary phase for affinity chromatography which may well allow for the isolation of an antibody from a complex mixture in a single step.

4) Compounds such as fat-soluble vitamins which are highly soluble in organic solvents are not always well suited to partition chromatography and are best resolved by adsorption chromatography.

5) A mixture of long-chain polymers will cover a wide range of molecular masses and are thus well suited to separation by size exclusion chromatography.

6) Organic acids could probably be separated by reverse phase HPLC but as they are ionisable they are probably better resolved by ion-pair chromatography. An alternative would be to esterify the acids to give volatile esters which can then be separated by GLC.

7) Geometric isomers of biliverdin methyl esters can be separated by reverse phase HPLC but, as mentioned above, adsorption chromatography is often the most efficient method of resolving such isomers.

If you had different answers to any of the above you are necessarily wrong - methods other than those mentioned could undoubtably be used to separate most of the mixtures. For example, the serum solution could also be subjected to gel filtration, dialysis and electrophoresis. In practice, the choice of separation methods often depends on more practical considerations such as apparatus available, amount of sample and whether the material needs to be recovered.

5.8

One explanation of the simple peak in the first chromatography system is that the two compounds had the same retention time and thus co-eluted.

It is possible that the solution did only contain a single solute which was degraded by the conditions of the second system, resulting in the two peaks of the breakdown products. Alternatively, one of the two compounds in the original mixture may have been destroyed by the conditions of the first system which then detected only one compound. The first system may not have employed a detection system which was capable of detecting both solutes, eg a TLC stain specific for only one type of material whereas the detection method of the second system recorded the presence of both solutes, eg a TLC plate stained in such a way as to visualize all organic material.

5.9

Your completed table of separation techniques should contain most of the following information.

Technique	Solute properties utilised for separation	Examples of mixtures which can be separated
Dialysis	Size	Proteins and buffer salts
Adsorption chromatography	Partitioning behaviour between chemically inert solid adsorbent and liquid mobile phase	Grass or leaf pigments
Reverse phase HPLC	Partitioning behaviour between a non-polar bonded liquid stationary phase and a polar liquid mobile phase	Drug metabolites
Ultrafiltration	Size	Proteins
Ultracentrifugation	Sedimentation properties (size and density)	Nuclear and mitochondrial DNA
Zone electrophoresis	Charge	Amino Acids
Polyacrylamide gel electrophoresis	Size	Proteins
Paper partition chromatography	Solubility, partition	Amino acids
Gas-Liquid chromatography (GLC)	Solubility, partition	Lipids
Size exclusion chromatography	Size	Proteins
Chiral HPLC	Absolute stereochemistry	(+) and (-) forms of enantiomers
Affinity chromatography	Biological specificity	Receptor proteins
Ion-exchange chromatography	Charge	Organic acids
Solvent extraction (counter current chromatography)	Solubility in immiscible solvents	Lipids, dyes

You may, however, have selected quite different examples of the mixture which can be separated by the various techniques.

Responses to Chapter 6 SAQs

6.1

An absorbency change of 0.1 per min is defined as one enzyme unit.

An absorbency change of 0.26 per min is therefore produced by 0.26/0.1 = 2.6 enzyme units.

Therefore the 50 μl aliquot contained 2.6 enzyme units.

Therefore in 1 cm^3 of the extract there are 1000/50 x 2.6 = 52.0 enzyme units.

The original extract therefore contained 52 enzyme units per cm^3.

6.2

Total enzyme units in original extract = 100 x 23 = 2300 units.

Total enzyme units in ammonium sulphate fraction = 30 x 21 = 630 units.

Thus:

Yield = 630/2300 x 100 = 27.4%

This is a very low yield for a single purification step and would not normally be acceptable.

6.3

You will see from the equations in Figures 6.10 and 6.11, that for every glyceraldehyde-3-phosphate molecule that is converted to dihydroxyacetone phosphate, one molecule of NADH is converted to NAD^+ when the dihydroxyacetone phosphate is converted to glyceraldehyde-1-phosphate. As we saw earlier, NADH has an absorption maximum at 340 nm whereas NAD^+ does not absorb at this wavelength. The conversion of NADH to NAD^+ is therefore observed as a decrease in absorbency at 340 nm, the rate of decrease being a measure of triosephosphate isomerase activity.

6.4

Include MDH and NADH in the substrate solution. As soon as oxaloacetate is formed it will be converted to malate, and at the same time, for every molecule that is converted, one NADH molecule is converted to NAD^+ This can be monitored by the decrease in absorbency at 340 nm, which in turn will be a measure of the rate of formation of oxaloacetate.

The example described here is a simple one-step linked assay. There are examples where as many as three enzymes are linked in an assay to generate a measurable product.

6.5

Either 350 nm or 420 nm could be used. Compounds A and B have near identical spectra at 220 and 250 nm so their interconversion would result in no obvious absorbency change at these wavelengths. However, the conversion of A into B results in a decrease in absorbency at 350 nm and an increase in absorbency at 4210 nm.

Therefore either of these wavelengths could be used to monitor the reaction.

6.6

1) Glucose and glucose-6-phosphate cannot be differentiated by their spectra. Glucose-6-phosphate dehydrogenase is therefore included with the glucose substrate to oxidise the glucose-6-phosphate produced. In the process, for every glucose-6-phosphate molecule produced, one $NADP^+$ molecule is converted to NADPH. The production of NADPH, and hence the enzyme rate, is measured by the rate of increase in absorbency at 340 nm. This is therefore an example of a linked reaction.

2) Glucose and gluconic acid cannot be differentiated by their spectra. However, by introducing the enzyme peroxidase into the substrate solution, together with the reduced form of o-dianisidine, each hydrogen peroxide molecule released is used to oxidise the o-dianisidine to a yellow colour which is recorded continuously on a spectrophotometer. This is another example of a linked assay.

6.7

1) The reason why the direct measurement of the product of an enzyme catalysed reaction might not be possible is because it does not absorb in either the UV or visible region of the spectrum.

2) A potential disadvantage of the use of a chemical derivatizing agent is that it is either inhibitory or affects the enzyme itself so that if added to the reaction system only a single measurement can be obtained. The method of discontinuous analysis will increase the error involved. Further many chemical derivatizing agents are non-specific.

3) By definition an artificial substrate is not the natural substrate of the enzyme; it is chosen because its disappearance or the appearance of a reaction product can be easily monitored. Further the enzyme is unlikely to react, either qualitatively or quantitatively, in the same manner as it would with its natural substrate. The use of artificial substrates, however, enables rapid monitoring of enzyme activity.

6.8 In principle we could assay this enzyme manometrically by measuring the volume of oxygen evolved. This is not, however, a particularly sensitive or convenient assay.

6.9 1) Direct spectroscopic method.

Advantages: simplest approach, specific, sensitive, can be used in a continuous mode, does not usually require the use of artificial substrates, useful for quick routine assays, temperature control of assay not necessary.

Disadvantages: only of use if the product(s) or substrate absorb in the UV or visible region of the spectrum, cannot be used with turbid suspensions.

2) Microcalorimetry.

Advantages: useful for coloured and turbid reaction mixtures, widespread applicability, very sensitive, can be used in continuous mode.

Disadvantages: expensive instrumentation required, long measurement time compared to spectroscopic methods, non-specific.

3) Conductance method.

Advantages: can be used in continuous mode, useful for coloured or turbid mixtures.

Disadvantages: limited use, only applicable to reactions in which there is a change in the number of ions, difficulty of detecting small changes of conductance in the presence of a background of high concentration of buffer ions, not very sensitive, requires good temperature control.

Responses to Chapter 7 SAQs

7.1 No. This has always been a problem. Each rabbit responds differently.

In the first place the antibodies against an antigen are different at the molecular level. This is the consequence of the fact that each animal has its own set of potential antibody-producing lymphocytes. Some epitopes will activate a lymphocyte in one rabbit but not in another.

Further the immunization factors play a role in determining the concentration of antibodies in the blood (titer): the number of times that a rabbit has been injected and the amount of injected antigen. Also the type of rabbit and its age are important factors.

7.2 The answer is 6. You should plot a graph of ratio of IgG: BSA (y axis) against quantitiy of BSA (x axis). We provide you with a description of how to calculate the ratio of IgG: BSA. Thus,

X mg of BSA yields an absorbance of X. 0.667 in 1 cm^3. This extinction value is subtracted from the total extinction, the difference is the absorbance caused by IgG.

The outcome is divided by 1.439 and yields the mgs of IgG in the precipitate. The molal ratio in the precipitate is yielded by the quotient (μg IgG/150): (μg BSA/68).

You should have calculated that: at 5 mg BSA the ratio is 5.9; at 25 mg BSA the ratio = 5.2; at 50 mg BSA the ratio = 4.2; at 100 mg BSA the ratio = 2.3. By extrapolation of a graph of these ratios against BSA to the y-axis the value of 6 is found.

7.3

1) Agar with a low electroendosmosis must be applied for rocket electrophoresis because the antibodies must move in the gel as little as possible. Hence, agar with as few charged groups as possible.

2) In the case of cross-over electrophoresis the IgG molecules must move as much as possible due to electrophoretic endosmosis. For this type of experiment, agar with comparatively many negatively charged groups is chosen.

3) The type of agar is not important for the Ouchterlony technique because no electric field is created here and thus electroendosmosis will not occur.

7.4

1) nitrocellulose

2) nylon

3) nitrocellulose

4) polyvinylidene difluoride

5) polyvinylidene difluoride

The answers to these are given in Tables 7.1 and 7.2

7.5

1) This experiment indicates that two of the fragments of bacterial DNA separated by electropheresis contained nucleotide sequences that were complementary to nucleotide sequences present in the bacteriophage. It could indicate (but not prove) that the original bacteria were 'contaminated' by bacteriophage.

2) This experiment would indicate that the bacterium carried a gene which coded for β-galactosidase. Thus, the ^{32}P-mRNA would hybridise with this sequence. This in turn would cause a blackening of the film. We could use the position of this blackened area on the film to locate the DNA fragement carying the β-galactosidase gene on the original electrophoretogram.

Responses to Chapter 8 SAQs

8.1

This question contains wavelengths given in a variety of units -to conform to SI units they must all be expressed in meters before using the equations.

Wavenumber, $\bar{v} = 1/\lambda$ (Equation 8.2)

Frequency, $v = c / \lambda$ (Equation 8.1)

Energy per quantum $E = h v$ (Equation 8.3)

Energy per mole $E_m = N_A h v$ (Equation 8.4)

a) $\lambda = 5 \times 10^{-4}$ nm $= 5 \times 10^{-13}$ m Thus

$\bar{v} = 1/5 \times 10^{-13} = 2 \times 10^{12}$ m^{-1} $= 2 \times 10^{10}$ cm^{-1}

$v = (2.998 \times 10^{8}$ ms$^{-1}) / (5 \times 10^{-13}$ m$) = 6 \times 10^{20}$ s^{-1}

$E = (6.625 \times 10^{-34}$ J s$)(6 \times 10^{20}$ s$^{-1}) = 4 \times 10^{-13}$ J

$E_m = (6.022 \times 10^{23}$ mol$^{-1})(4 \times 10^{-13}$ J$) = 2.4 \times 10^{8}$ J mol^{-1}

b) $\lambda = 25\,00$ Å $= 25\,00 \times 10^{-10}$ m $= 2.5 \times 10^{-7}$ m

$\bar{v} = 1/2.5 \times 10^{-7} = 4 \times 10^{6}$ m^{-1} $= 4 \times 10^{4}$ cm^{-1}

$v = (2.998 \times 10^{8}$ m s$^{-1})/(2.5 \times 10^{-7}$ m$) = 1.2 \times 10^{15}$ s^{-1}

$E = (6.625 \times 10^{-34}\ J\ s)(1.2 \times 10^{15}\ s^{-1}) = 7.95 \times 10^{-19}\ J$

$E_m = (6.022 \times 10^{23}\ mol^{-1})(7.95 \times 10^{-19}\ J) = 4.78 \times 10^{5}\ J\ mol^{-1}$

c) $\lambda = 30\ \mu m = 30 \times 10^{-6}\ m$

$\bar{v} = 1/30 \times 10^{-6} = 33.33 \times 10^{3}\ m^{-1} = 333\ cm^{-1}$

$v = (2.998 \times 10^{8}\ m\ s^{-1})/(30 \times 10^{-6}\ m) = 10^{13}\ s^{-1}$

$E = (6.625 \times 10^{-34}\ J\ s)(10^{13}\ s^{-1}) = 6.625 \times 10^{-21}\ J$

$E_m = (6.022 \times 10^{23}\ mol^{-1})(6.625 \times 10^{-21}\ J) = 3.98 \times 10^{3}\ J\ mol^{-1}$

8.2

For this calculation we can use equation 8.9 $A = \varepsilon cl$.

Thus, after converting the concentration units from mol dm^{-3} to mol m^{-3}, we have; $\varepsilon = 0.88/(7.5 \times 10^{-2} \times 0.02)$

Therefore $\varepsilon = 586$ and answer (d) is correct.

8.3

It is necessary to draw up the following Table - in this instance we will keep to pre SI units (ie practical units which are in every day use), $l = 0.2$ cm.

$[Br_2]/(mol\ dm^{-3})$	0.001	0.005	0.010	0.015	0.02
T(%)	81.4	35.6	12.7	4.5	1.6
$A = -\log T$	0.089	0.449	0.896	1.346	1.796
$\varepsilon = A/cl$	445	449	448	449	449

Thus mean extinction coefficient = 448 $dm^3\ mol^{-1}\ cm^{-1}$ and the molar extinction coefficient = 44.8 $m^2\ mol^{-1}$

As an alternative you could have plotted a calibration graph of A against concentration:

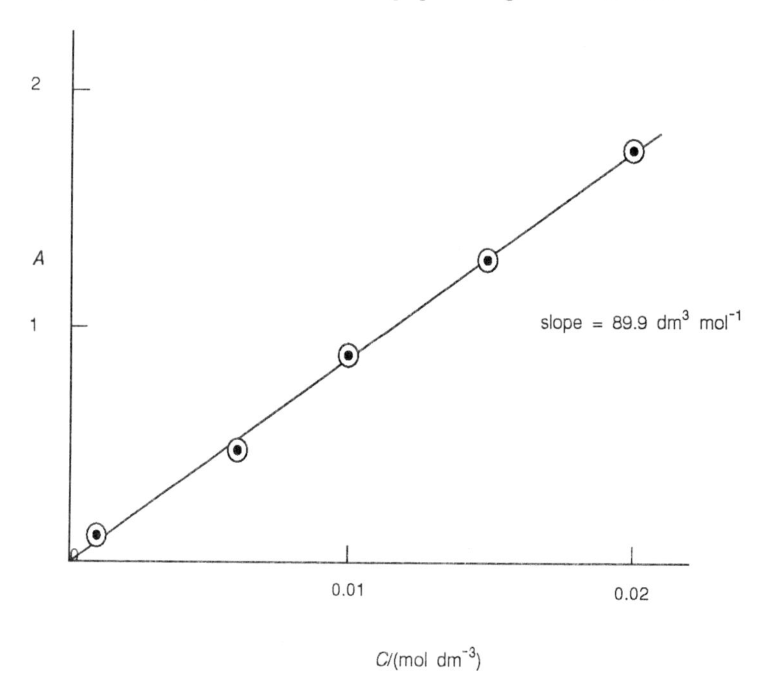

From Equation 8.9

Slope = $A/c = \varepsilon l$

ε = slope/l = 89.8/0.2 = 449 $dm^3\ mol^{-1}\ cm^{-1}$

= 44.9 $m^2\ mol^{-1}$

One advantage of plotting such a graph is that the validity of Beer-Lambert law can be tested - ie a linear plot passing through the origin.

8.4 The answer is c). If you chose d) you were in fact describing fluorescence or phosphorescence; the energy released as electrons fall back to a lower energy level may be emitted as light resulting in fluorescence or phosphorescence. These phenomena can also be used as biochemical tools.

8.5 The right response is b). You could, of course, just turned the page to find that the absorbance of a sample is given by the expression; A = log (I_o/I). If you chose (c) you were getting confused with transmittance.

8.6 There are two types of protons on the molecule - those on the CH_2 group and those on the CH_3 group. There is no spin-spin coupling between the protons of the two ethyl groups as they are too far apart. The triplet must correspond to the protons of the methyl group(s) as this is coupled to two adjacent protons which have a total of three different spin states. The quartet must correspond to the protons of the CH_2 group(s) as this group is coupled to three protons with a total of four different spin states.

8.7 The correct answer is d). A doublet and a quartet have these relative intensities as we can predict using Pascal's triangle.

8.8 Statement (a) is false. Only a certain few nuclei with an odd number of protons such as hydrogen nuclei (protons), ^{13}C, ^{31}P, ^{19}F, can undergo NMR. These are nuclei which are spin-active, that is, in a strong magnetic field, they can exist in more than one spin-state, each of a different energy.

Statement (b) is false though this might have been slightly confusing. The actual chemical shift measured in Hz, will increase as the operating frequency increases but in practice we express the chemical shift in terms of δ ppm which takes this into account, and is therefore independent of the operating frequency.

Statement (c) is also false. Spin-spin couplings are due to interactions with only nearby protons - not with all of the protons on the molecule. If protons are separated by more than 3 σ bonds they will not generally interact.

Statement (d) is true. The signal from a proton coupled to a methyl group will be split into a quartet with relative intensities 1:3:3:1.

8.9 About 6.8%. You probably calculated this in the following way. Hexane contains 6 C atoms and 14 H atoms and so 6 x 1.1% of its C atoms and 14 x 0.015% of its H atoms will be present in the form of ^{13}C and ^{2}H, respectively. This means that about 6.8% of hexane molecules will have a molecular mass of 87 and therefore the intensity of the M+1 peak relative to the M^+ peak will be about 6.8%.

This answer is not however quite correct. Some molecules may have more than one ^{13}C or more than one ^{2}H atoms. In such caes they will give peaks at M+x (where x is greater than 1). The calculation done above assumes that only one atom is replaced by the heavy isotope per molecule.

8.10

1) Infrared spectroscopy involves absorption of energy which gives rise to vibrations within a molecule, about covalent bonds. Therefore **infrared spectroscopy** is a form of vibrational spectroscopy.
2) **Mass spectrometry** involves analysis of ionised molecules and fragments.
3) X-ray diffraction patterns can only be obtained from crystals and so **X-ray crystallography** is the correct answer.
4) Both **visible spectroscopy** and **ultraviolet spectroscopy** are forms of electronic spectroscopy, ie they involve the transition of electrons between different energy levels.

8.11 Statement 1) is true. Unlike the other techniques mentioned, mass spectrometry differentiates molecules on the basis of their ionization and fragmentation by other particles.

Statement 2) is false. Infrared radiation is of lower energy and therefore, longer wavelength and lower frequency than visible or ultraviolet radiation.

Statement 3) is false. Ultraviolet spectroscopy is not capable of determining molecular formulae; however, the presence of different chromophores may give an indication as to the type of molecule under study.

Statement 4) is true. Like hydrogen nuclei, ^{13}C nuclei resonate at different frequencies depending on the local magnetic environment which ultimately depends on the structure of the molecule.

8.12

1) One of the best means of obtaining the molecular mass of a compound is by subjecting it to mass spectrometry. From this it is often possible to obtain a molecular mass (from the molecular ion) and often the molecular formula.

2) To obtain a picture of the molecule with detailed atomic resolution, it is necessary to carry out X-ray crystallography.

3) A useful method of identifying the presence of particular functional groups is by infrared spectroscopy. However, other methods such as nmr may also provide a great deal of structural information, including the presence of particular groups.

4) Mass spectrometry is a very valuable technique for the study of biosynthesis as it can be used to indicate the incorporation of an isotopically labelled precursor - such as ^{13}C- or ^{18}O-labelled acetate - into a particular molecule. NMR can also be used to follow the fate of labelled precursors.

Responses to Chapter 9 SAQs

9.1

1) 1 Curie = $3.7 . 10^{10}$ desintegrations per second

   10^{-3} Curie = $3.7.10^{7}$ dps

2) The half life of ^{32}P is about 2 weeks, thus after 8 weeks $1/2^4 = 1/16$ part of the radioactivity will be present, which is about 2.10^{6} dps.

9.2 ^{3}H is a much softer (lower energy) beta emitter than ^{32}P. Therefore, a smaller number of beta particles emitted by ^{3}H will penetrate the window of the counter and can be registered. Thus at the same number of disintegrations ^{32}P will yield more counts per second than ^{3}H.

9.3 The best method is to use a lot of ^{3}H and little ^{14}C (if other experimental boundary conditions allow). For a proper adjustment of the counter no correction in the ^{14}C channel (channel B) is needed because only the ^{14}C desintegrations detected in this channel). The correction in the ^{3}H channel for the ^{14}C count will be small when we use (relatively) little ^{14}C.

9.4 A counting error of 1% is obtained at 10000 counts (error is $\sqrt{10000} = 100$); thus in 10 minutes at least 10000 counts must be collected. Of these 10.100 = 1000 are originating from the background. So there must be at least 9000 counts per 10 minutes attributed to the ^{14}C radioactivity in the sample. At an efficiency of 50% this corresponds to 18000 disintegrations per 10 minutes = 30 disintegrations per second; thus the sample must contain 30 Becquerel. (Remember 1 Becquerel = 1 disintegration per second).

Responses to Chapter 10 SAQs

10.1 Your diagram of a batch calorimeter should be similar to that shown in Figure 10.3.

The basic components of a batch calorimeter include a heavy metal block (aluminium or steel) as a heat sink, capable of rotation in an air thermostat. Located in the block are two box-shaped calorimeter vessels, each is surrounded by semi-conducting thermocouple plates. The reaction vessels are two-compartmented stoppered glass containers; the contents of the separate compartments are mixed on rotation.

10.2 Batch calorimetry is not commonly used to measure the heat produced by growing cultures because:

- normal growth is not achieved when the cells sediment;
- aerobic cells will be in an enclosure of ever decreasing oxygen tension leading to the possibility of different metabolic processes taking over;
- it is not possible to monitor increase in biomass and other parameters as growth proceeds.

10.3 When growth ceases in a medium containing excess glucose there will be an immediate decrease in the power; however, this will not fall to the base line but to a low value where it will remain constant until all the residual carbon sources, including breakdown products formed during growth, have been broken down. The carbon dioxide-time profile is identical to the power-time plot.

10.4 To establish the mass balance it is necessary to determine the total amount of carbon available initially in the glucose and that in the biomass produced and carbon dioxide evolved.

Total carbon initially available	=	$3.3 \times 10^{-3} \times (620/1000) \times 6$
	=	12.28×10^{-3} mol
[glucose contains 6 C atoms]		
Total carbon stored in the biomass	=	$0.1843 \times 0.451/12$
[Atomic mass of (C) = 12]	=	6.928×10^{-3} mol
Total carbon in the carbon dioxide	=	5.351×10^{-3} mol

Total carbon recovered = 12.279×10^{-3} mol = total carbon available
Thus all of the carbon is accounted for.

10.5 To establish the energy balance it is necessary to calculate the total energy available from the oxidation of glucose and then find out how this is partitioned between energy stored in the biomass and the energy liberated as heat.

Total energy available from the oxidation of glucose

$= 3.3 \times 10^{-3} \times 28903 \times (620/1000) = 5.73$ kJ

The energy stored in the biomass is obtained from the ratio of mol of carbon stored in the cells to that originally present (from SAQ 10.4 this ratio is 6.928/12.28 = 0.564).

Hence total energy stored in biomass = $0.564 \times 5.73 = 3.23$ kJ

This is 56.4% of the available energy.

The energy liberated as heat = 1.89 kJ; ie 33% of available energy.

The total energy which can be accounted for is 5.12 kJ, ie 89.4% of the available energy.

The residual 10.6% of the energy is that required for biosynthetic and maintenance processes. Similar figures have been reported for other organisms.

10.6 For the calculation of ΔG^o and ΔS^o values you need to use the following thermodynamic relationships:

$$\Delta G^o = -RT \ln K_c$$

and

$$\Delta S^o = (\Delta H^o - \Delta G^o)/T$$

Using these equations the values obtained are given in the table (R = 8.314 J K^{-1} mol^{-1} and T = 298 K):

Macromolecule	Ligand	ΔG^o (kJ mol^{-1})	ΔS^o (J K-1 mol^{-1})
α–chymotrypsin	indole-3-propionic acid	-18.44	10.7
human serum albumin	L-tryptophan	-23.65	-13.3
(HSA)	benzodiazepin	-23.47	-61.5
HSA/L-tryptophan complex	benzodiazepin	-17.1	20.8
Troponin C	CA^{2+}	-34.22	81.3
Myosin	ADP	-34.68	-128.9

All the binding reactions are exothermic and all are spontaneous, ie they have negative values for ΔG^o. The binding of indole-3-propionic acid to α-chymotrypsin and benzodiazepin to the HSA/tryptophan complex are weak as indicated by the low values of the formation constant and the corresponding low values of ΔG^o. The binding of calcium ions to troponin C and ADP to myosin are, however, very strong. In the reaction of calcium ions with troponin C the entropy term is an important factor in deciding the sign and value of the free energy change.

Responses to Chapter 11 SAQs

11.1

Using Equation 11.1, ΔG = - nFE. For this cell E = 1.1 V, n = 2, and F = 96 500 C mol^{-1}, hence:

ΔG = 2 x 1.1 x 96 500

= - 212 300 J mol^{-1} = -212.3 kJ mol^{-1}

If you are confused about units: V = J A^{-1} s^{-1} where A = ampere and C (Coulomb) = A s so the units of the ΔG from these is J A^{-1} s^{-1} x A s mol^{-1} = J mol^{-1}.

11.2

To write their electrode reactions you must remember the general equation:

$$\text{ox} + n\,e \rightleftarrows \text{red}$$

and then decide on the reduced and oxidized states; the reduction reactions and the electrode formulations are:

1)	Mg^{2+} + 2e	→	Mg	Mg^{2+} | Mg
2)	H^+ + e	→	0.5 H_2(g)	H^+ | H_2, Pt
3)	0.5 Cl_2(g) + e	→	Cl^-	Pt, Cl_2 | Cl^-
4)	AgCl(s) + e	→	Ag(s) + Cl^-	AgCl, Ag | Cl^-
5)	Fe^{3+} + e	→	Fe^{2+}	Fe^{3+}, Fe^{2+} | Pt
6)	CH_3CHO + $2H^+$ + 2e	→	C_2H_5OH	CH_3CHO, C_2H_5OH | Pt

Note that in all cases, except (a) and (d) where the metals also form the electronic conductor, platinum is included as the inert electronic conductor. You will also note that for (f) it is necessary to include hydrogen ions; this is a common feature of biological redox systems.

11.3

For the cell:

- Pt, H_2 (g, 1 atmos) | H^+ Cl^- | AgCl, Ag +

1) The electrode reactions are:

negative electrode	$0.5\ H_2(g)$	→	$H^+ + e$	(oxidation)
positive electrode	$AgCl(s) + e$	→	$Ag(s) + Cl^-$	(reduction)
Cell	$0.5\ H_2(g) + AgCl(s)$	→	$Ag(s) + H^+ + Cl^-$	

2) To obtain the overall emf of the cell we must first write down the separate electrode potentials ($p_{H2} = 1$ atmos):

$$E(H^+,H_2) = E^o(H^+,H_2) + (RT/F)\ln a_{H+}/(p_{H2})^{1/2} = (RT/F)\ln a_{H^+} \text{ since } E^o(H^+,H_2) = 0$$

$$E(AgCl,Ag,Cl^-) = E^o(AgCl,Ag,Cl^-) - (RT/F)\ln a_{Cl^-}$$

$$\text{Hence } E\text{ cell} = E(AgCl,Ag,Cl^-) - E(H^+,H_2)$$

$$= E^o(AgCl, Ag, Cl^-) - (RT/F)\ln a_{Cl^-} - (RT/F)\ln a_{H^+}$$

$$= E^o(AgCl, Ag, Cl^-) - (RT/F)\ln a_{Cl^-}\, a_{H^+}$$

3) The standard emf of this cell is the same as the standard electrode potential of the silver chloride electrode, since by definition

$E^o(H^+, H_2) = 0.$ thus E^o for the silver electrode $= 0.2224$ V

11.4 From the data in Table 11.1 it is possible to construct the hypothetical cell:

$- \text{Pt} \mid NAD^+, NADH, H^+, CH_3CHO, CH_3CH_2OH \mid \text{Pt} +$

for which the electrode reactions are:

$CH_3CHO + 2H^+ + 2e \longrightarrow CH_3CH_2OH$ and $NAD^+ + 2H^+ + 2e \longrightarrow NADH + H^+$

which on addition gives the required equation:

$$CH_3CHO + NADH + H^+ \rightleftarrows CH_3CH_2OH + NAD^+$$

Therefore the standard emf of the hypothetical cell is given by:

$$E^{o\prime} = -0.197 - (-0.320) = 0.123\text{ V}$$

Whence $\Delta G^o = -2 \times 96\,500 \times 0.123 = -23\,739$ J

Since from thermodynamics:

$$\Delta G^o = -RT \ln K \text{ or } K = \exp(-\Delta G^o/RT)$$

Using the value of ΔG^o, $K = \exp(23\,739/8.314 \times 298) = 1.45 \times 10^4$.

11.5 1) In its full form Equation 11.7 is:

$$E = (RT/F)\ln(a_{H^+,1}/a_{H^+,2}) + (RT/F)\ln(a_{Cl^-1}/a_{Cl^-,2})$$

If we assume ideal behaviour then we can replace activity values by molalities and the equation becomes:

$$E = (RT/F)\ln(m_{H^+,1}/m_{H^+,2}) + (RT/F)\ln(m_{Cl^-,1}/m_{Cl^-,1}/m_{Cl^-,2})$$

substituting the ratio $m_1/m_2 = 10$ in this equation gives:

$E = (RT/F) \ln 10 + (RT/F) \ln 10$

The quantity (RT/F) at 298 K = 8.314 x 298/96 500 = 0.0257 V.

Hence the emf of the cell E = 2 x 0.0257 x 2.303 = 0.1184 V.

Since the emf depends on the ratio of the molalities, the emf is independent of the actual values of m_1 and m_2.

2) Using the values of activity coefficients given and remembering that $a = m\ \gamma$ we can substitute directly into Equation 11.7:

a) $m_1 = 0.1$, $m_2 = 0.01$

$E = (RT/F) \ln (0.1 \times 0.83/0.01 \times 0.91) + (RT/F) \ln (0.1 \times 0.76/0.01 \times 0.90)$

$= 0.0257 \ln 9.12 + 0.0257 \ln 8.44$

$= 0.0568 + 0.0548 = 0.1116$ V

b) $m_1 = 0.01$, $m_2 = 0.001$

$E = (RT/F) \ln (0.01 \times 0.91/0/001 \times 0.97) + (RT/F) \ln (0.01 \times 0.90/0.001 \times 0.96)$

$= 0.0257 \ln 9.38 + 0.0257 \ln 9.375$

$= 0.0575 + 0.575 = 0.115$ V

It is thus apparent that as the two solutions become more dilute the value of the emf approaches the ideal value.

11.6

The basic reaction in both the electrodes is:

$$\text{glucose} + O_2 + H_2O \xrightarrow{\text{glucose oxidase}} H_2O_2 + \text{gluconicacid}$$

a) A glass electrode coated with a thin film of glucose oxidase can be used to detect small changes of pH resulting from the generation of gluconic acid.

b) Glucose oxidase is immobilized in a layer over the iodide electrode. The hydrogen peroxide liberated by the glucose oxidase reaction oxidizes the iodide ions (added to the sample solution):

$$H_2O_2 + 2I^- + 2H^+ \longrightarrow I_2 + 2H_2O$$

and the local changes in the iodide ion concentration are detected.

Suggestions for further reading

General texts

Covering essential biological chemistry

'Molecular Fabric of Cells'. BIOTOL series Buttersworth-Heinemann 1991

'Infrastructure and Activities of Cells' BIOTOL series. Butterworth-Heinemann 1991

D Friefelder 'Physical Biochemistry', Applications to Biochemistry and Molecular biology. Freeman 1976

L Stryer 'Biochemistry' (3rd Edition) Freeman 1988.

General texts covering a range of techniques

D J Holme & H Peck 'Analytical Biochemistry' Longmans 1983.

J Steesh Experimental Biochemistry Allyn and Bacon 1984

K H Wilson and K L Goulding 'A Biologiest's Guide to Principles and techniques of Practical Biochemistry'. 3rd Edition Arnold 1986.

T S Work and E Work, 'Laboratory Techniques in Biochemistry and Molecular Biology', North Holland 1979.

Texts - relating to specific techniques

We recommend texts from the ACOL (Analytical Chemistry by Open Learning) series published by John Wiley & Sons (Chichester, New York, Brisbane, Toronto and Singapore). These texts are prepared in open/distance learning style and are of particular value ot distance learners. Important titles are:

P A Sewel, V B Clarke, 'Chromatographic Separations', ACOL series Wiley 1987.

S Lindsay, 'High Performance Liquid Chromatography', ACOL Series Wiley 1987.

M Melvin 'Electrophoresis' ACOL Series, Wiley 1987.

R C Denney and R Sinclair, 'Visible and Ultraviolet Spectroscopy', ACOL series Wiley 1987.

B George and P McIntyre, 'Infra Red Spectroscopy', ACOL Series Wiley 1987.

D Williams 'Nuclear Magnetic Resonance Spectroscopy' ACOL series Wiley 1986.

W J Geary 'Radiochemical methods', ACOL series, Wiley 1987.

D Hawcroft, T Hector and F Rowell 'Quantitative Bioassay', ACOL series, Wiley 1987.

J Dodd and K Tonge, 'Thermal Methods', ACOL series, Wiley 1987.

T Riley and C Tomlinson, 'Principles of Electroanalytical Methods', ACOL series, Wiley 1987.

A Evans 'Potentiometry and Ion Selective Electrodes' ACOL series Wiley 1987.

In addition, the following texts provide additional information on specific techniques

C M Collins and P M Lyne 'Microbiological Methods' Butterworth-Heinemann, (1987).

W B Jakoby and I H Pastam, 'Cell Culture' Academic Press (1988).

A E Beezer 'Biological Microcalorimetry' Academic Press (1980).

A M James, 'Thermal and Energetic Studies of Cellular Biological Systems'. Wright Publishers (1987)

D B Hibbert and A M James 'Dictionary of Electrochemistry' 2nd Edition MacMillan (1984).

R L Solsky, 'Ion-Selective Electrodes in Biomedical Analyses' CRC Cnt Rev. Anal.Chem. 14.1 (1983).

D Rickwood, 'Centrifugation, A practical approach', Information Retreval Ltd 1978.

Texts dealing with the analysis of specific groups of compounds.

'Analyses of Amino Acids, Proteins and Nucleic Acids' (1992) BIOTOL Series Butterworth-Heinemann Oxford.

'Analysis of Carbohydrates and Lipids' (in press) BIOTOL series Butterworth-Heinemann, Oxford.

Texts dealing with underpinning issues.

B Woodget and D Cooper, 'Samples and Standards' ACOL series Wiley 1987

D McCormic and A Roach ' Measurement, Statistics and Computation' ACOL series Wiley (1987).

Text dealing with the application of the techniques covered by this text on large scale processes.

'Product Recovery' in Bioprocess Technology (1992) BIOTOL series Butterworth-Heinemann, Oxford

J Krijgsman 'Release of Intracellular Components' in Advanced Course in Downstream Processing. J Krigsman, Delft Institute of Technology, The Netherlands (1991)

L Svrosky 'Filtration Fundamentals in Solid Liquid Separation' 2nd Edition, L Svarovsky Ed. Butterworth-Heinemann, (1981)

T C Lo, M M I Bourd and C Hanson 'Handbook of Solvent Extraction' (1983)

F J Dechow 'Separation and Purification Techniques in Biotechnology' Noyes Publ. Pork Ridge N.Y (1989)

G Jagchies 'Process scale Chromatography' Ullmann's Encyclopaedia of Industrial Chemistry. Vol B3 10 1-44 (1988).

Appendix 1 - Enzyme assays and enzyme kinetics

In Chapter 6, we examined enzyme assays especially from the point of view of monitoring the purification of an enzyme. The discussion was based upon the assumption that the reader has some knowledge of enzymes especially in relation to their reaction kinetics. This appendix provides a revision of some of the important aspects of enzyme catalysed reaction kinetics. It also includes some additional practical considerations that need to be borne in mind when we assay enzymes, particularly the need to ensure that the rate of product formation is proportional to the enzyme concentration and the need to measure initial reaction rates.

A.1 Ensure that the rate of product formation is proportional to the enzyme concentration

Enzyme assays should enable us to quantitatively compare how much enzyme is present in various solutions. To be able to do this, the above condition must be satisfied, such that results similar to those shown in Figure A.1 are obtained experimentally.

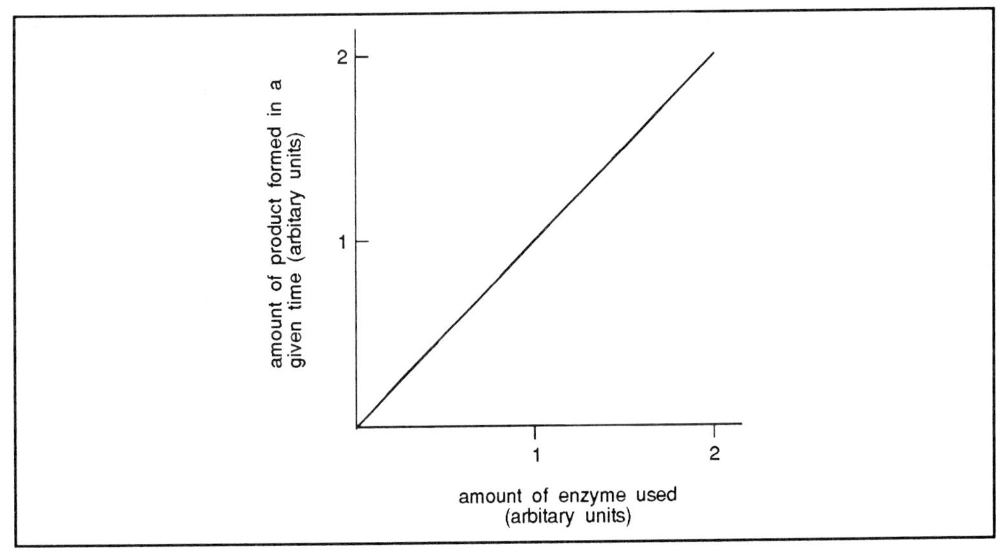

Figure A.1 Effect of enzyme concentration on product formation. For an accurate assay of enzyme activity, doubling the volume of enzyme added must result in double the amount of product being formed in a given time. Rate of reaction should be proportional to the volume of enzyme solution used.

If double the amount of enzyme is present, then the rate of product formation should be doubled. This requirement needs to be stressed because several other factors influence enzyme activity eg pH, temperature. To assay an enzyme, we should ensure, as far as possible, that these other factors are not limiting the rate and that it is the enzyme concentration which is determining the rate. In practice, this means checking that rate is proportional to the volume of enzyme solution used (Figure A.1).

Π Whilst attempting to assay an enzyme, the following data were obtained. Does the volume of enzyme solution used influence the result? If so, which volume do you recommend to best assay the enzyme. Why was incubation (A) included, in which no enzyme was present?

Incubation	Volume of enzyme added (cm^3)	Amount of product formed (μmol), after 5 min incubation
(A)	0	0
(B)	0.05	0.325
(C)	0.1	0.65
(D)	0.25	1.4
(E)	0.5	1.95

The purpose of an enzyme assay is to allow us to measure how much enzyme is present. To do this, the rate of the reaction must be proportional to the amount of enzyme present. This is most easily seen if the amount formed is normalised, for example to 1.0 cm^3 of enzyme. This gives the following:

Incubation	Vol enz used (cm^3)	μmol product formed	μmol product formed per cm^3 enzyme, assuming proportionality
(A)	0	0	0
(B)	0.05	0.325	6.5
(C)	0.1	0.65	6.5
(D)	0.25	1.4	5.6
(E)	0.5	1.95	3.9

For incubations (B) and (C) there is proportionally between rate and volume of enzyme: (C) has formed 2.0 x as much as (B). (D) should have formed 5 x as much as (B) but has not do so (5 x that formed in (B) is 1.63); this under-production is shown up in the normalised figure in the table above. This assay condition would lead to a slight underestimate. (E) is more seriously flawed, giving only just over half of the true value.

Assuming the other assay conditions remain the same, it does not matter which volume is used: a suitable volume is 0.05 or 0.1 cm^3. On a more general note, too much enzyme present for too long will lead to poor results. What was the purpose of incubation (A)? This is a control, to allow any product formed in the absence of enzyme to be allowed for. This will be discussed further shortly.

A.2 Ensure that the initial rate is measured.

rate often declines with time

If we examine the amount of product formed with increasing time, we often observe that the rate declines with time (Figure A.2).

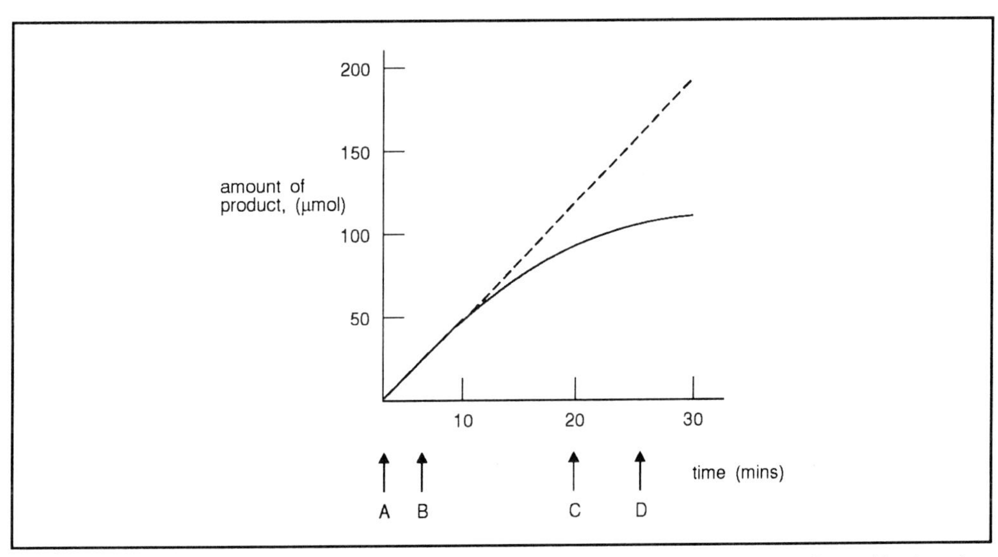

Figure A.2 Progress of an enzyme-catalysed reaction with time. The amount of product formed is plotted on the Y axis.

The dashed line indicates a 'theoretical' result, in which product formation is maintained at a steady rate. In practice, as shown by the continuous line, the reaction usually slows down. If we estimate the rate at different stages in the case above, we get very different results:

rate estimated over period A to B = 5 μmol min^{-1}

rate estimated over period C to D = 2 μmol min^{-1}

Of these, the higher value is the better estimate of enzyme activity; the second is seriously misleading and is a consequence of changes in the reaction mixture. There are various possible causes for this: any one of the following would be sufficient to produce the result shown.

[S] must be saturating

- Substrate is no longer saturating. As already noted, only the enzyme concentration should be rate-limiting in an enzyme assay. Consequently, we try to ensure that a high and saturating substrate concentration is used; with time, the substrate concentration will drop so that a lower reaction rate results.

- The product formed in the reaction may build up to inhibitor levels. We can visualise the reaction as follows:

$$E + S \rightleftarrows ES \rightleftarrows EP \rightleftarrows E + P$$

- If product is allowed to build up, its binding to enzyme molecules may increase, which will then not be available (however transiently) for reaction with substrate molecules. Presence of EP means there is less free E available, hence a lower rate occurs.

- The reverse reaction is occurring and equilibrium is being approached. Looking at the reaction scheme above, if we start with only enzyme and substrate, reaction can only be towards the right-hand side. If product concentration builds up, the reverse reaction may begin to occur; eventually an equilibrium would result, in which there was no net formation of product (reaction to the right proceeds at the same rate as reaction towards the left): the enzyme may be catalysing many millions of reactions per minute but there would be no more product detected.

- Some change in reaction conditions may have occurred, leading to lessening of enzyme activity: this could include change of pH, such that the pH was no longer optimal (see Chapter 6), or the enzyme may have been denatured, so that fewer active enzyme molecules are present.

must measure initial rate

How does the experimenter cope with these problems? The solution is to avoid them by making measurements before the rate slows down. This is known as measuring the initial velocity, when conditions are still those determined by the experimenter (buffered at a known pH, with a known substrate concentration, no product present, controlled temperature) If this is done, all of the above problems will be avoided: in practice, one should check that the rate is linear over the chosen period of assay.

Π More data to analyse! Attempts at assaying an enzyme yielded the data shown below. Recommend an assay duration for routine use (several hundred samples may have to be assayed per day!).

	Incubation	Amount of product formed (μmol) after time					
		0	1	2	5	10	20 (min)
(A)	No enzyme	0	0	0	0	0	0
(B)	0.1 cm^3 enzyme	0	0.2	0.4	1.0	2.0	4.0
(C)	0.2 cm^3 enzyme	0	0.4	1.2	2.0	3.9	6.3
(D)	0.5 cm^3 enzyme	0	0.8	1.9	4.2	6.1	7.0

Hint: If in doubt how to analyse this, plot a graph!

The most important thing is to ensure that the initial velocity is being measured. Plotting a graph of amount of product formed (Y axis) against time (X axis) will show for how long the initial rates are maintained. This is shown here:

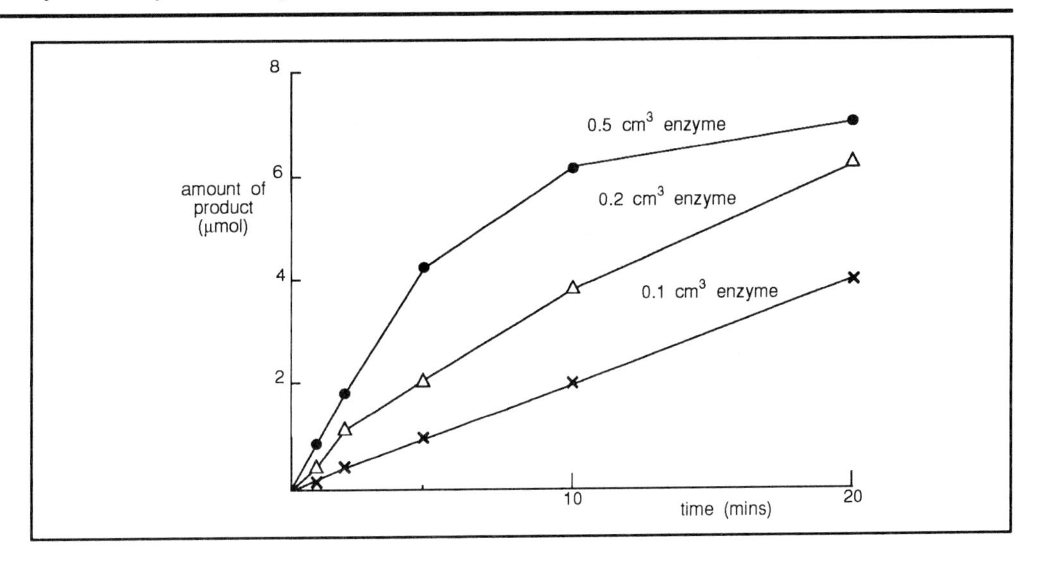

For incubation (B) there is no problem: the amount of product increases linearly with time. The duration of the incubation is a matter of convenience - perhaps 5 or 10 minutes, because with shorter times, errors in timing will become more serious. In (C), the reaction is beginning to slow down, as seen by the curve between 10 and 20 minutes. For (D), this slowdown has occurred much earlier and net formation of product has been reduced dramatically by 10 minutes.

Where does this lead us? Assays of 5 or 10 minutes, with 0.1 or 0.2 cm^3 enzyme solution, look suitable, providing the amount of product formed does not exceed 2-3 μmol. If it does, we should either shorten the assay period or use a smaller volume of enzyme.

A.3 Conduct appropriate controls

In assaying an enzyme, we must ensure that we distinguish between the enzyme-catalysed reaction and any non-enzymic product formation. For example, invertase hydrolyses sucrose to glucose and fructose; at low pH, acid hydrolysis will also occur. When measuring the amount of glucose formed, we will over estimate the enzyme activity if we do not correct for any non-enzyme reaction (Figure A.3).

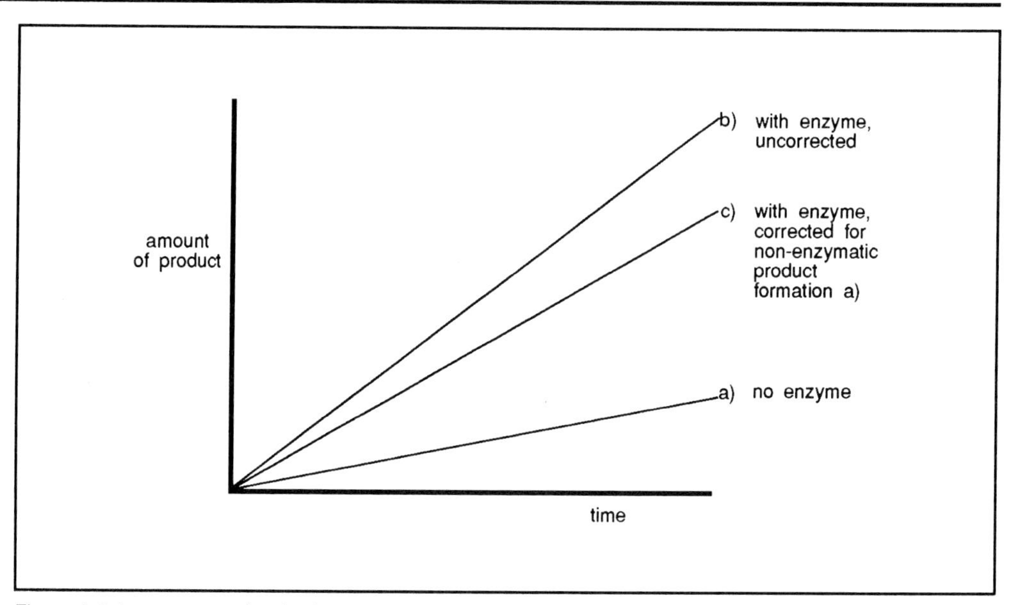

Figure A.3 In an enzyme incubation, some product may arise non-enzymically (line a); to obtain the true enzyme activity this must be deducted from that formed in the presence of enzyme (line b), giving the net product formation which is attributable to enzyme (line c).

non-enzymic reaction

Various controls are conducted to allow non-enzymic product formation to be deducted. The commonest is to use enzyme which has been denatured by boiling to test for the occurrence of the non-enzymic reaction: whatever product is found in this incubation is deducted from that found with the active enzyme. This approach ensures that any product present at the start of the assay (as a contaminant of the substrate or the enzyme extract), together with any formed non-enzymically during the assay, is allowed for. We will now do an exercise to illustrate the importance of this point.

Π Calculate the enzyme activity, as μmol product formed per incubation per cm^3 enzyme, for the following data,

1 Using incubation (A) alone.

2 Then taking into account the additional information which incubation (B) provides.

		μmol product at end of incubation
(A)	0.5 cm^3 enzyme	15
(B)	0.5 cm^3 boiled enzyme	8

(NB boiling normally destroys enzyme activity)

1 If just incubation A had been conducted, we should have concluded that 1 cm^3 of enzyme would make 30 μmol product in the incubation. This gives an activity of 30 μmol product per cm^3 enzyme.

2 The boiled enzyme control (incubation B) shows also that some non-enzymic product formation occurred. We must assume that this also occurred in incubation (A). We thus deduce that, of the 15 μmol formed in the presence of

active enzyme in incubation (A), 8 μmol were formed non-enzymically. Hence the 0.5 cm^3 active enzyme produced 15-8 = 7 μmol product, and its true activity was 14 μmol product per cm^3 enzyme.

A.4 Use defined conditions

define the conditions

Ideally, enzyme assays are conducted at optimal conditions for the enzyme. This includes providing a substrate concentration which is saturating, and any other compounds necessary for activity of the enzyme in question. The pH should be optimal and controlled by a suitable buffer. The temperature should also be held constant, preferably at a temperature which is unlikely to denature the enzyme - such as 30°C or 37°C. These factors are discussed in Chapter 6.

A.5 Effect of substrate concentration on enzyme activity

It is well established that substrate concentration influences the rate of enzyme-catalysed reactions. In this section the precise relationship between these parameters will be explored in detail.

first order reaction

If the rate of a chemical reaction (non-enzyme) is analysed, it is found to be proportional to substrate concentration (Figure A.4); this is referred to as a first order reaction meaning that the rate is proportional to the concentration of a single component (the substrate).

$$\text{ie rate} \propto [S]$$

In contrast, experiments with enzyme-catalysed reactions revealed a different relationship (Figure A.4).

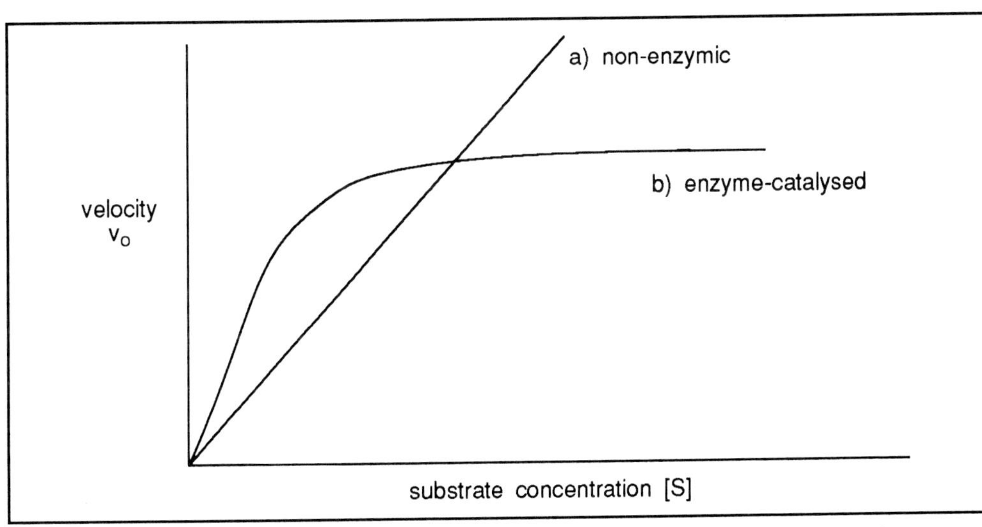

Figure A.4 Relationship between rate and reactant concentration for a) Non-enzymic reaction; b) Enzyme-catalysed reaction.

zero order reaction

Here, rate increased in proportion to substrate concentration only at low substrate concentrations; at higher concentrations, the increase in rate lessened for unit increase in substrate concentration, eventually levelling off to a plateau. At high substrate concentration the rate is said to be zero order as rate is no longer affected by a change in substrate concentration. Between these two regions is a mixture of first and zero order kinetics.

enzyme-substrate complex

The mathematical relationship between initial velocity and substrate concentration was analysed in the early part of the twentieth century by two scientists, Michaelis and Menten. They analysed the enzyme distribution on the basis of the following reaction scheme:

$$E + S \underset{k_2}{\overset{k_1}{\rightleftharpoons}} ES \underset{k_4}{\overset{k_3}{\rightleftharpoons}} E + P$$

where k_1, k_2 etc are rate constants. Catalysis was assumed to require the combination of enzyme (E) with substrate (S) to form an enzyme-substrate complex (ES). This reaction was reversible, but ES could also break down to form free enzyme (E) and product (P). In deriving an equation to describe the observed relationship between initial velocity and substrate concentration, they made several assumptions:

- that the concentration of enzyme, (E), was low, relative to that of substrate. This meant that formation of ES did not significantly alter the concentration of free substrate, and hence simplified the analysis. Typically, this assumption is acceptable;

- that product was absent; as a consequence, the possible recombination of E and P to form ES could be ignored (thus any term involving rate constant k_4 was eliminated);

- that the concentration of ES was constant and was in equilibrium with E and S; this is referred to as the equilibrium assumption. Subsequent analysis showed this to be a restricted case: a wider generalisation nonetheless remains that (ES) is constant: a steady state exists between its formation ($E + S \rightarrow ES$) and its breakdown ($ES \rightarrow E + S$ and $ES \rightarrow E + P$) and is established very soon after enzyme and substrate are mixed. The rate of the overall reaction is given by k_3 (ES) and thus velocity is proportional to (ES). The existence of ES was initially deduced by Michaelis, on the basis of the hyperbolic relationship between rate and substrate concentration; it has since been observed by electron microscopy and confirmed by spectroscopic techniques.

Michaelis-Menten equation

These assumptions are used in the derivation of the Michaelis-Menten equation. At this stage you do not need worry about deriving the relationship (which is therefore omitted) but you should know and understand the Michaelis-Menten equation. This is usually given as:

$$v = \frac{V_{max}\ [S]}{K_M + [S]}$$

where v - initial velocity of the reaction

[S] = substrate concentration in mol dm^{-3}

V_{max} = maximum velocity, seen at (infinitely) high substrate concentration; V_{max} is a constant

K_M = Michaelis constant, in mol dm^{-3}.

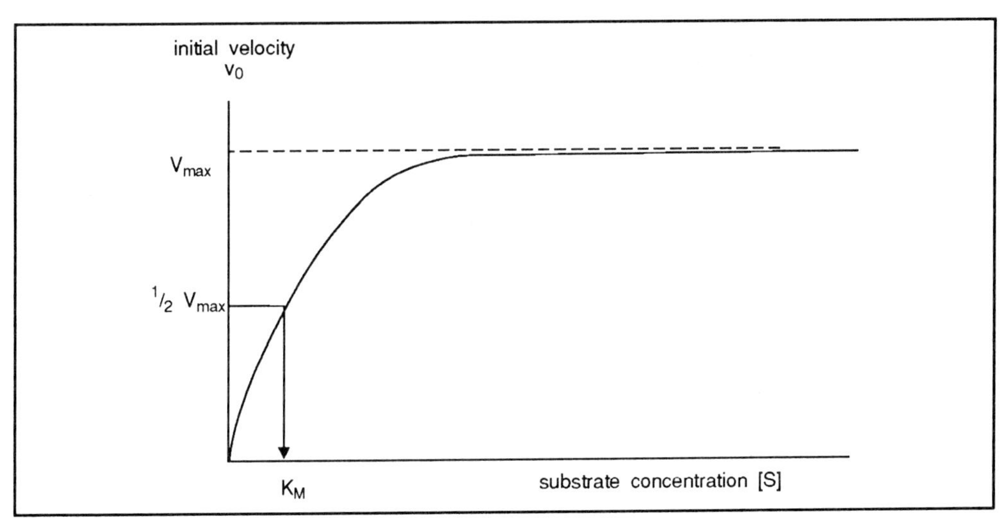

Figure A.5 Relationship between initial velocity, Vo and substrate concentration. [S], for an enzyme which obeys the Michaelis-Menten equation. The Michaelis constant, K_M is the [S] at which half maximum velocity ($\frac{1}{2} V_{max}$) is displayed.

We can now analyse the observed relationship between initial velocity and substrate concentration (Figure A.5) in terms of the Michaelis-Menten equation.

- At low substrate concentration where $[S] < K_M$ the denominator approximates to K_M (because the contribution of [S] to $K_M + [S]$ is negligible). Under these conditions, the equation becomes

$$v = \frac{V_{max} \cdot [S]}{K_M}$$

ie $v \propto [S]$, which is consistent with our earlier comments that at low substrate concentrations first order kinetics are observed.

- At high substrate concentration, the contribution of K_M to the denominator can be ignored.

The equation then becomes

$$v = \frac{V_{max} \cdot [S]}{[S]}$$

ie velocity is maximal and zero order kinetics (with respect to substrate) will be displayed. Thus both features which we described qualitatively are consistent with the mathematical analysis.

meaning of K_M

What is the meaning and significant of K_M, the Michaelis constant? We can answer this by analysing a third situation, when [S] is made equal to the K_M. The Michaelis Menten equation then becomes

$$v = \frac{V_{max}}{2} \quad (\text{via } v = \frac{V_{max} \cdot [S]}{[S] + [S]})$$

and the Michaelis constant, K_M can be seen to be the substrate concentration which gives half-maximal velocity. This will be when half the enzyme active sites have substrate bound and are in the ES form (remember that we said that rate was proportional to (ES)). This is a useful operational definition for K_M: other definitions are obtained from the derivation of Michaelis Menten equation; the interested reader will find these (together with the derivation of the equation) in standard biochemistry textbooks.

The significance of the Michaelis constant is that knowledge of it enable us to predict over what substrate concentration range the activity of an enzyme will vary. The next ITA will amplify this point. Knowledge of the K_M of an enzyme also facilitates optimisation of assay conditions.

Π What is the initial velocity of an enzyme-catalysed reaction at the following substrate concentrations, if $K_M = 1\ \mu mol\ dm^{-3}$ and $V_{max} = 100\ \mu mol\ min^{-1}$. Assume that the enzyme obeys Michaelis-Menten kinetics.

	[S] μmol dm^{-3}
(A)	0.1
(B)	0.2
(C)	1.0
(D)	2.0
(E)	10.0
(F)	100.0

When you have calculated the initial velocities for each [S] it would be useful to plot a graph of v against [S].

The initial velocities at the various concentrations are obtained by substitution into the Michaelis-Menten equation. For case (A), this gives

$$v = \frac{V_{max} \cdot [S]}{K_M + [S]} = \frac{100 \times 0.1}{1 + 0.1}\ \mu mol.min^{-1}$$

$$v = \frac{10}{1.1} = 9.1\ \mu mol.min^{-1}$$

corresponding calculations give:

	[S], μmol dm^{-3}	v, μmol min^{-1}
(A)	0.1	9.1
(B)	0.2	16.7
(C)	1.0	50.0
(D)	2.0	66.7
(E)	10.0	90.9
(F)	100.0	99.0

If plotted as a graph, this will give the characteristic hyperbolic plot (Figure A.5), with near proportionally between rate and [S] at low [S], and diminishing proportionally at higher [S], particularly above the K_M. Notice that the rate shows only a very small response to a 10-fold increase in [S] when it is substantially (10x) above the K_M value. Note also that at very low [S], whilst the rate may change nearly proportionally with any change in [S] (look at (A) and (B)), it is nonetheless a small percentage of the maximum activity of the enzyme.

A low K_M means that an enzyme will bind substrate at low substrate concentrations. It is said to have a high affinity for its substrate. For example, if the K_M is 10^{-5} mol dm^{-3} the rate will be 91% of V_{max} by a substrate concentration of 10^{-4} mol dm^{-3}.

A high K_M, in contrast, indicates that a relatively high substrate concentration will be required to achieve (near) saturation. This is described as low affinity. As an example, a K_M of 10^{-2} mol dm^{-3} would require a substrate concentration of 10^{-1} mol dm^{-3} to display 91% of its maximum rate. At a substrate concentration of 10^{-4} mol dm^{-3}, it will only be working at 1% of its maximum rate!

A.6 Determination of K_M and V_{max}

direct plot of v and [S]

When enzymes are being characterised and described, K_M and V_{max} are routinely determined. How may this best be done? Experimentally, one measures initial velocity at a variety of substrate concentrations (having fixed other parameters such as pH and temperature). The data could then be plotted as in Figure A.5. From this direct plot, V_{max} can, in principle, be obtained. K_M is then the substrate concentration which gives half maximal velocity. This strategy is not a good one! Whilst plotting the data in this way it is useful as a check that Michaelis-Menten kinetics are occurring , satisfactory values for K_M and V_{max} are unlikely to be obtained. Look back at the previous ITA. If the substrate concentration is 10x higher than the K_M, the rate is only 10/11th of V_{max}, even when the substrate concentration is 100x that of K_M, the rate is still only 99% of V_{max}! It may thus be extremely difficult to experimentally obtain V_{max}. Any error in estimating V_{max} will result in error in estimating.

linear plots of v and [S]

These problems are avoided by various mathematical transformations of the data which result in linear plots (providing Michaelis-Menten kinetics are obeyed). These are obtained by taking reciprocals of the Michaelis-Menten equation, giving:

$$\frac{1}{v} = \frac{K_M + [S]}{V_{max} \cdot [S]} = \frac{K_M}{V_{max}\ [S]} + \frac{1}{V_{max}}$$

Lineweaver-Burk

This is the equation of a straight line ($y = mx + c$). Although various ways of plotting data have been devised, the most widely used plot is the Lineweaver-Burk plot (Figure A.6), in which 1/v is plotted against 1/[S]. For an enzyme which conforms to Michaelis-Menten kinetics, this will give a straight line, with slope of K_M / V_{max}.

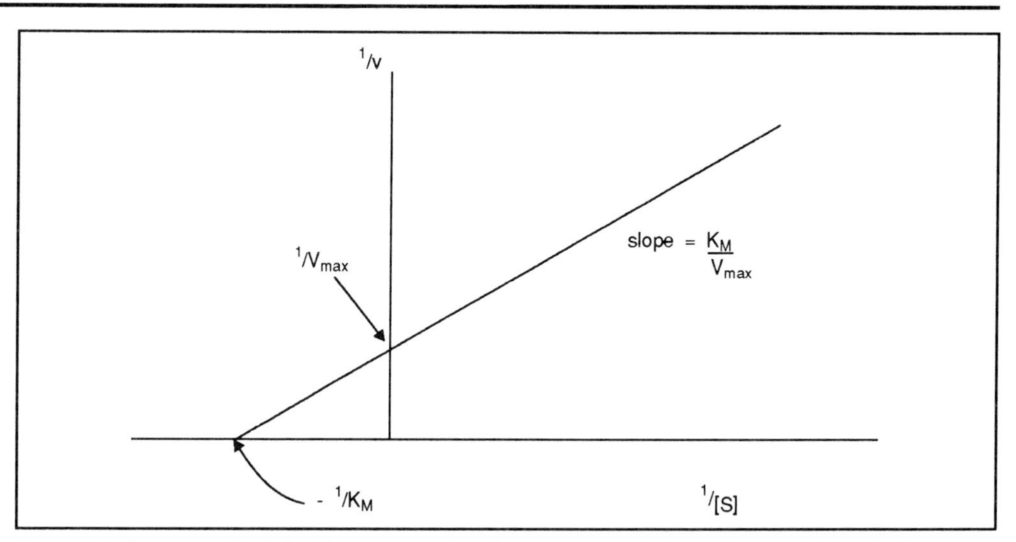

Figure A.6 Lineweaver-Burk double reciprocal plot. For an enzyme which conforms to Michaelis-Menten kinetics, a straight line will be obtained. The intercept on the 1/v axis is $1/V_{max}$, whilst that on the negative side of the 1/[S] axis equals - $1/K_M$..

The intercept on the 1/v axis equals $1/V_{max}$, that on the 1/[S] axis equals - $1/K_M$. Since a linear relationship has been established, it is much easier and more accurate to extrapolate and the line of best fit can be obtained by regression analysis, this avoids the need to experimentally obtain V_{max}. Indeed, data for a Lineweaver-Burk plot is generally most satisfactorily gained over the substrate concentration range 0.2-5.0xK_M.

A.6 Enzymes showing other kinetics properties.

We conclude this appendix by pointing out that some enzymes,the so called allostearic enzymes, do not follow Michaelis Menten kinetics but exhibited quite different relationships between velocity and substrate concentration. You should also remember that enzymes may also be subject to various forms of inhibition by a variety of chemicals. These include competitive, non-competitive and un-competitive inhibition. For a treatment of allostearic enzymes and of affects of various types of inhibitors on the kinetics of enzyme catalysed reactions, we recommend the BIOTOL text "Principles of Enzymology for Technological Application". Simpler treatments of enzyme kinetics are given in the BIOTOL texts "The Molecular Fabric of Cells" and "Principles of Cell Energetics".

Appendix 2

Units of measurement

For historical reasons a number of different units of measurement have evolved. The literature reflects these different systems. In the 1960s many international scientific bodies recommended the standardisation of names and symbols and a universally accepted set of units. These units, SI units (Systeme Internationale de Unites) were based on the definition of: metre (m), kilogram (kg); second (s); ampare (A); mole (mol) and candela (cd). Although, in the intervening period, these units have been widely adopted, their adoption has not been universal. This is especially true in the biological sciences.

It is, therefore, necessary to know both the SI units and the older systems and to be able to interconvert between both sets.

The BIOTOL series of texts predominantly uses SI units. However, in areas of activity where their use is not common, other units have been used. Tables 1 and 2 below provides some alternative methods of expressing various physical quantities. Table 3 provides prefixes which are commonly used.

Mass (S1 unit: kg)	Length (S1 unit: m)	Volume (S1 unit: m^3)	Energy (S1 unit: J = kg m^2 s^{-2})
g = 10^{-3} kg	cm = 10^{-2} m	l = dm^3 = 10^{-3} m^3	cal = 4.184 J
mg = 10^{-3} g = 10^{-6} kg	Å = 10^{-10} m	dl = 100 ml = 100 cm^3	erg = 10^{-7} J
µg = 10^{-6} g = 10^{-9} kg	nm = 10^{-9} m = 10Å	ml = cm^3 = 10^{-6} m^3	eV = 1.602 x 10^{-19} J
	pm = 10^{-12} m = 10^{-2} Å	µl = 10^{-3} cm^3	

Table 1 Units for physical quantities

Concentration (SI units: mol m^{-3})

a) M = mol l^{-1} = mol dm^{-3} = 10^3 mol m^{-3}
b) mg1^{-1} = μg cm^{-3} = ppm = 10^{-3} g dm^{-3}
c) μg g^{-1} = ppm = 10^{-6} g g^{-1}
d) ng cm^{-3} = 10^{-6} g dm^{-3}
e) ng dm^{-3} = pg cm^{-3}
f) pg g^{-1} = ppb = 10^{-12} g g^{-1}
g) mg% = 10^{-2} g dm^{-3}
h) μg% = 10^{-5} g dm^{-3}

Table 2 Units for concentration

Fraction	Prefix	Symbol	Multiple	Prefix	Symbol
10^{-1}	deci	d	10	deka	da
10^{-2}	centi	c	10^2	hecto	h
10^{-3}	milli	m	10^3	kilo	k
10^{-6}	micro	μ	10^6	mega	M
10^{-9}	nano	n	10^9	giga	G
10^{-12}	pico	p	10^{12}	tera	T
10^{-15}	femto	f	10^{15}	peta	P
10^{-18}	atto	a	10^{18}	exa	E

Table 3 Prefixes for S1 units

Appendix 3

Chemical Nomenclature

Chemical nomenclature is quite a difficult issue especially in dealing with the complex chemicals of biological systems. To rigidly adhere to a strict systematic naming of compounds such as that of the International Union of Pure and Applied Chemistry (IUPAC) would lead to a cumbersome and overly complex text. BIOTOL has adopted a pragmatic approach by predominantly using the names or acronyms of chemicals most widely used in biologically-based activities. It is recognised however that there remains some potential for confusion amongst readers of different background. For example the simple structure CH_3COOH can be described as ethanoic acid or acetic acid depending on the environment or industry in which the compound is produced or used. To reduce such confusion, the BIOTOL series makes every effort to provide synonyms for compounds when they are first mentioned and to provide chemical structures where clarity and context demand.

Appendix 4

Abbreviations used for the common amino acids

Amino acid	Three-letter abbreviation	One-letter symbol
Alanine	Ala	A
Arginine	Arg	R
Asparagine	Asn	N
Aspartic acid	Asp	D
Asparagine or aspartic acid	Asx	B
Cysteine	Cys	C
Glutamine	Gln	Q
Glutamic acid	Glu	E
Glutamine or glutamic acid	Glx	Z
Glycine	Gly	G
Histidine	His	H
Isoleucine	Ile	I
Leucine	Leu	L
Lsyine	Lys	K
Methionine	Met	M
Phenylalanine	Phe	F
Proline	Pro	P
Serine	Ser	S
Threonine	Thr	T
Tryptophan	Trp	W
Tyrosine	Tyr	Y
Valine	Val	V

Index

A

Index

A

B

C

D